From Genes to Genomes

Second Edition

Concepts and Applications of DNA Technology

Jeremy W. Dale and **Malcolm von Schantz**

University of Surrey, UK

John Wiley & Sons, Ltd

Telephone (+44) 1243 779777

Email (for orders and customer service enquiries): cs-books@wiley.co.uk
Visit our Home Page on www.wileyeurope.com or www.wiley.com

Reprinted with corrections November 2008

Other Wiley Editorial Offices

John Wiley & Sons Inc., 111 River Street, Hoboken, NJ 07030, USA

Jossey-Bass, 989 Market Street, San Francisco, CA 94103-1741, USA

Wiley-VCH Verlag GmbH, Boschstr. 12, D-69469 Weinheim, Germany

John Wiley & Sons Australia Ltd, 33 Park Road, Milton, Queensland 4064, Australia

John Wiley & Sons (Asia) Pte Ltd, 2 Clementi Loop #02-01, Jin Xing Distripark, Singapore 129809

John Wiley & Sons Canada Ltd, 6045 Freemont Blvd, Mississauga, Ontario, L5R 4J3

Wiley also publishes its books in a variety of electronic formats. Some content that appears in print may not be available in electronic books.

Library of Congress Cataloging-in-Publication Data

Dale, Jeremy.
From genes to genomes : concepts and applications of DNA technology / Jeremy W. Dale and Malcolm von Schantz. – 2nd ed.
p. ; cm.
Includes bibliographical references and index.
ISBN 978-0-470-01733-3 (cloth)
1. Genetic engineering. 2. DNA. 3. Genes. 4. Genomes. I. Schantz, Malcolm von. II. Title.
[DNLM: 1. Genetic Engineering. 2. Cloning, Molecular. 3. DNA, Recombinant. 4. Genome. QU 450 D139f 2007]
QH442.D35 2007
660.6′5 | – dc22 2007025153

British Library Cataloguing in Publication Data

A catalogue record for this book is available from the British Library

ISBN 978-0-470-01733-3 (HB) 978-0-470-01734-0 (PB)

Typeset in 10.5 on 13 pt Times by SNP Best-set Typesetter Ltd., Hong Kong
Printed and bound in Spain by Graphos SpA, Barcelona
This book is printed on acid-free paper.

From Genes to Genomes

Second Edition

Contents

Preface

The first edition of this book was published in 2002. At that time the revolution in molecular biology, with *gene cloning* at its heart, had already progressed to encompass a variety of other techniques and advances, especially genome sequencing, the polymerase chain reaction and microarray technology. This has now developed still further, to the extent that many researchers no longer need sophisticated techniques to clone and identify specific genes. They can simply screen the genome sequence for the relevant gene, and (if needed) amplify it, using PCR. The changes we have made in this edition represent an attempt to reflect this development – especially by enhancing the treatment of bioinformatics. At the same time, we are conscious of the need to keep the book accessible to our target audience, which means not allowing it to expand much beyond its original size. Therefore, some topics have had to go. However, we have tried not to throw out the baby with the bathwater. Many of the older techniques still have a role to play – especially for those who are working with organisms whose genomes have not yet been sequenced. We have also retained a description of some techniques which, although now used less frequently, are either significant in understanding the development of the subject or play a useful role in understanding underlying concepts.

The main title of the book, *From Genes to Genomes*, is derived from the progress of this revolution. It signifies the move from the early focus on the isolation and identification of specific genes to the exciting advances that have been made possible by the sequencing of complete genomes. This has in turn spawned a whole new range of technologies (*post-genomics*) that are designed for genome-wide analysis of gene structure and expression. In this edition, we have strengthened this theme, by dealing first with methods that are applicable to individual genes before moving on to genome-wide concepts.

Dealing only with the techniques, without the applications, would be rather dry. Some of the applications are obvious – recombinant product formation, genetic diagnosis, transgenic plants and animals, and so on – and we have attempted to introduce these to give you a flavour of the advances that continue to be made, but at the same time without burdening you with excessive detail. Equally important, possibly more so, are the contributions made to the advance of fundamental knowledge in areas such as developmental studies and molecular phylogeny.

The purpose of this book is to provide an introduction to the concepts and applications of this rapidly moving and fascinating field. In writing it, we had in mind its usefulness for undergraduate students in the biological and biomedical sciences (who we assume will have a basic grounding in molecular biology). However, it will also be relevant for many others, ranging from research workers and teachers who want to update their knowledge of related areas to anyone who would like to understand rather more of the background to current controversies about the applications of some of these techniques.

Finally, we would like to thank Celia Carden, Andy Slade and other staff at John Wiley, for their encouragement and patience during the preparation of this book.

Jeremy W. Dale
Malcolm von Schantz

1 Introduction

The classical approach to genetics starts with the identification of variants which have a specific *phenotype*, i.e. they are altered in some way that can be seen (or detected in other ways) and defined. For Mendel, this was the appearance of his peas (e.g. green *vs* yellow, or round *vs* wrinkled). One of the postulates he arrived at was that these characteristics assorted independently of one another. For example, if you cross a strain that produces yellow round peas with another strain that produces green wrinkled peas, the first generation (F_1) are all round and yellow (because round is dominant over wrinkled, and yellow is dominant over green). Of course Mendel did not know why this happened. We now know that if two genes are located on different chromosomes, which will segregate independently during meiosis, the genes will thus be distributed independently amongst the progeny. The same can happen if the two genes are on the same chromosome, but are so far apart that the recombination between the homologous chromosomes will be sufficient to reassort them independently. On the other hand, if they are quite close together, they will tend to remain associated during meiosis, and will therefore be inherited together. We refer to genes that do *not* segregate independently as *linked*; the closer they are, the greater the degree of linkage, i.e. the more likely they are to stay together during meiosis. Measuring the degree of linkage (*linkage analysis*) is a central tool in classical genetics, in that it provides a way of mapping genes, i.e. determining their relative position on the chromosome. Furthermore, it provides us with an important method for correlating genetic and physical maps.

Bacteria and yeasts provide much more convenient systems for genetic analysis, because they grow quickly, as unicellular organisms, and on defined media. You can therefore use chemical or physical mutagens (such as ultraviolet irradiation) to produce a wide range of mutations, and can select

From Genes to Genomes, Second Edition, Jeremy W. Dale and Malcolm von Schantz

specific mutations from very large pools of organisms – remembering that an overnight culture of *E. coli* will contain some 10^9 bacteria per millilitre. These mutations may simply affect the ability to produce a specific amino acid, manifested as a requirement for that amino acid to be added to the growth medium, or to use a particular carbon source such as lactose. Or it may be a more complex phenotype, such as loss of motility, or inability to divide into two cells, leading to the production of filaments. Therefore we can use genetic techniques to investigate detailed aspects of the physiology of such cells, including identifying the relevant genes by mapping the position of the mutations. Although the techniques in bacteria differ from those in higher organisms, forms of linkage analysis still play a major role.

For multicellular organisms, the range of phenotypes is even greater, as there are then questions concerning the development of different parts of the organism. However animals have much longer generation times than bacteria, and using millions of animals (especially mammals) to identify the mutations you are interested in is either cumbersome, impossible or indefensible. Human genetics is even more difficult as you cannot use selected breeding to map genes; you have to rely on the analysis of real families who have chosen to breed with no consideration for the needs of science. Nevertheless, classical genetics has contributed extensively to the study of developmental processes, notably in the fruit fly *Drosophila melanogaster*, where it is possible to study quite large numbers (although nothing like the numbers that can be used in bacterial genetics), and to use mutagenic agents to enhance the rate of variation.

These methods suffered from a number of limitations. In particular, they could only be applied, in general, to mutations that gave rise to a phenotype that could be defined by morphological, physiological, biochemical or behavioural means. Furthermore, there was no easy way of characterizing the nature of the mutation. The situation changed radically in the 1970s, with the development of techniques that enabled DNA to be cut precisely into specific fragments, and to be joined together, enzymatically – techniques that became known variously as genetic manipulation, genetic modification, genetic engineering or recombinant DNA technology. The term 'gene cloning' is also used, since joining a fragment of DNA with a vector such as a plasmid that can replicate in bacterial cells enabled the production of a bacterial strain (a clone) in which all the cells contained a copy of this specific piece of DNA. For the first time, it was possible to isolate and study specific genes. Since this applied equally to human genes, the impact on human genetics was particularly marked.

The revolution also depended on the development of a variety of other molecular techniques. The earliest of these (actually predating gene cloning) was *hybridization* which enabled the identification of specific DNA sequences on the basis of their sequence similarity. Later on came methods for determining the sequence of these DNA fragments, and the polymerase chain reaction (PCR), which provided a powerful way of amplifying specific DNA sequences. Combining those advances with automated techniques and the concurrent advance in computer power has led to the determination of the full genome sequence of many organisms, including the human genome, as well as enormous advances in understanding the roles of genes and their products.

This revolution does not end with understanding how genes work and how the information is inherited. Genetics, and especially modern molecular genetics, underpins all the biological sciences. By studying, and manipulating, specific genes, we develop our understanding of the way in which the products of those genes interact to give rise to the properties of the organism itself. This could range from, for example, the mechanism of motility in bacteria to the causes of human genetic diseases and the processes that cause a cell to grow uncontrollably, giving rise to a tumour. In many cases, we can identify precisely the cause of a specific property. We can say that a change in one single base in the genome of a bacterium will make it resistant to a certain antibiotic, or that a change in one base in human DNA could cause debilitating disease. Understanding the cause of a genetically determined disease leads firstly to genetic diagnosis, and ultimately to remedying it, using gene therapy.

Furthermore, since these techniques enabled the cloning and expression of genes from any one organism (including humans) into a more amenable host, such as a bacterium, they allowed the use of genetically modified bacteria (or other hosts) for the production of human gene products, such as hormones, for therapeutic use. This principle was subsequently extended to the genetic modification of plants and animals – both by inserting foreign genes and by knocking out existing ones – to produce plants and animals with novel properties.

As you are probably aware, the construction and use of genetically modified organisms (GMOs) is not without controversy. In the early days, *E. coli* strains carrying recombinant DNA molecules were treated with extreme caution. *E. coli* is a bacterium which lives in its billions within our digestive system, and those of other mammals, and which will survive quite easily in our environment, unfortunately including our food and our beaches. So there was a lot of concern that the introduction of foreign DNA into *E. coli* would generate bacteria with dangerous properties. Fortunately, this is one fear that has been shown to be unfounded. Some natural *E. coli* strains *are* pathogenic – in particular the O157:H7 strain, which can cause severe disease or death. In contrast, the strains used for genetic manipulation are harmless disabled laboratory strains that will not even survive in the gut. Working with genetically modified *E. coli* can therefore be done very safely (although work with *any* bacterium has to follow some basic safety rules). However, the most commonly used type of vector, plasmids, is shared readily between bacteria; the transmission of plasmids between bacteria is behind much of the natural spread of antibiotic resistance. What if our recombinant plasmids were transmitted to other *bacterial* strains that *do* survive on their own? This, too, has turned out not to be a worry in the majority of cases. The plasmids themselves have been manipulated so that they cannot be readily transferred to other bacteria. Furthermore, expressing a gene such as that coding for, say, dogfish insulin, or carrying an artificial chromosome containing 100 000 bases of human genomic DNA, is a great burden to an *E. coli* cell, and confers no reward whatsoever. In order to make them accept it, we have to create conditions that will kill all bacterial cells *not* carrying the foreign gene. Although we have to recognize the possibility that some unscrupulous individual might evade the regulations and produce a harmful bacterium, in practice it is not that easy. In general, genetically modified bacteria are not well able to cope with life outside the laboratory, and nature is quite capable of producing pathogenic organisms without our assistance.

The debate has now largely moved on to issues relating to genetically modified plants and animals. It is important to distinguish the *genetic modification* of plants and animals from *cloning* plants and animals. The latter simply involves the production of genetically identical individuals; it does not involve any genetic modification whatsoever. (The two technologies can be used in tandem, but that is another matter.) There are ethical issues to be considered, but cloning plants and animals is not the subject of this book.

The debate largely revolves around two factors: food safety and environmental impact. The first thing to be clear about is that there is no imaginable reason why genetic modification, *per se*, should make a foodstuff hazardous in any way. There is no reason to suppose that cheese made with rennet from a genetically modified bacterium is any more dangerous than similar cheese made with 'natural' rennet. It is possible to imagine a risk associated with some genetically modified foodstuffs, due to unintended stimulation of the production of natural toxins – remembering for example that potatoes are related to deadly nightshade. However, this can happen equally well (or perhaps is even more likely) with conventional procedures for developing new strains, which are not subject to the same degree of rigorous testing for safety.

The potential environmental impact is more difficult to assess. The main issue here is the use of genetic modification to make plants resistant to herbicides or to insect attack. When such plants are grown on a large scale, it is difficult to be certain that the gene in question will not spread to related wild plants in the vicinity (although measures can be taken to reduce this possibility). However, this may be an exaggerated concern. As with the bacterial example above, these genes will not spread significantly unless there is an evolutionary pressure favouring them. We would not expect widespread resistance to weedkillers unless the plants are being sprayed with those weed-killers. There might be an advantage in becoming resistant to insect attack, but the insects concerned have been around for a long time, so the wild plants have had plenty of time to develop natural resistance anyway. We also have to balance the use of genetically modified plants against the use of chemicals. If genetic modification of the plants means a reduction in the use of environmentally damaging chemicals, then that must be a good thing.

Nonetheless, this issue is by no means as clear-cut as that of genetically modified bacteria. We cannot test these organisms in a contained laboratory. They take months or years to produce each generation, not 20 min as *E. coli* does. Thus, this is an important and complicated issue, and to understand it fully you need to know about evolution, ecology, food chemistry, nutrition and molecular biology. We hope that reading this book will be of some help for the last of these. We also hope that it will convey some of the wonder, excitement, and intellectual stimulation that this science brings to its practitioners. What better way to reverse the boredom of a long journey than to indulge in the immense satisfaction of constructing a clever new screening algorithm? Who needs jigsaw and crossword puzzles when you can figure out a clever way of joining two DNA fragments together? And how can you ever lose the fascination you feel about the fact that the drop of enzyme that you are adding to your test tube is about to manipulate the DNA molecules in it with surgical precision?

2 Basic Molecular Biology

In this book, we assume you already have a working knowledge of the basic concepts of molecular biology. This chapter serves as a reminder of the key aspects of molecular biology that are especially relevant to this book.

2.1 Nucleic acid structure

2.1.1 The DNA backbone

Manipulation of nucleic acids in the laboratory is based on their physical and chemical properties, which in turn are reflected in their biological function. Intrinsically, DNA is a remarkably stable molecule. Scientists routinely send plasmid DNA samples in the post without worrying about refrigeration. Indeed, DNA of high enough quality to be analysed has been recovered from frozen mammoths and mummified pharaohs thousands of years old. This stability is provided by the robust repetitive phosphate–sugar backbone in each DNA strand, in which the phosphate links the 5′ position of one sugar to the 3′ position of the next (Figure 2.1). The bonds between these phosphorus, oxygen and carbon atoms are all *covalent bonds*. Controlled degradation of DNA requires enzymes (nucleases) that break these covalent bonds. These are divided into *endonucleases*, which attack internal sites in a DNA strand, and *exonucleases*, which nibble away at the ends. We can for the moment ignore other enzymes that attack, for example, the bonds linking the bases to the sugar residues. Some of these enzymes are non-specific, and lead to a generalized destruction of DNA. It was the discovery of *restriction endonucleases* (or *restriction enzymes*), which cut DNA strands at specific positions, coupled with *DNA ligases*, which can join two double-stranded DNA molecules together, that opened up the possibility of *recombinant DNA technology* (‘*genetic engineering*’).

From Genes to Genomes, Second Edition, Jeremy W. Dale and Malcolm von Schantz

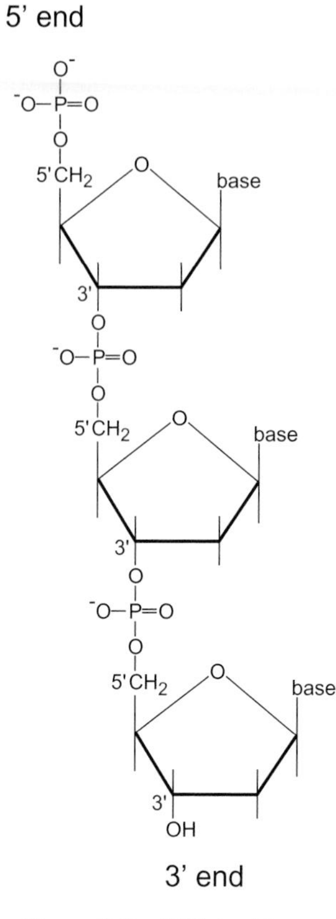

Figure 2.1 DNA backbone

RNA molecules, which contain the sugar ribose (Figure 2.2), rather than the deoxyribose found in DNA, are less stable than DNA. They show greater susceptibility to attack by nucleases (*ribonucleases*), and are also more susceptible to chemical degradation, especially by alkaline conditions.

2.1.2 The base pairs

In addition to the sugar (2′deoxyribose) and phosphate, DNA molecules contain four nitrogen-containing bases (Figure 2.3): two pyrimidines, thymine (T) and cytosine (C), and two purines, guanine (G) and adenine (A). (Other bases can be incorporated into synthetic DNA in the laboratory, and sometimes other bases occur naturally.) Because the purines are bigger than the pyrimidines, a regular double helix requires a purine in one strand to be matched by a pyrimidine in the other. Furthermore, the regularity of the double helix requires specific hydrogen bonding between the bases so that they fit together, with an A opposite a T, and a G opposite a C (Figure 2.4). We refer to these pairs of bases as *complementary*, and hence to each strand as the *complement* of the other.

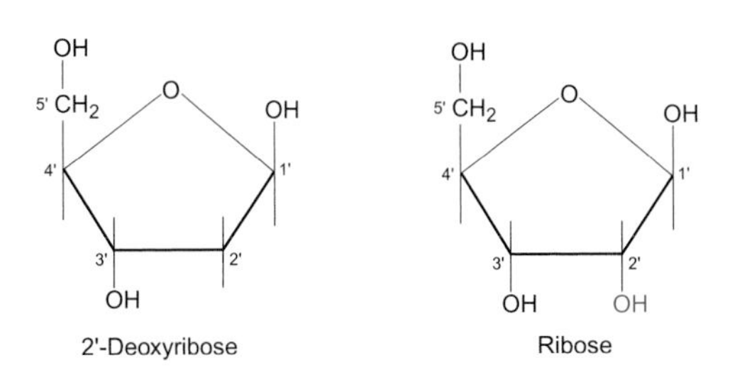

Figure 2.2 Nucleic acid sugars

Purines **Pyrimidines**

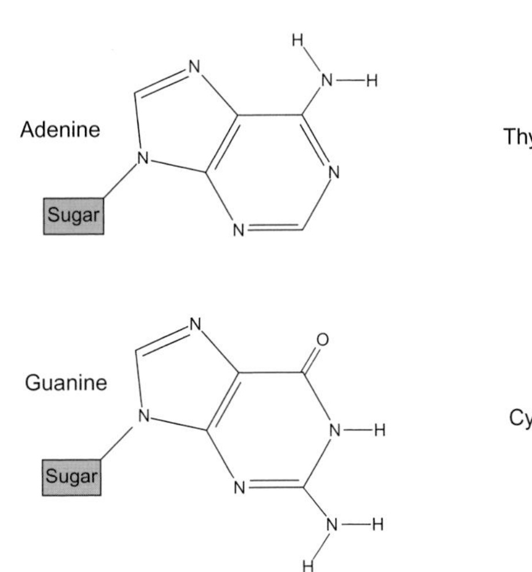

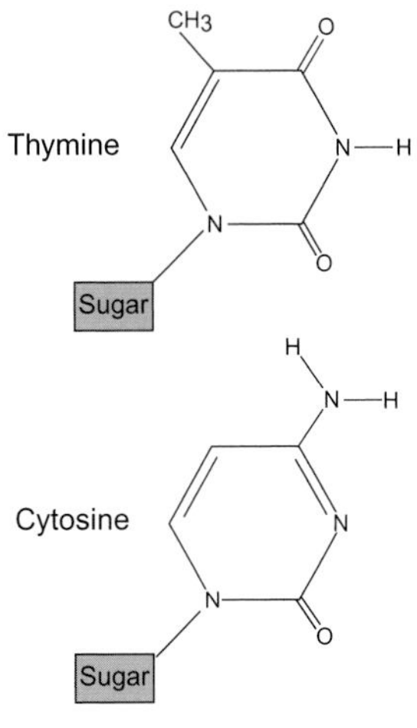

Figure 2.3 Nucleic acid bases

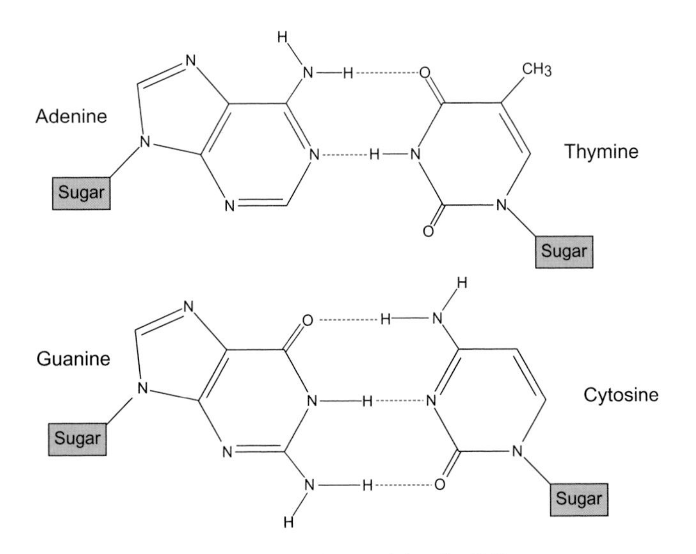

Figure 2.4 Base pairing in DNA

Box 2.1 Complementary sequences

DNA sequences are often represented as the sequence of just one of the two strands, in the 5′ to 3′ direction, reading from left to right. Thus the double-stranded DNA sequence

```
5'-AGGCTG-3'
3'-TCCGAC-5'
```

would be shown as AGGCTG, with the orientation (i.e. the position of the 5′ and 3′ ends) being inferred.

To get the sequence of the other (complementary) strand, you must not only change the A and G residues to T and C (and vice versa), but you must also reverse the order.

In this example, the complement of AGGCTG is CAGCCT, reading the lower strand from right to left (again in the 5′ to 3′ direction).

Note that the two DNA strands run in opposite directions. In a conventional representation of a double-stranded sequence the 'top' strand has a 5′ hydroxyl group at the left-hand end (and is said to be written in the 5′ to 3′ direction), while the 'bottom' strand has its 5′ end at the right-hand end. Because the two strands are complementary, there is no information in one strand that cannot be deduced from the other one. Therefore, to save space, the convention is to represent a double-stranded DNA sequence by showing the sequence of only one strand, in the 5′ to 3′ direction. The sequence of the second strand is inferred from that, and you have to remember that the second strand runs in the opposite direction. Thus a single-strand sequence written as AGGCTG (or more fully 5′AGGCTG3′) would have as its complement CAGCCT (5′CAGCCT3′) (see Box 2.1).

Thanks to this base pairing arrangement, the two strands can be separated intact – both in the cell and in the test tube – under conditions which disrupt the hydrogen bonds between the bases but are much too mild to pose any threat to the covalent bonds in the backbone. Such a separation is referred to as *denaturation* of DNA, and unlike the denaturation of many proteins, it is reversible. Because of the complementarity of the base pairs, the strands will easily join together again and *renature* by reforming their hydrogen bonds. In the test tube, DNA is readily denatured by heating, and the denaturation process is therefore often referred to as *melting* even when it is accomplished by means other than heat (e.g. by NaOH). Denaturation of a double-stranded DNA molecule occurs over a short, specific, temperature range, and the midpoint of that range is defined as the *melting temperature* (T_m). This is influenced by the base composition of the DNA. Since guanine:cytosine (GC) base pairs have three hydrogen bonds, they are stronger (i.e. melt less easily) than adenine:thymine (AT) pairs, which have only two hydrogen bonds. It is therefore possible to estimate the melting temperature of a DNA fragment if you know the sequence (or the base composition and length). These considerations are important in understanding the technique known as *hybridization*, in which *gene probes* are used to detect specific nucleic acid sequences. We will look at hybridization in more detail in Chapter 7.

In addition to the hydrogen bonds, the double-stranded DNA structure is maintained by *hydrophobic interactions* between the bases. The hydrophobic nature of the bases means that a single-stranded structure, in which the bases are exposed to the aqueous environment, is unstable. Pairing of the bases enables them to be removed from interaction with the surrounding water. In contrast to the hydrogen bonding, hydrophobic interactions are relatively non-specific. Thus, nucleic acid strands will tend to stick together even in the absence of specific base pairing, although the specific interactions make the association stronger. The specificity of the interaction can therefore be increased by the use of chemicals (such as formamide) that reduce the hydrophobic interactions.

What happens if there is only a single nucleic acid strand? This is normally the case with RNA, but single-stranded forms of DNA (ssDNA) also exist. In some viruses, the genetic material is single-stranded DNA. A single-stranded nucleic acid molecule will tend to fold up on itself to form localized double-stranded regions, including structures referred to as hairpins or stem-loop structures. This has the effect of removing the bases from interaction with the surrounding water. At room temperature, in the absence of denaturing agents, a single-stranded nucleic acid will normally consist of a complex set of such localized secondary structure elements, which is especially evident with RNA molecules. This can also happen to a limited extent with double-stranded DNA, where short sequences sometimes tend to loop out of the regular double helix. Because this makes it easier for enzymes to unwind the DNA, and to separate the strands, these sequences can play a role in the regulation of gene expression, and in the initiation of DNA replication.

A further factor to be taken into account is the negative charge on the phosphate groups in the nucleic acid backbone. This works in the opposite direction to the hydrogen bonds and hydrophobic interactions; the strong negative charge on the DNA strands causes electrostatic repulsion that tends to make the two strands repel each other. In the presence of salt, this effect is counteracted by a cloud of counterions surrounding the molecule, neutralizing the negative charge on the phosphate groups. However, if you reduce the salt concentration, any weak interactions between the strands will be disrupted by electrostatic repulsion. Hence, at low ionic strength, the strands will only remain together if the hydrogen bonding is strong enough, and therefore we can use low salt conditions to increase the specificity of hybridization (see Chapter 7).

2.1.3 RNA structure

Chemically, RNA is very similar to DNA. The fundamental chemical difference is that the RNA backbone contains ribose rather than the 2′deoxyribose (i.e. ribose without the hydroxyl group at the 2′ position) present in DNA (Figure 2.5). However, this slight difference has a powerful effect on some properties of the nucleic acid, especially on its stability. Thus, RNA is unstable under alkaline conditions. Under these conditions, DNA is stable: although the strands will separate, they will remain intact and capable of renaturation when the pH is lowered again. A further difference between RNA and DNA is that the former contains uracil rather than thymine (Figure 2.5).

Generally, while most of the DNA we encounter is double stranded, most of the RNA we encounter consists of a single polynucleotide strand. However, DNA can also exist as a single-stranded molecule, and RNA is able to form double-stranded molecules. Thus, this distinction between RNA and DNA is not an inherent property of the nucleic acids themselves, but is a reflection of the natural roles of RNA and DNA in the cell, and of the method of production. In all *cellular* organisms (i.e. excluding viruses), DNA is the inherited material responsible for the genetic composition of the cell, and the replication process that has evolved is based on a double-stranded molecule. In contrast, the roles of RNA in the cell do not require a second strand, and indeed the presence of a second, complementary, strand would preclude its

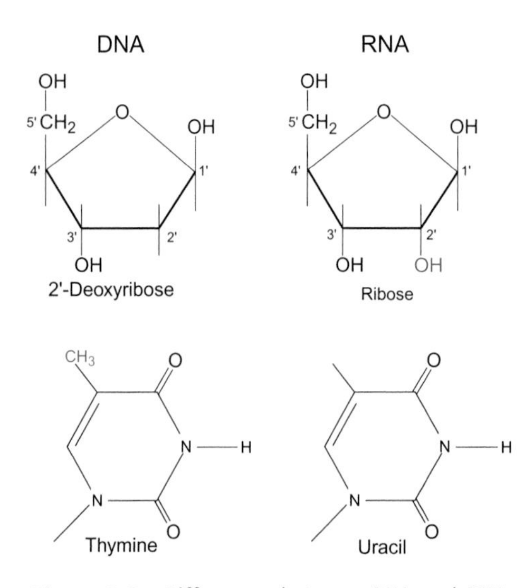

Figure 2.5 Differences between DNA and RNA

role in protein synthesis. However there are some viruses that have double-stranded RNA (dsRNA) as their genetic material, as well as some viruses with single-stranded RNA, and some viruses (as well as some plasmids) replicate via single-stranded DNA forms.

2.1.4 Nucleic acid synthesis

We do not need to consider here all the details of how nucleic acids are synthesized. The fundamental features that we need to remember are summarized in Figure 2.6, which shows the addition of a nucleotide to the growing end (3′-OH) of a DNA strand. The substrate for this reaction is the relevant deoxynucleotide triphosphate (dNTP), i.e. the one that makes the correct base pair with the corresponding residue on the template strand. The DNA strand is always extended at the 3′-OH end. For this reaction to occur it is essential that the existing residue at the 3′-OH end, to which the new nucleotide is to be added, is accurately base paired with its partner on the other strand.

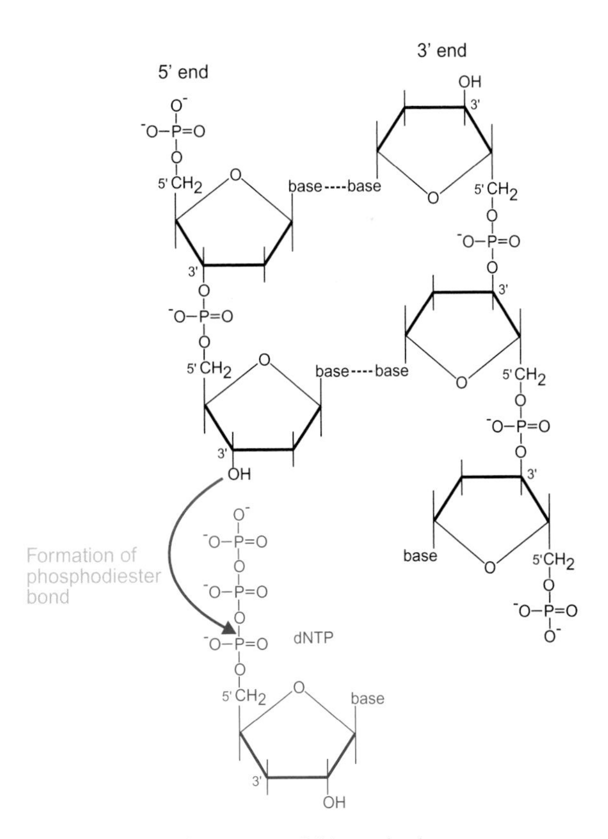

Figure 2.6 DNA synthesis

RNA synthesis occurs in much the same way, as far as this description goes, except that of course the substrates are nucleotide triphosphates (NTPs) rather than the deoxynucleotide triphosphates (dNTPs). There is one very important difference though. DNA synthesis only occurs by extension of an existing strand – it always needs a *primer* to get it started to make a complementary copy of its template. RNA polymerases on the other hand are capable of starting a new RNA strand, complementary to its template, from scratch, given the appropriate signals.

2.1.5 Coiling and supercoiling

DNA can be denatured and renatured, deformed and reformed, and still retain unaltered function. This is a necessary feature, because as large a molecule as DNA will need to be packaged if it is to fit within the cell that it controls. The DNA of a human chromosome, if it were stretched out into an unpackaged double helix, would be several centimetres long. Thus, cells are dependent on the packaging of DNA into modified configurations for their very existence.

Double-stranded DNA, in its relaxed state, normally exists as a right-handed double helix with one complete turn per 10 base pairs; this is known as the *B form* of DNA. Hydrophobic interactions between consecutive bases on the same strand contribute to this winding of the helix, as the bases are brought closer together, enabling a more effective exclusion of water from interaction with the hydrophobic bases.

The DNA double helix can exist in other forms, notably the *A form* (also right-handed but more compact, with 11 bases per turn) and *Z-DNA*, which is a left-handed double helix with a more irregular appearance (a zigzag structure, hence its designation). However, that is not the complete story. There are higher orders of conformation. The double helix is in turn coiled on itself – an effect known as *supercoiling*. There is an interaction between the coiling of the helix and the degree of supercoiling. As long as the ends are fixed, changing the degree of coiling will alter the amount of supercoiling, and *vice versa*. DNA *in vivo* is constrained; the ends are not free to rotate. This is most obviously true of circular DNA structures such as (most) bacterial plasmids, but the rotation of linear molecules (other than very short oligonucleotides) is also constrained within the cell. The net effect of coiling and supercoiling (a property known as the *linking number*) is therefore fixed, and cannot be changed without breaking one of the strands. In nature, there are enzymes known as topoisomerases (including DNA gyrase) that do just that: they break the DNA strands, and then in effect rotate the ends and reseal them. This alters the degree of winding of the helix, and thus affects the supercoiling of the DNA. Topoisomerases also have an ingenious use in the laboratory, which we will consider in Chapter 4.

The plasmids that we will be referring to frequently in later pages are naturally supercoiled when they are isolated from the cell. However, if one of the strands is broken at any point, the DNA is then free to rotate at that point and can therefore relax into a non-supercoiled form. This is known as an *open circular* form (in contrast to the *covalently closed circular* form of the native plasmid).

2.2 What is a gene?

The definition of a 'gene' is rather imprecise. Its origins go back to the early days of genetics, when it was used to describe the unit of inheritance of an observable characteristic (a *phenotype*). This meaning persists in non-scientific usage, rather loosely, as the 'gene for blue eyes', or 'the gene for red hair'. As it became realized that many characteristics were determined by the presence or properties of individual proteins, the definition became refined to relate to the chromosomal region that carried the information for that protein, leading to the concept of 'one gene, one protein'. As the study of genetics and biochemistry progressed further, it was realized that many proteins consist of several distinct polypeptides, and that the chromosomal regions coding for the different polypeptides could be distinguished genetically. Therefore, the definition was refined further to mean a piece of DNA containing the information for a single specific polypeptide ('one gene, one polypeptide'). With the advent of DNA sequencing, it became possible to consider a gene in molecular terms. So we now often use the term 'gene' as being synonymous with 'open reading frame' (ORF), i.e. the region between the start and stop codons. In bacteria, this is (usually) a simple uninterrupted sequence, but in eukaryotes, the presence of introns (see below) makes this definition more difficult, since the region of the chromosome that contains the information for a specific polypeptide may be many times longer than the actual coding sequence. We also have to be careful as we may want to refer to the whole transcribed region, which will be longer than the translated open reading frame, or indeed to include the control regions that are necessary for the start of transcription.

Furthermore, this definition, by focusing solely on the regions that code for proteins (or polypeptides), is too limited in its scope. It ignores many regions of DNA that, although not coding for proteins, are nevertheless important for the viability of the cell, or influence the phenotype in other ways. The most obvious of these are DNA sequences that code for RNA molecules. Most well known of these are ribosomal and transfer RNA, although later in the book we will encounter other RNA molecules that play significant roles in gene regulation and other activities. Other DNA regions are important in gene regulation because they act as binding sites for regulatory proteins.

In organisms with small genomes, such as bacteria, a high proportion of the genome is accounted for by these coding and regulatory regions (together with elements that may be described as 'parasitic' such as integrated viruses and insertion sequences). There is relatively little DNA to which we can ascribe no likely function, compared with eukaryotic cells, and especially animal and plant cells with much larger genomes, where there is a much higher proportion of non-coding DNA. However, we should not be too hasty in writing these sequences off as 'junk'. Increasingly more of it is recognized as having a function, such as enabling the DNA to be folded correctly and in ensuring that the coding regions are available for expression under the appropriate conditions.

Thus, we have to accept that it is not possible to produce an entirely satisfactory definition of the word 'gene'. However, this is rarely a serious problem. We just have to be careful as to how we use it depending on whether we are discussing only the coding region (ORF), the length of sequence that is transcribed into mRNA (including untranslated regions), or whether we wish to include DNA regions with regulatory functions as well as coding sequences.

2.3 Information flow: gene expression

The way in which genes are expressed is so central to the subsequent material in this book that it is worth reviewing briefly the salient features. The basic dogma (Figure 2.7) is that, while DNA is the fundamental genetic material

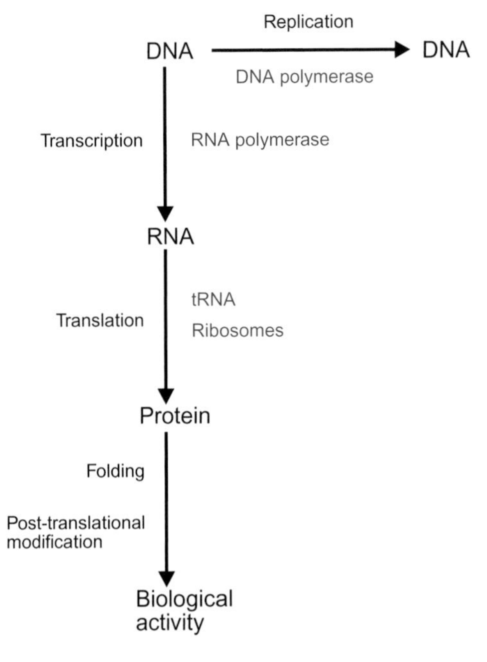

Figure 2.7 Information flow

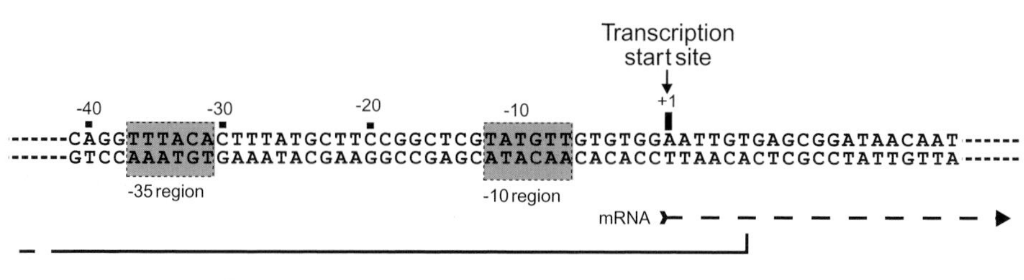

Figure 2.8 Structure of the promoter region of the *lac* operon. Note that the –35 and –10 regions of the *lac* promoter do not correspond exactly to the consensus sequences TTGACA and TATAAT, respectively

(ignoring RNA viruses) that carries information from one generation to the next, its effect on the characteristics of the cell requires firstly its copying into RNA (*transcription*), and then the *translation* of the mRNA into a polypeptide by ribosomes. Further processes are required before its proper activity can be manifested: these include the folding of the polypeptide, possibly in association with other subunits to form a multi-subunit protein, and in some cases modification, e.g. by glycosylation or phosphorylation. It should be remembered that, in some cases, RNA rather than protein is the final product of a gene (ribosomal and transfer RNA molecules for example).

2.3.1 Transcription

Transcription is carried out by RNA polymerase. RNA polymerase recognizes and binds to a specific sequence (the *promoter*), and initiates the synthesis of mRNA from an adjacent position.

A typical bacterial promoter carries two *consensus* sequences (i.e. sequences that are closely related in all genes): TTGACA centred at position –35 (i.e. 35 bases before the transcription start site) and TATAAT at –10 (Figure 2.8). It is important to understand the nature of a consensus: few bacterial promoters have exactly the sequences shown, but if you line up a large number of promoters you will see that at any one position a large number of them have the same base (see Box 2.2). The RNA polymerase has higher affinity for some promoters than others – depending not only on the exact nature of the two consensus sequences but to a lesser extent on the sequence of a longer region. The nature and regulation of bacterial promoters, including the existence of alternative types of promoters, is considered further in Chapter 9.

In eukaryotes, by contrast, the promoter is a considerably larger area around the transcription start site, where a number of *trans*-acting transcription factors (i.e. DNA-binding proteins encoded by genes in other parts of the genome) bind to a number of *cis*-acting promoter elements (i.e. elements that affect the expression of the gene next to them) in a considerably more

Box 2.2 Examples of *E. coli* promoters

```
              -35                                     -10         1
TGGCGGTGTTGACATAAATA  CCACTGGCGGTGATACTGAGCA  CA    Lambda P_L
CGTGCGTGTTGACTATTTTA  CCTCTGGCGGTGATAATGGTTG  CA    Lambda P_R
TGCCGAAGTTGAGTATTTTT  GCTGTATTTGTCATAATGACTCCTG    Lambda P_O
ATGAGCTGTTGACAATTAAT  CATCGAACTAGTTAACTAGTACGCA    trp
CATCGAATGGCGCAAAACCTTTCGCGGTATGGCATGATAGCGCCCG    lacI
CCCCAGGCTTTACACTTTATGCTTCCGGCTCGTATGTTGTGTGG  A    lacZ
CGTAACACTTTACAGCGGCG    CGTCATTTGATATGATGCGCC  CG    tyr tRNA
        TTGACA                        TATAAT              consensus
```

Bases matching the –10 and –35 consensus sequences are boxed. Spaces are inserted to optimize the alignment. Note that the consensus is derived from a much larger collection of characterized promoters.
Position 1 is the transcription start site.

complex scenario. The need for this added complexity can easily be imagined; if cells carrying the same genome are differentiated into a multitude of cell types fulfilling very different functions, a very sophisticated control system is needed to provide each cell type with its specific repertoire of genes, and to fine-tune the degree of expression for each one of them. Nonetheless, the promoter region, however simple or complex, gives rise to different levels of transcription of various genes.

Eukaryotes have three different RNA polymerases. Only one of these, RNA polymerase II, is involved in the transcription of protein-coding transcripts and small RNAs, while RNA polymerase I is responsible for the synthesis of large ribosomal RNAs, and RNA polymerase III makes small RNAs such as tRNA and 5S ribosomal RNA.

In eukaryotes, the primary transcript, produced by RNA polymerase II, *heterogeneous* or *heteronuclear RNA* (hnRNA), is very short-lived as such, as it is processed in a number of steps called *maturation*. A specialized nucleotide *cap* is added to the 5′ end; this is the site recognized by the ribosomes in protein synthesis (see below). The precursor mRNA is cleaved at a specific site towards the 3′ end and a *poly-A tail*, consisting of a long sequence of adenosine residues, is added to the cut end. This is a specific process, governed by polyadenylation recognition sequences in the 3′ untranslated region. Nature's 'tagging' of mRNA molecules comes in very useful in the laboratory for the isolation of eukaryotic mRNA (see Chapter 6). This transcript contains intervening sequences (*introns*) between the *exons* which carry the coding information (see below). The final step is the process of *splicing*, by which the introns are removed and the exons are joined together. This process is quite complex; some introns are removed by specific proteins known as splicing factors, whereas other introns are removed independently through autocatalysis. To complicate things further, in some cases the transcript is edited, leading for example to the introduction of a tissue-specific earlier stop codon that is not encoded in the corresponding DNA.

A recently discovered mechanism is known as *RNA interference* or *RNAi*. RNAi consists of the degradation of mRNA initiated by double-stranded RNA fragments. This mechanism has probably developed to protect the cell from infection by RNA viruses, which replicate via a double-stranded RNA intermediate. It may also be exploited as a way of silencing gene expression in the laboratory that demands far less resources and is much more amenable to high-throughput screening than other methods (see Chapters 12 and 15).

In bacteria, the processes of transcription and translation take place in the same compartment and simultaneously. In other words, the ribosomes translating the mRNA follow closely behind the RNA polymerase, and polypeptide production is well under way long before the mRNA is complete. In eukaryotes, by contrast, the mature mRNA molecule is transported out of the nucleus to the cytoplasm where translation takes place.

The resulting level of protein production is dependent on the amount of the specific mRNA available, rather than just the rate of production. The level of an mRNA species will be affected by its rate of degradation as well as by its rate of synthesis. In bacteria, most mRNA molecules are degraded quite quickly (with a half-life of only a few minutes), although some are much more stable. The instability of the majority of bacterial mRNA molecules means that bacteria can rapidly alter their profile of gene expression by changing the transcription of specific genes. By contrast, the lifespans of most eukaryotic mRNA molecules are measured in hours or days rather than minutes. Again, this is a reflection of the fact that an organism (or a cell) that is able to control its own environment to a substantial extent is subjected to less radical environmental changes. Consequently, mRNA molecules tend to be more stable in multicellular organisms than in, for example, yeast. Nonetheless, the principle remains; the level of an mRNA is a function of its production and degradation rates. We will discuss how to study and disentangle these parameters in Chapter 10.

2.3.2 Translation

In bacteria, translation starts when ribosomes bind to a specific site (the *ribosome binding site, RBS*), which is adjacent to the start codon. The sequence of the ribosome binding site (also known as the *Shine–Dalgarno sequence*) has been recognized as being complementary to the 3′ end of the 16S rRNA (Figure 2.9). The precise sequence of this site, and its distance from the start codon, does affect the efficiency of translation, although in nature this is less important than transcriptional efficiency in determining the level of gene expression. Translation efficiency will also depend on the codon usage, i.e. the match between synonymous codons and the availability of tRNA that will recognize each codon. This concept is explored more fully in Chapter 11.

In bacterial systems, where transcription and translation occur in the same compartment of the cell, ribosomes will bind to the mRNA, a process known

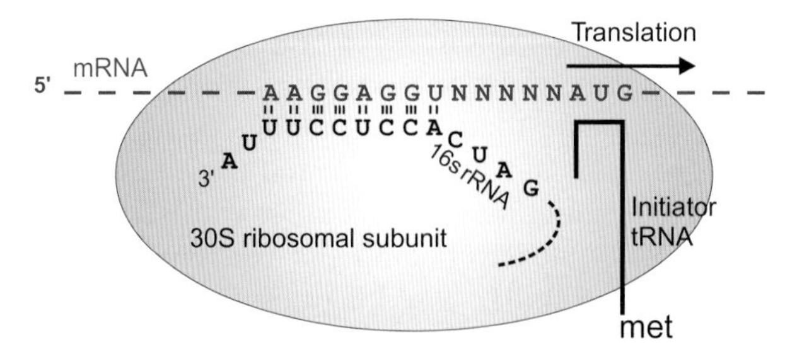

Figure 2.9 Bacterial ribosome binding site

as *initiation*, as soon as the RBS has been synthesized. Thus there will be a procession of ribosomes following close behind the RNA polymerase, translating the mRNA in the process or *elongation* as and when it is being produced. So, although the mRNA may be very short-lived, the bacteria are capable of producing substantial amounts of the corresponding polypeptide in a short time. Translation stops when a termination codon is reached.

In eukaryotes, the mechanism (as usual) is much more complicated. Instead of binding just upstream of the initiation codon, the ribosome binds at the 5′ end of the mRNA to the cap, and reads along the 5′ untranslated region (UTR) until it reaches an initiation codon. The sequence AUG may be encountered on the way without initiation, because the surrounding sequence is also important to define the start of protein synthesis. The fact that the 5′ UTR is scanned in its full length by the ribosome makes it an important region for specifying translation efficiency, and different secondary structures can have either a positive or a negative effect on the amount of protein that is produced. There are some exceptions to this, in that sometimes translation may be initiated at an internal ribosome entry site (IRES); the best studied of these occur in some viruses, but they also occur in some transcripts encoded naturally by the cell.

2.4 Gene structure and organization

2.4.1 Operons

In bacteria, it is quite common for a group of genes to be transcribed from a single promoter into one long RNA molecule; this group of genes is known as an *operon* (Figure 2.10). If we are considering protein-coding genes, the transcription product, messenger RNA (mRNA), is then translated into a

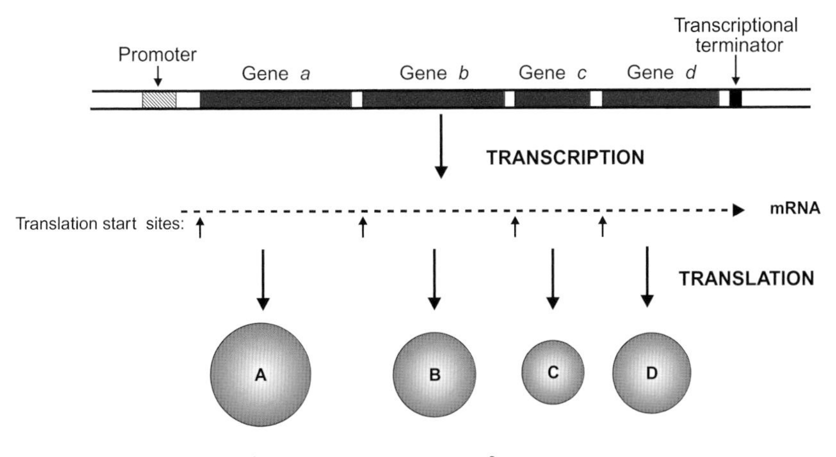

Figure 2.10 Structure of an operon

number of separate polypeptides. This can occur by the ribosomes reaching the stop codon at the end of one polypeptide-coding sequence, terminating translation and releasing the product before re-initiating (without dissociation from the mRNA). Alternatively, the ribosomes may attach independently to internal ribosome binding sites within the mRNA sequence. Generally, the genes involved are responsible for different steps in the same pathway, and this arrangement facilitates the co-ordinate regulation of those genes, i.e. expression of each gene in the operon goes up (or down) together in response to changing conditions.

In eukaryotes, in contrast, the way in which ribosomes initiate translation is different, which means that they cannot usually produce separate proteins from a single mRNA in this way. There are occasions when a single mRNA can give rise to different proteins, but these work in different ways, such as alternative processing of the mRNA (see below) or by producing one long polyprotein or precursor which is then cleaved into different proteins (as occurs in some viruses).

2.4.2 Exons and introns

In bacteria there is generally a simple one-for-one relationship between the coding sequence of the DNA, the mRNA and the protein. This is usually not true for eukaryotic cells, where the initial transcript is many times longer than that needed for translation into the final protein. It contains blocks of sequence (*introns*) which are removed by processing to generate the final mRNA for translation (Figure 2.11). One important consequence of the presence

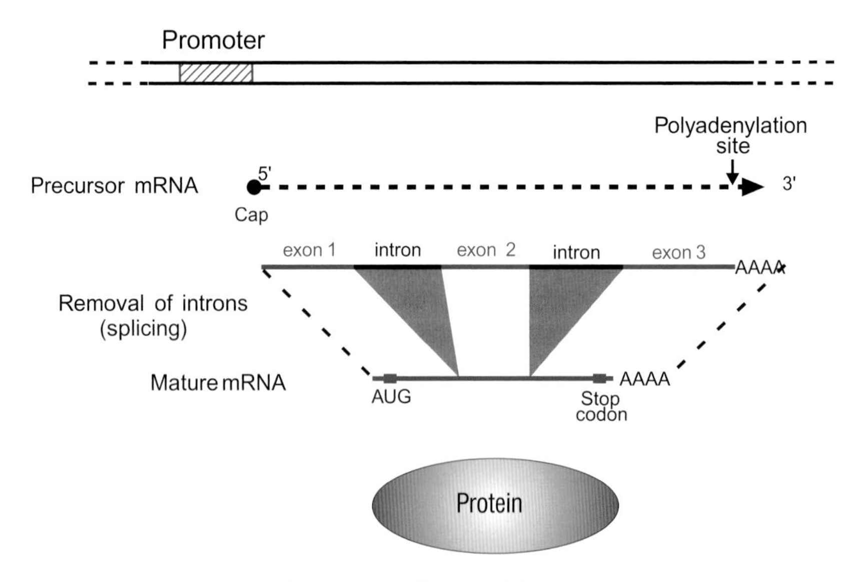

Figure 2.11 Exons and introns

of introns is that several different variants of a protein can be produced from the same coding region of the genome, by differential processing of the mRNA.

Introns do occur in bacteria, but quite infrequently. This is partly due to the need for economy in a bacterial cell; the smaller genome and generally more rapid growth provides an evolutionary pressure to remove unnecessary material from the genome. A further factor arises from the nature of transcription and translation in a bacterial cell. As the ribosomes are translating the mRNA while it is being made, there is less opportunity for sections of the RNA to be removed before translation.

3 How to Clone a Gene

3.1 What is cloning?

Cloning means using asexual reproduction to obtain organisms that are genetically identical to one another, and to the 'parent'. This contrasts with sexual reproduction, where the offspring are not usually identical. It is worth stressing that clones are only identical *genetically*; the actual appearance and behaviour of the clones will be influenced by other factors such as their environment. This applies to all organisms, from bacteria to humans.

Despite the emotive language that increasingly surrounds the use of the word 'cloning', this is a concept that will be surprisingly familiar to many people. In particular, anyone with an interest in gardening will know that it is possible to propagate plants by taking cuttings, and that in this way you will produce a number of plants that are identical to the original, and to each other. These are clones. Similarly, the routine bacteriological procedure of purifying a bacterial strain by picking a single colony for inoculating a series of fresh cultures is also a form of cloning.

The term *cloning* is also applied to genes, by extension of the concept. If you introduce a foreign gene into a bacterium, or any other type of cell, in such a way that it will be copied when the cell replicates, then you will produce a large number of cells all with identical copies of that piece of DNA – you have *cloned* the gene (Figure 3.1). By producing a large number of copies in this way, you can sequence it or label it as a probe to study its expression in the organism it came from. You can express its protein product in bacterial or eukaryotic cells. You can mutate it and study what difference that mutation makes to the properties of the gene, its protein product or the cell that carries it. You can even purify the gene from the bacterial clone and inject it into a mouse egg, and produce a line of transgenic mice that express

From Genes to Genomes, Second Edition, Jeremy W. Dale and Malcolm von Schantz

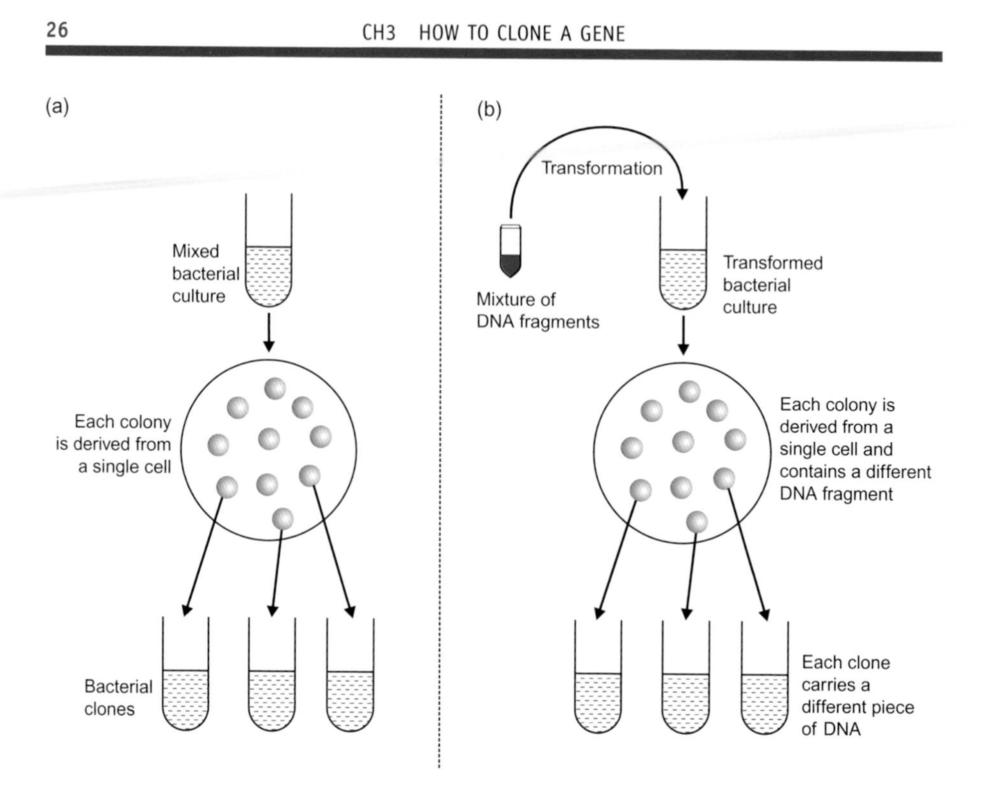

Figure 3.1 (a) Bacterial cloning and (b) gene cloning

it. Behind all these applications lies a cloning process with the same basic steps.

In subsequent chapters, we will consider how this process is achieved, initially with bacterial cells (mainly *E. coli*) as the host and later extending the discussion to alternative host cells. The purpose of this chapter is to present an overview of the process, with further details of the various steps being considered in subsequent chapters.

3.2 Overview of the procedures

Some bacterial species will naturally take up DNA by a process known as *transformation*. However, most bacteria have to be subjected to chemical or physical treatments before DNA will enter the cells. In all cases, the DNA will not be replicated by the host cell unless it either recombines with (i.e. is inserted into) the host chromosome, or alternatively is incorporated into a molecule that is recognized by enzymes within the host cell as a substrate for replication. For most purposes the latter strategy is the relevant one. We use *vectors* to carry the DNA and allow it to be replicated. There are many types of vectors for use with bacteria. Some of these vectors are *plasmids*, which are naturally occurring pieces of DNA that are replicated independently of the chromosome, and are inherited by the two daughter cells when the cell divides.

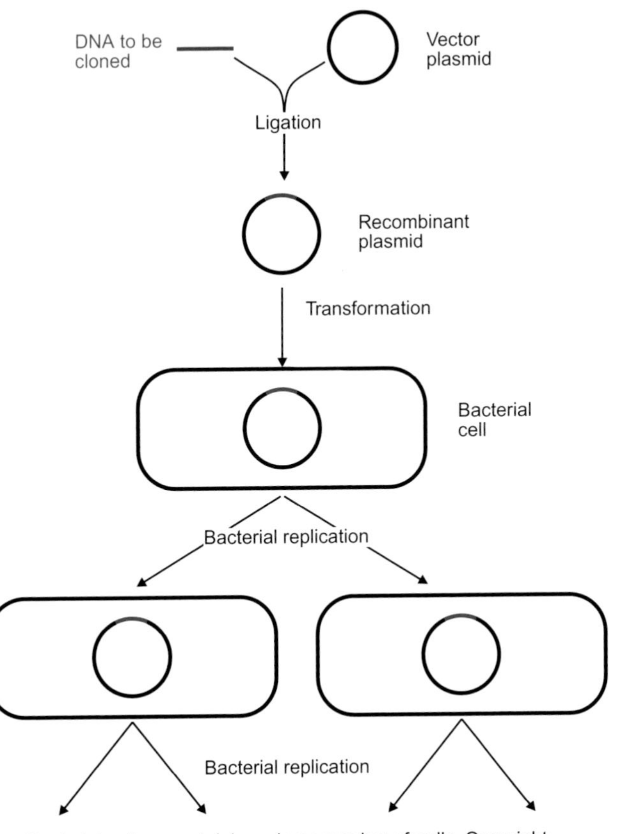

Figure 3.2 Basic outline of gene cloning

(In Chapter 5 we will encounter other types of vectors, including viruses that infect bacteria; these are known as *bacteriophages,* or phages for short.)

The DNA that we want to clone is inserted into a suitable vector, producing a *recombinant molecule* consisting of vector plus insert (Figure 3.2). This recombinant molecule will be replicated by the bacterial cell, so that all the cells descended from that initial transformant will contain a copy of this piece of recombinant DNA. A bacterium like *E. coli* can replicate very rapidly under laboratory conditions, doubling its population size every 20 min or so. This *exponential growth* gives rise to very large numbers of cells; after 30 generations (10 h), there will, in theory, be 1×10^9 (one billion) descendants of the initial transformant. Each one of these cells carries a copy of the recombinant DNA molecule, so we will have produced a very large number of copies of the cloned DNA.

Of course, exponential growth does not continue indefinitely; after a while, the bacteria start to run out of nutrients, and stop multiplying. (The reasons for growth stopping are actually rather more complex than that, but depletion of nutrients, including limitations on the diffusion of oxygen, is the main factor.) With *E. coli*, this commonly occurs with about 1×10^9 bacteria per millilitre of culture. However, if we take a small sample and add it to fresh medium, exponential growth will resume. The clone can thus be propagated, and in this way we can effectively produce unlimited quantities of the cloned DNA. So long as we can get the bacteria to express the cloned gene, we can also get very large amounts of the product of that gene.

In order to carry out this procedure, we require a method for joining pieces of DNA to such a vector, as well as a way of cutting the vector to provide an opportunity for this joining to take place. The key to the development of gene cloning technology was the discovery of enzymes that would carry out these reactions in a very precise way. The main enzymes needed are *restriction endonucleases*, which break the sugar–phosphate backbone of DNA molecules at precise sites, and *DNA ligases*, which are able to join together the fragments of DNA that are generated in this way (Figure 3.3). These enzymes, and the ways in which they are used, are described in more detail in Chapter 4.

Once a piece of DNA has been inserted into a plasmid (forming a *recombinant* plasmid), it then has to be introduced into the bacterial host by a

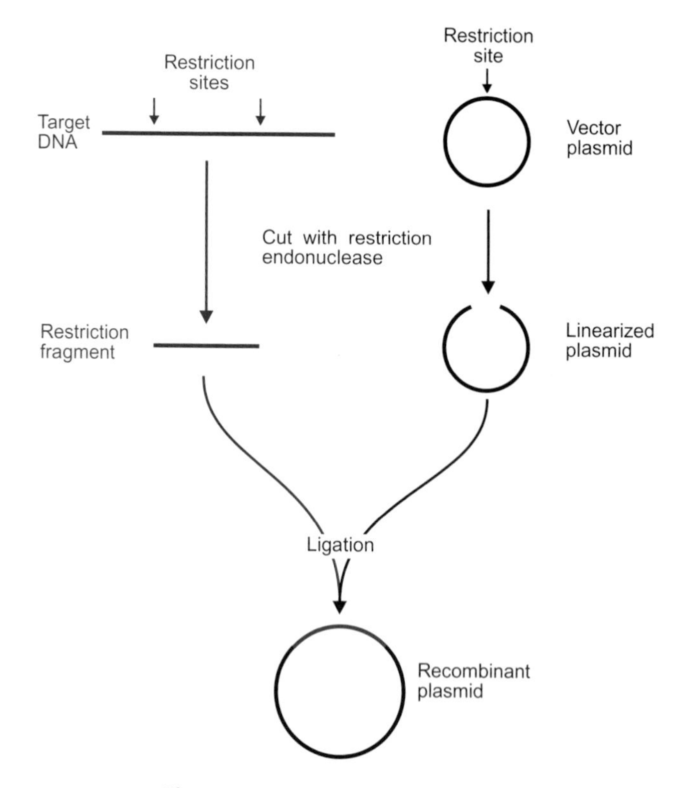

Figure 3.3 Cutting and joining DNA

transformation process. Generally, this process is not very efficient, so only a small proportion of bacterial cells actually take up the plasmid. However, by using a plasmid vector that carries a gene coding for resistance to a specific antibiotic, we can simply plate out the transformed bacterial culture onto agar plates containing that antibiotic, and only the cells which have received the plasmid will be able to grow and form colonies.

This description does not consider how we get hold of a piece of DNA carrying the specific gene that we want to clone. Even a small and relatively simple organism like a bacterium contains thousands of genes, and they are not arranged as discrete packets, but are regions of a continuous DNA molecule. We have to break this molecule into smaller fragments, which we can do specifically (using restriction endonucleases) or non-specifically (by mechanical shearing). However we do it, we will obtain a very large number of different fragments of DNA with no easy way of reliably purifying an individual fragment, let alone isolating the specific fragment that carries the required gene. The only way of separating the fragments is by size, but there are will be a lot of different pieces of DNA that are so similar in size that they cannot be separated.

3.3 Gene libraries

Fortunately, it is not necessary to purify specific DNA fragments. One of the strengths of gene cloning is that it provides another, much more powerful, way of finding a specific piece of DNA. Rather than attempting to separate the DNA fragments, we take the complete mixture and use DNA ligase to insert the fragments into the prepared vector. Under the right conditions, only one fragment will be inserted into each vector molecule. In this way, we produce a mixture of a large number of different recombinant vector molecules, which is known as a *gene library* (or more specifically, in this case, a *genomic library*, to contrast it with other forms of gene library that will be described in Chapter 6). On transforming a bacterial culture with this library, each cell will only take up one molecule. When we then plate the transformed culture, each colony, which arises from a single transformed cell, will contain a large number of bacteria all of which carry the same recombinant plasmid, with a copy of the same piece of DNA from our starting mixture. So instead of a mixture of thousands (or millions, or tens of millions) of different DNA fragments, we have a large number of bacterial colonies, each of which carries one fragment only (Figure 3.4). The production and screening of gene libraries is considered in Chapters 6 and 7, where we will see that a variety of different vectors, other than simple plasmids, are used for constructing genomic libraries.

We still have a very complex mixture, but whereas purifying an individual DNA fragment is extremely difficult, it is simple to isolate individual bacterial

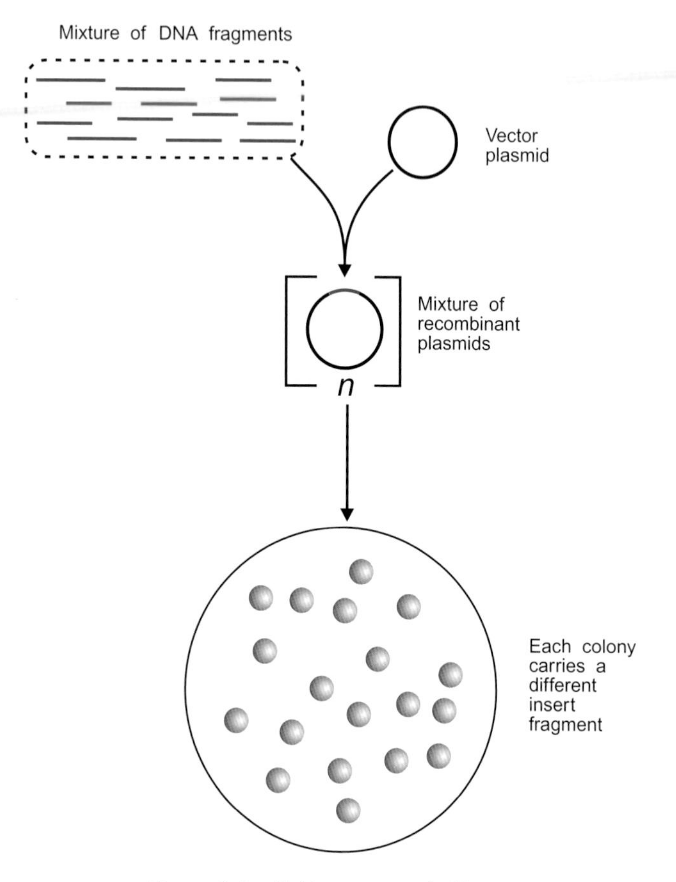

Figure 3.4 Making a genomic library

colonies from this mixture – we just pick them from a plate. In addition, each individual bacterial colony will carry a different piece of DNA from our original complex mixture. Therefore, if we can identify which bacterial colony carries the gene that we are interested in, purifying it becomes a simple matter. We just have to pick the right colony and inoculate it into fresh medium. However, we still have the problem of knowing which of these thousands/millions of bacterial colonies actually carry the gene that we want. This is considered more fully in Chapter 7, but one commonly used and very powerful method can be introduced here as an example. This depends on the phenomenon of *hybridization*.

3.4 Hybridization

If a double-stranded DNA fragment is heated, the non-covalent bonds holding the two strands together will be disrupted, and the two strands will separate. This is known as *denaturation*, or less formally (and less accurately)

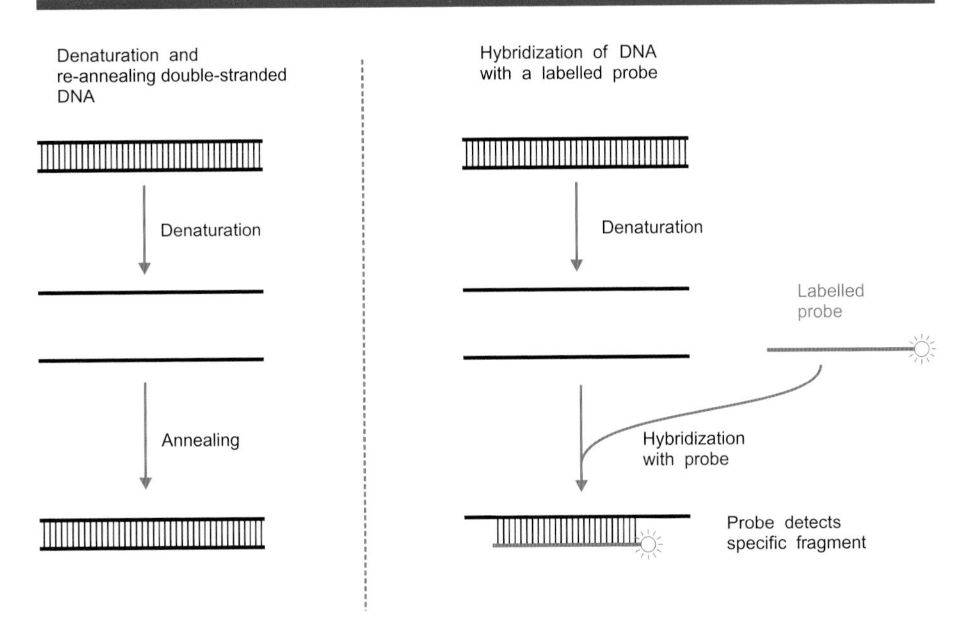

Figure 3.5 Hybridization and gene probes

as '*melting*'. When the solution is allowed to cool again, these bonds will reform and the original double-stranded fragment will be reformed (the two strands are said to *anneal*).

We can use this phenomenon to identify a particular piece of DNA in a complex mixture by labelling a specific DNA sequence (the *probe*), and mixing the labelled probe with the denatured mixture of fragments. When the mixture is cooled down, the probe will tend to hybridize to any related DNA fragments (Figure 3.5), which enables us to identify the specific DNA fragments that we want.

When screening a gene library, the labelled probe will hybridize to DNA from any colony that carries the corresponding gene or part of it; we can then recover that colony and grow up a culture from it, thus producing an unlimited amount of our cloned gene.

Of course it is not quite as simple as that. We cannot hybridize the probe to the colonies on an agar plate. However, it is easy to transfer a part of each colony onto a membrane by replication, and then lyse the colony so that the DNA it contains is fixed to the membrane. This produces a pattern of DNA spots on the membrane in positions corresponding to the colonies on the original plate (Figure 3.6), which can then be hybridized to the labelled probe to enable identification, and recovery, of the required colony. Hybridization, using labelled DNA or RNA probes, is an important part of many other techniques that we will encounter in subsequent chapters.

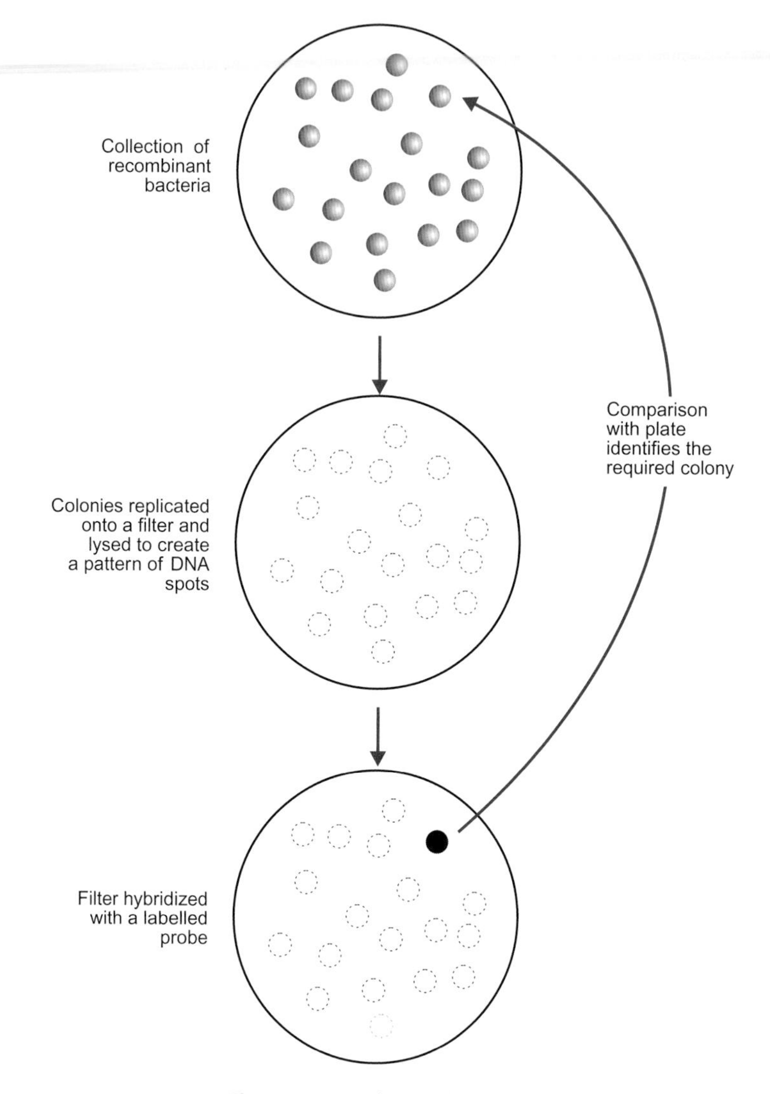

Figure 3.6 Colony hybridization

3.5 Polymerase chain reaction

The technique known as the polymerase chain reaction (PCR), usually combined with the information available in sequence databanks, provides an alternative to the above approaches for obtaining a cloned gene (in addition to many other applications). PCR requires the use a pair of primers that will anneal to sites at either side of the required region of DNA (Figure 3.7). DNA polymerase action will then synthesize new DNA strands starting from each primer. Denaturation of the products, and re-annealing of the primers,

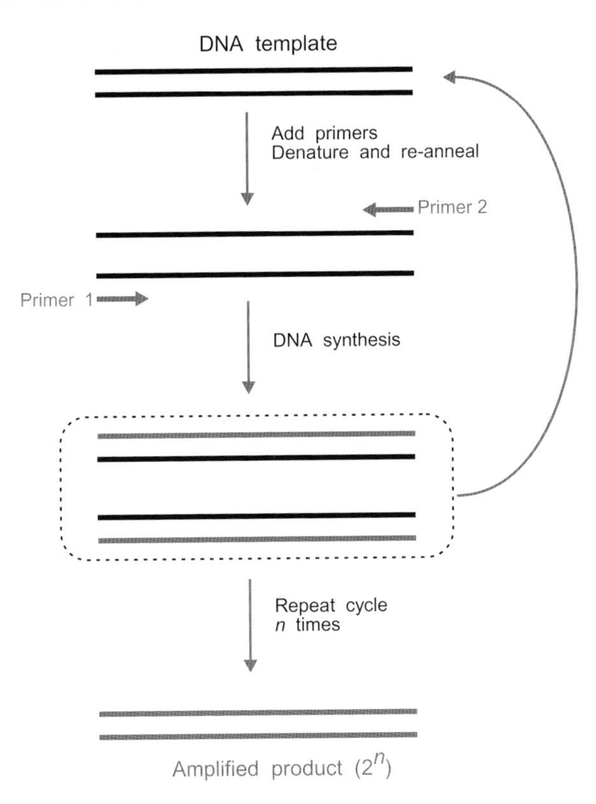

Figure 3.7 Polymerase chain reaction

will allow a second round of synthesis. Repeated cycles of denaturation, annealing and extension will give rise to an exponential amplification of the DNA sequence between the two primers, with the amount of product doubling in each cycle, so that after, say, 20 cycles there will (theoretically) be a million-fold increase in the amount of product. This enables the amplification of a specific region of the DNA, and the product can then be cloned directly. The polymerase chain reaction, and some of the many applications, is described more fully in Chapter 8.

3.6 Extraction and purification of nucleic acids

The first step for most of the procedures referred to in this book is to extract the DNA (or for some purposes RNA) from the cell and to purify it by separating it from other cellular components. Although recent technological advances have made this much less of a challenge than it once was, the quality of the starting material remains a crucially important factor for most purposes. For some applications, such as PCR (see Chapter 8) or hybridization analyses (see Chapter 7), less pure material may be acceptable. In this chapter, we review the concepts underlying the most commonly used methods of purifying and fractionating nucleic acids; for further experimental details, you will need to consult a laboratory manual (see the Bibliography).

3.6.1 Breaking up cells and tissues

The starting material could be a culture of bacterial or eukaryotic cells, which would simply need to be separated from the growth medium (for example by centrifugation), or a more complex tissue sample, which first needs to be homogenized so that the individual cells can be lysed. Wherever possible, the material should be freshly harvested or frozen until ready to use, to avoid degradation by enzymes present in the cell extract.

The cells then need to be lysed to release their components. Bacterial cell walls can often be broken using *lysozyme*, usually in conjunction with EDTA and a detergent such as SDS (sodium dodecyl sulfate). EDTA eliminates divalent cations and thus destabilizes the outer membrane in bacteria such as *E. coli*, and also inhibits DNases that would otherwise tend to degrade the DNA, while the detergent will solubilize the membrane lipids.

Plant and fungal cells have cell walls that are different from those in bacteria, and require alternative treatments, either mechanical or enzymatic, while animal cells (which lack a cell wall) can usually be lysed by more gentle treatment with a mild detergent.

The crude extract contains a complex solution of DNA, RNA, proteins, lipids and carbohydrates. Note that the sudden lysis of the cell will usually result in some fragmentation of chromosomal DNA. Where it is necessary to obtain very large (even intact) chromosomal DNA, more gentle lysis conditions are necessary (see the description of pulsed field gel electrophoresis in Chapter 13). Bacterial plasmids, however, are readily obtained in their native, circular, state by standard lysis conditions.

The next step in the procedure is to separate the desired nucleic acid from these other components. Removal of RNA from a DNA preparation is easily achieved by treatment with ribonuclease (RNase). Since RNase is a very heat-stable enzyme, it is easy to ensure that it is free of traces of deoxyribonuclease (DNase) that would otherwise degrade your DNA, simply by heating the enzyme before use. Removal of DNA from RNA preparations requires the use of RNase-free DNase.

Removal of proteins is particularly important as the cell contains a number of enzymes that will degrade nucleic acids, as well as other proteins that will interfere with subsequent procedures by binding to the nucleic acids. Although much of the protein contamination can be removed by affinity chromatography or digestion with a proteolytic enzyme such as proteinase K, the most effective way of removing proteins is by extraction with a mixture of liquefied phenol and chloroform. When the mixture is vigorously agitated, the proteins will be denatured and precipitated at the interphase, and the

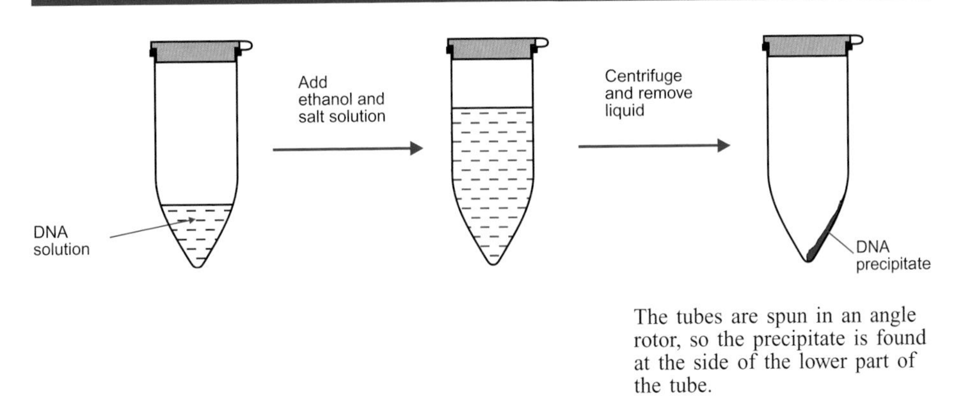

Figure 3.8 Ethanol precipitation

nucleic acids can be recovered from the aqueous layer. If you carry out the extraction with untreated (and therefore acidic) phenol, DNA will partition into the organic phase, allowing you to recover pure RNA from the aqueous phase. However, if (as is more often the case) you wish to recover DNA from the extraction, it is essential that the phenol is first equilibrated with a neutral, or even alkaline, buffer so that the DNA will partition into the aqueous phase.

Note: phenol is highly toxic by skin absorption, and gloves *must* be worn.

Following phenol extraction, you will have a protein-free sample of your nucleic acid(s). However, it will probably be more dilute than you want it to be, and furthermore it will contain traces of phenol and chloroform. The answer is normally to concentrate (and further purify) the solution by precipitating the nucleic acid. This is done by adding either isopropanol or (more frequently) ethanol. This process requires the presence of monovalent cations (Na^+, K^+ or NH_4^+) for the formation of a nucleic acid precipitate, which can be collected at the bottom of the test tube by centrifugation (Figure 3.8). Some of the salt will precipitate as well, but can be removed by washing with 70 per cent ethanol.

3.6.2 Alkaline denaturation

This is a widely used procedure for separating plasmids from chromosomal DNA in bacterial cell extracts. Chromosomal DNA is broken into linear fragments during cell lysis. Raising the pH disrupts the hydrogen bonds and allows the linear strands to separate. Plasmids are much less prone to breakage and are not disrupted by cell lysis; they remain as intact supercoiled circular DNA. Although the high pH will disrupt the hydrogen bonds, the two circular strands will not be able to separate physically, and will remain interlinked. When the pH is reduced, the interlinked plasmid strands will snap back to reform the double-stranded plasmid (Figure 3.9). On the other hand, the

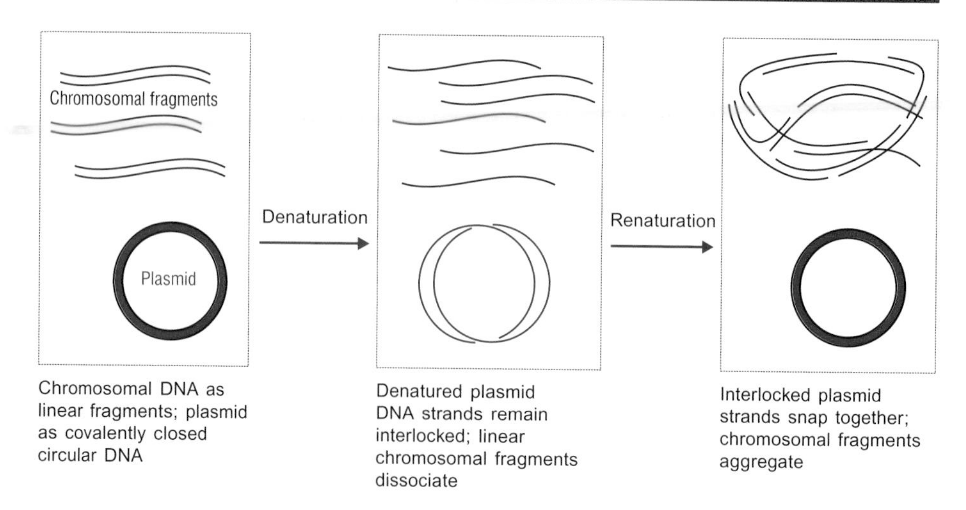

Figure 3.9 Alkaline denaturation procedure for plasmid purification

separated linear chromosomal fragments will aggregate into an insoluble network that can be removed by centrifugation, leaving the plasmids in solution. Other cell components, including cell wall debris and many proteins, are also removed by this procedure, avoiding the need for phenol extraction.

3.6.3 Column purification

Two types of column purification are frequently used when purifying nucleic acids. In *size-selection chromatography*, a sample is passed through a matrix of small porous beads. Smaller molecules, such as salts and unincorporated nucleotides, will enter the beads, whereas larger ones such as longer nucleic acid chains will pass right through the column. This type of purification is a fast and simple alternative to purification by alcohol precipitation.

In *affinity chromatography* purification, the macromolecules in your sample bind to the resin in the column. This could be an anionic resin, which binds to the negatively charged phosphate groups in the nucleic acid backbone, or more sophisticated ones such as resins coated with oligo-dT sequences, which specifically bind to the poly-A tails of eukaryotic mRNA molecules (see Chapter 6). In both cases, undesirable molecules can be washed from the column, after which the stringency conditions are changed and the bound nucleic acids eluted.

3.7 Detection and quantitation of nucleic acids

If your DNA preparation is reasonably pure [i.e. free of other materials that absorb ultraviolet radiation (UV), including RNA, free nucleotides, and proteins], then you can estimate the DNA concentration by measuring the absorbance of the solution in a UV spectrophotometer at 260 nm. This is convenient, but not very sensitive: a solution of 50 μg/ml of double-stranded DNA will have an absorbance of 1. However, equipment is now available for determining the UV absorbance of very small samples of nucleic acids. Note that the presence of proteins or phenol will affect this estimate, and that UV absorbance gives you no check on the integrity of your DNA; it can be completely degraded and still give you a reading.

Dyes such as ethidium bromide are commonly used for both detecting and quantitating nucleic acids. Ethidium bromide has a flat ring structure which is able to stack in between the bases in nucleic acids; this is known as *intercalation*. The dye can then be detected by its fluorescence (in the red-orange region of the spectrum) when exposed to UV. This is the most widely used method for staining electrophoresis gels, and can also be used for estimating the amount of DNA (or RNA) in each band on the gel, by comparing the intensity of the fluorescence with a sample of known concentration on the same gel.

Note that ethidium bromide is mutagenic, and precautions must be taken to eliminate health risks. Alternative, less hazardous, dyes are increasingly being used for this purpose.

3.8 Gel electrophoresis

Gel electrophoresis is a crucial technique for both the analysis and the purification of nucleic acids. When a charged molecule is placed in an electric field, it will migrate towards the electrode with the opposite charge; DNA and RNA, being negatively charged, will move towards the positive pole (anode). In a gel, which consists of a complex network of pores, the rate at which a nucleic acid molecule moves will be determined by its ability to penetrate through this network. For linear fragments of double-stranded DNA within a certain size range, this will reflect the size of the molecule (i.e. the length of the DNA). We do not have to consider the amount of charge that the molecule carries (unlike some other applications of electrophoresis), since all nucleic acids carry the same amount of charge per unit size.

The effective size range of a gel is determined by its composition. We can use agarose gels for separating nucleic acid molecules greater than a few hundred base pairs, reducing the agarose concentration to obtain effective separation of larger fragments, or increasing it for small fragments. For even smaller molecules, down to only a few tens of base pairs, we would use polyacrylamide gels. These are capable of distinguishing DNA chains with only a single base difference in their length, which is important for sequence analysis (see Chapter 9).

3.8.1 Analytical gel electrophoresis

We can therefore use agarose gel electrophoresis for analysing the composition and quality of a nucleic acid sample. In particular, it is invaluable for determining the size of DNA fragments from a restriction digest or the products of a PCR reaction (see Chapter 8). For this purpose it is necessary to calibrate the gel by running a standard marker containing fragments of known sizes; in Figure 3.10, the standard marker is provided by a *Hin*dIII digest of DNA from the lambda bacteriophage. It can be seen that, over much of the size range, there is a linear relationship between the logarithm of the fragment size and the distance it has moved. From this calibration graph, you can then estimate the size of your unknown fragment(s) – assuming they are linear double-stranded DNA (see below).

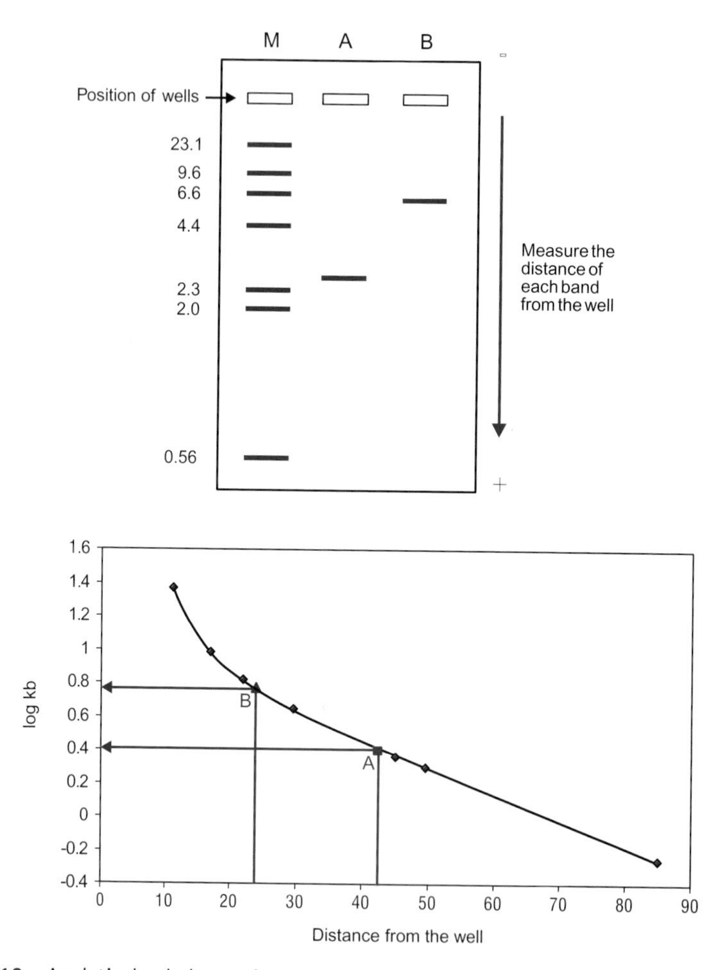

Figure 3.10 Analytical gel electrophoresis. Using the standard marker to provide a calibration curve, the size of fragment A is estimated as 2.5 kb and fragment B as 6.0 kb

The reason for the nonlinearity of the curve with larger DNA molecules is that these molecules move through the gel in a different way. Although these molecules are large, they are also very thin, and they can in effect slither through the gel end-on. It takes some time for them to become lined up, but once they are, then the rate at which they move is largely independent of their size. So for a particular gel, all molecules above a certain size will have virtually the same mobility. In the gel shown, all DNA molecules larger than about 20 kb will not be separated, but the use of gels with a lower agarose concentration will extend the size range. Special techniques, involving frequent switching of the direction of the electric field, are available for separating very large DNA molecules – see the description of pulsed-field gel electrophoresis (PFGE) in Chapter 13. If a more precise confirmation of the nature of the sample is required, the gel is *blotted* onto a membrane support, and *hybridized* with a nucleic acid probe (see the section on Southern blotting in Chapter 7).

Polyacrylamide gel electrophoresis offers a much sharper size-separation of nucleic acid molecules, down to the separation of fragments that only differ in size by one single base. This is used in methods such as *primer extension* analysis and *gel retardation* (or *band shift*) assays (see Chapter 10), and has formed the basis for the development of DNA sequencing methods (Chapter 9).

Not all DNA molecules are linear. Native plasmids are supercoiled circular molecules, but if a plasmid is nicked (i.e. one of the strands is broken) the loose ends are free to rotate, and it adopts a relaxed, open circular form. A double strand break will produce a linearized plasmid (see Figure 3.11). It is therefore quite normal for a purified plasmid preparation to show two or three bands in a gel; it does not necessarily mean that there is more than one

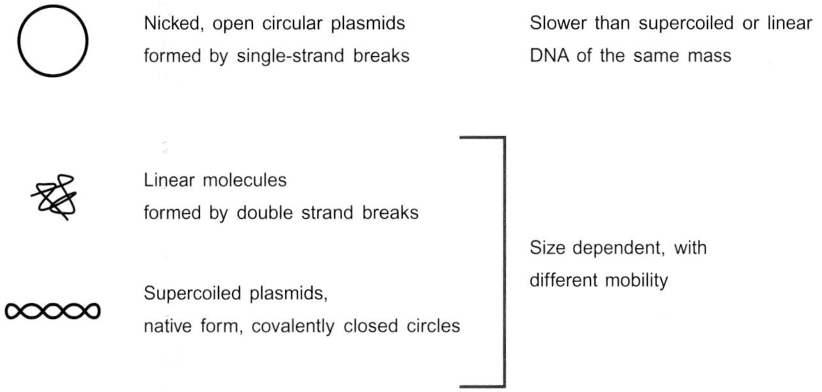

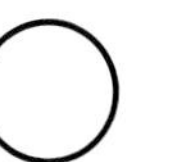

Figure 3.11 Electrophoretic mobility of forms of plasmid DNA

plasmid. (A further complication is that there may be dimeric or multimeric forms of the plasmid present.) Relaxed and supercoiled plasmids show different relative mobilities, and both are different from linear molecules. Unless plasmids are linearized before electrophoresis by digestion with restriction enzymes, a special supercoiled molecular weight marker has to be used.

Most RNA molecules you will encounter are single-stranded, and in some cases (such as in DNA sequencing) we will come across single-stranded DNA as well. These need special consideration. As described in Chapter 2, single-stranded nucleic acids tend to fold up into complex secondary structures, to remove the hydrophobic bases from the aqueous environment. The migration of the molecule will be greatly influenced by the way it folds. If we want to get a true picture of its size, we have to make sure that it remains in an unfolded state. To do this, we use *denaturing gels*, in which denaturing agents such as urea or formaldehyde are included.

3.8.2 Preparative gel electrophoresis

Gel electrophoresis is also an important tool in the purification of a specific nucleic acid fragment from a complex mixture. In this case, using a low-melting-point preparation of agarose makes it easier to recover the DNA fragment. After separating the sample, it is visualized with ethidium bromide. With the gel still on the transilluminator, and using suitable equipment to protect yourself from the UV radiation, the band(s) that need to be purified are excised using a razor blade or scalpel. The DNA can then be recovered and purified from the gel fragment, using standard DNA purification procedures.

In this chapter, we have provided a brief overview of the principal methods used in gene cloning. These procedures, and some of the main alternative strategies, are described more fully in subsequent chapters.

4 Cutting and Joining DNA

4.1 Restriction endonucleases

Restriction endonucleases derive their name from the phenomenon of host-controlled restriction and modification. This can occur if a bacteriophage preparation that has been grown using one bacterial host strain is used to infect a different strain. It may be found that infection is extremely inefficient (phage growth is *restricted* by the new host) compared with infection using the same strain for both propagation and assay. The reason for this restriction of phage growth is that many bacterial strains produce an endonuclease, which is therefore known as a *restriction endonuclease*, that cuts DNA into pieces, so that the incoming phage DNA is rapidly broken down and only occasionally escapes to produce phage progeny (Figure 4.1). The host DNA is protected against the action of the endonuclease by a second enzyme, which modifies DNA by methylation, so that it is not attacked by the endonuclease. Although this phenomenon was initially demonstrated with bacteriophage infection, it can also occur whenever DNA is transferred between one bacterial strain and another. If the receiving strain possesses a restriction/modification system that is not present in the originating host, then the yields of transformants or recombinants will be very low.

The restriction enzymes that were initially studied in this way (type I restriction endonucleases) recognize specific sequences in the DNA but do not cut there. Instead they track along the DNA for a variable distance, sometimes as far as 5 kb, before breaking the DNA strand. This is not very useful for our purposes, since it does not generate specific fragments. The enzymes that are commonly used for gene cloning (type II restriction endonucleases) typically recognize and cut within (or immediately adjacent to) specific target sequences and therefore generate specific fragments.

From Genes to Genomes, Second Edition, Jeremy W. Dale and Malcolm von Schantz

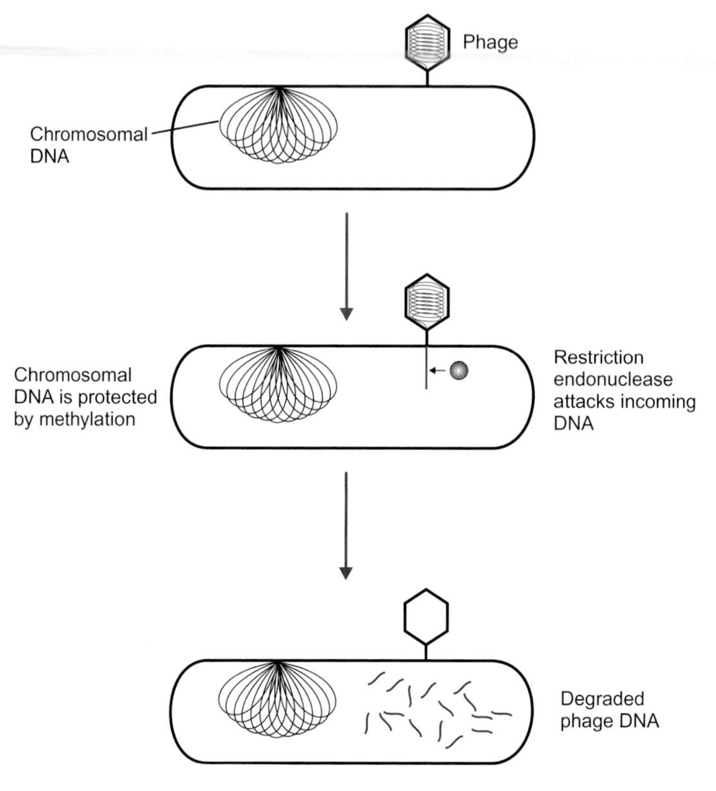

Figure 4.1 Bacteriophage restriction

4.1.1 Specificity

A very large number of type II restriction endonucleases have been characterized and used in DNA manipulations. These are identified by the name of the organism from which they are obtained, using the first letter of the genus and the first two letters of the species name, together with a suffix indicating the specific enzyme from that species. Thus *Pst*I indicates a specific enzyme obtained from the bacterium *Providencia stuartii*, and *Hae*I, *Hae*II and *Hae*III indicate three different enzymes, with different specificities, from *Haemophilus aegyptius*. The convention is to write the first part (derived from the name of the source organism) in italics, just as you would the species name.

Some examples are shown in Box 4.1, together with their recognition sequences and the position of the site at which they break the DNA. One of the key parameters that influences how we use these enzymes is the length of the recognition site, which affects the frequency with which they cut DNA, and hence the average size of the fragments generated. For example, take

Box 4.1 Examples of restriction endonucleases

Enzyme	Recognition site	Number of bases	Ends generated	Original source of enzyme
*Eco*RI	G/AATTC	6	5′ Sticky	*Escherichia coli* RY13
*Bam*HI	G/GATCC	6	5′ Sticky	*Bacillus amyloliquefaciens* H
*Bgl*II	A/GATCT	6	5′ Sticky	*Bacillus globigii*
*Pst*I	CTGCA/G	6	3′ Sticky	*Providencia stuartii*
*Xma*I[a]	C/CCGGG	6	5′ Sticky	*Xanthomonas malvacearum*
*Sma*I[a]	CCC/GGG	6	Blunt	*Serratia marcescens*
*Acc*65I[a]	G/GTACC	6	5′ Sticky	*Acinetobacter calcoaceticus* 65
*Kpn*I[a]	GGTAC/C	6	3′ Sticky	*Klebsiella pneumoniae*
*Sau*3A	/GATC	4	5′ Sticky	*Staphylococccus aureus* 3A
*Alu*I	AG/CT	4	Blunt	*Arthrobacter luteus*
*Not*I	GC/GGCCGC	8	5′ Sticky	*Nocardia otitidis-caviarum*
*Pac*I	TTAAT/TAA	8	3′ Sticky	*Pseudomonas alcaligenes*

Only one strand of the recognition site is shown, with a slash (/) showing the position of the cleavage site. All the examples shown are palindromic, so the sequence of the second strand, read as the reverse complement, and the position of the cleavage site will be the same as that shown. Thus the reverse complement of 5′GAATTC3′ is also 5′GAATTC3, and both strands are cut by *Eco*RI between G and A.

[a]*Xma*I is an isoschizomer of *Sma*I, and *Acc*65I is an isoschizomer of *Kpn*I.

the enzyme *Sau*3A, which has a four-base recognition site, GATC. If we assume, for the moment, that there is an equal proportion of all four bases, and also that they are randomly distributed, then at any position on the chromosome there is a 1 in 4 chance that it is a G. Then there is a 1 in 4 chance that the next base is an A; the chance of having the sequence GA is the product of the two, or 1 in 4^2 (1 in 16). Extending the argument, the chance of the sequence GAT occurring is 1 in 4^3, and that of the four-base sequence GATC is 1 in 4^4, or 1 in 256. So this enzyme (and any others with a four-base recognition site) would cut DNA (on these assumptions) on average every 4^4 bases, and so would generate a set of fragments with an *average* size of 256 bases. The actual sizes of the fragments will be distributed quite widely on either side of this average value. A six-base site such as that recognized by *Eco*RI (GAATTC) would occur every 4^6 bases (or 4096 bases). To give some perspective to this calculation, a moderately sized protein might have about 300 amino acids and would therefore be coded for by a DNA sequence 900 bases in length (ignoring the possible presence of introns). Therefore a four-base cutting enzyme would often give fragments much smaller than a whole gene. On the other hand a bacterial genome of perhaps 4×10^6 bases (4 Mb) would be expected (given the same assumptions) to be cut into about 1000 fragments by an enzyme such as *Eco*RI.

Of course the assumptions in this calculation are not necessarily valid. Firstly, we assumed an equal proportion of all four bases. If we consider both strands of a double-stranded DNA molecule, the number of G and C residues is equal, since for every G on one strand there is a C on the other, and vice versa. Similarly, the number of A residues is the same as the number of T residues. (This equality of G and C, and of A and T, does not apply in the same way to each strand taken individually – but they are in fact usually nearly balanced.) Therefore we can simplify the examination of base composition by comparing the number of G + C bases to the number of A + T residues. This is commonly expressed as the percentage of (G + C), or GC per cent. This parameter can vary widely from one species to another. For *E. coli*, our assumption is valid, as the GC content *is* actually 50 per cent, but other fully sequenced bacterial genomes range quite widely, for example from 37 per cent G + C (*Staphylococcus aureus*) to 72 per cent G + C (*Streptomyces coelicolor*). For an organism with a high proportion of G + C bases in its DNA, the predicted number of sites for an enzyme such as *Eco*RI, with an AT-rich recognition site (GAATTC), will be lower than in the calculation above.

Furthermore, the sequence of bases is not random. For various reasons some combinations of bases occur much less often than would be expected. Some sites are therefore significantly under-represented in the genome as a whole.

A further factor that affects the ability of restriction endonucleases to cut DNA is methylation of the DNA. We have already seen that each restriction endonuclease, in its original host, is accompanied by a modifying enzyme that protects the chromosomal DNA against endonuclease cleavage, by methylation of the same sequence. In addition, most laboratory strains of *E. coli* contain two site-specific DNA methyltransferases (methylases). The Dam methylase catalyses methylation of the adenine residues in the sequence GATC, while the Dcm methylase modifies the internal cytosine in the sequences CCAGG and CCTGG. Methylation at these positions may make the DNA resistant to attack by restriction enzymes that cut the DNA at sites containing these sequences (see Box 4.1). Methylases also occur in eukaryotes, especially the CpG methylase, which modifies the cytosine in some sites containing the dinucleotide CpG. Although these are not known to be involved in restriction/modification systems, they can affect the ability of restriction enzymes to cut certain sites in DNA from higher eukaryotes. A further complication is that many *E. coli* strains have restriction systems (McrA, McrBC, Mrr) that attack DNA that is methylated at CpG sites, and the recovery of cloned DNA from mammals and higher plants can be reduced in such strains. The full implications of methylation of DNA are too complex to be dealt with here, beyond saying that it may be necessary to consider these factors when selecting restriction enzymes and *E. coli* host strains.

Although most of the commonly used enzymes recognize either four-base or six-base sites, there are also important roles for enzymes that cut even less frequently, such as *Not*I and *Pac*I (see Box 4.1). These both have eight base recognition sites, which might be expected to occur (on average) every 4^8 bases (about 65 kb), but the sequences are far from random (being composed entirely of G + C or A + T, respectively), so a given DNA sequence may contain very few sites (or even none at all) for such an enzyme.

4.1.2 Sticky and blunt ends

Another important parameter that affects the use we make of restriction enzymes is the position of the cut site within the recognition sequence. In this context it should be noted that most (but not all) restriction endonuclease recognition sites are said to be *palindromic*, although the term is not strictly accurate. A verbal palindrome, for example 'radar' reads the same from left to right as from right to left. A restriction site 'palindrome', such as GAATTC, is slightly different. At first glance it does not seem to be the same when read in the other direction, but you have to remember that the two strands of DNA lie in opposite directions. When we write GAATTC we are looking at the sequence of the 'top' strand, which runs 5′ to 3′ when read from left to right – so we should write the sequence as 5′GAATTC3′. On the 'bottom' (complementary) strand, we have to read from right to left to see the 5′ to 3′ sequence, which would also be 5′GAATTC3′; see Figure 4.2.

Within this sequence, the restriction enzyme *Eco*RI will cut the DNA between the G and the A on each strand, and will hence produce fragments with four bases unpaired at the 5′ end (Figure 4.3). These four bases (AATT) are the same on all fragments generated with this enzyme, and these ends are complementary to one another. They will thus tend to form base pairs, and so help to stick the fragments together. They are therefore referred to as *cohesive* or *sticky* ends. Note that the base pairing formed with a sequence of four bases is very weak and would not be stable, so we need to use DNA ligase to finally join the fragments together with a covalent bond to form a recombinant molecule (see below). Nevertheless, the limited cohesiveness of these ends does make the ligation process much more efficient.

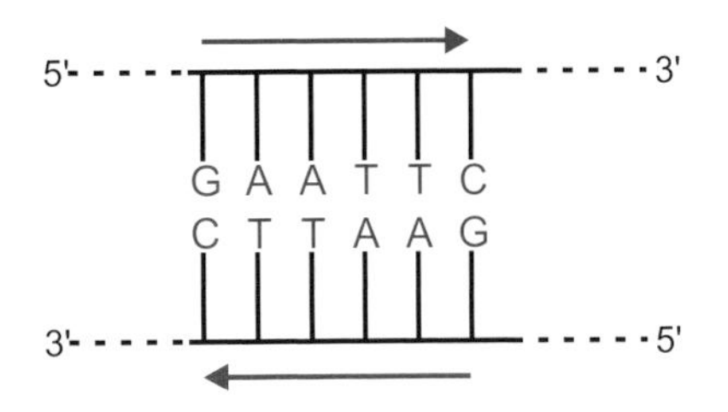

Figure 4.2 Reading a palindromic sequence. The top strand, read from left to right (5′ to 3′), is the same as the bottom strand when also read 5′ to 3′ (right to left)

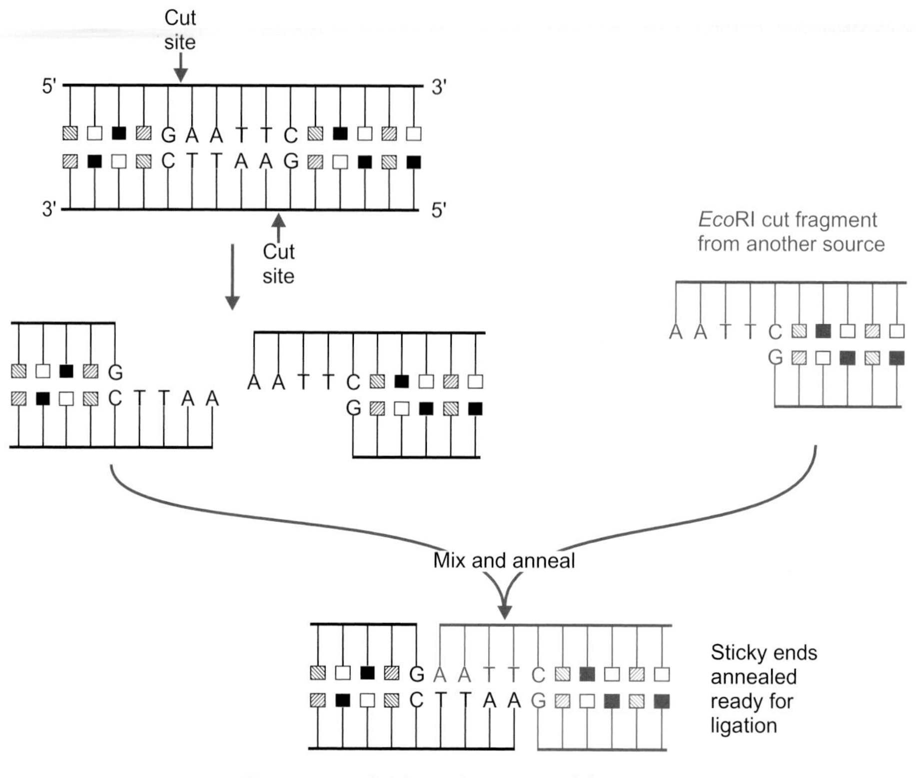

Figure 4.3 Sticky ends generated by *Eco*RI

Some enzymes, such as *Pst*I, also form cohesive ends, but by cutting asymmetrically within the right-hand part of the recognition site; they thus generate sticky ends with unpaired single-strand sequences at the 3′ ends of the fragment (Figure 4.4).

Although enzymes that generate cohesive ends are useful for gene cloning because of the efficiency of the ligation process, there is a limitation. You can only join together fragments with compatible ends. So two *Eco*RI fragments can be ligated to each other, or two *Bam*HI fragments can be joined, but you cannot ligate an *Eco*RI fragment directly to a *Bam*HI fragment. However, there are circumstances in which compatible ends can be generated by different restriction enzymes. For example the restriction enzymes *Bam*HI and *Bgl*II recognize different sequences (see Box 4.1), but they both generate the same sticky ends (with unpaired GATC sequences) and these can be joined.

More flexibility can be achieved by using an enzyme such as *Sma*I, which recognizes the six base sequence CCCGGG and cuts symmetrically at the centre position (see Figure 4.4); it generates blunt ends, which although much

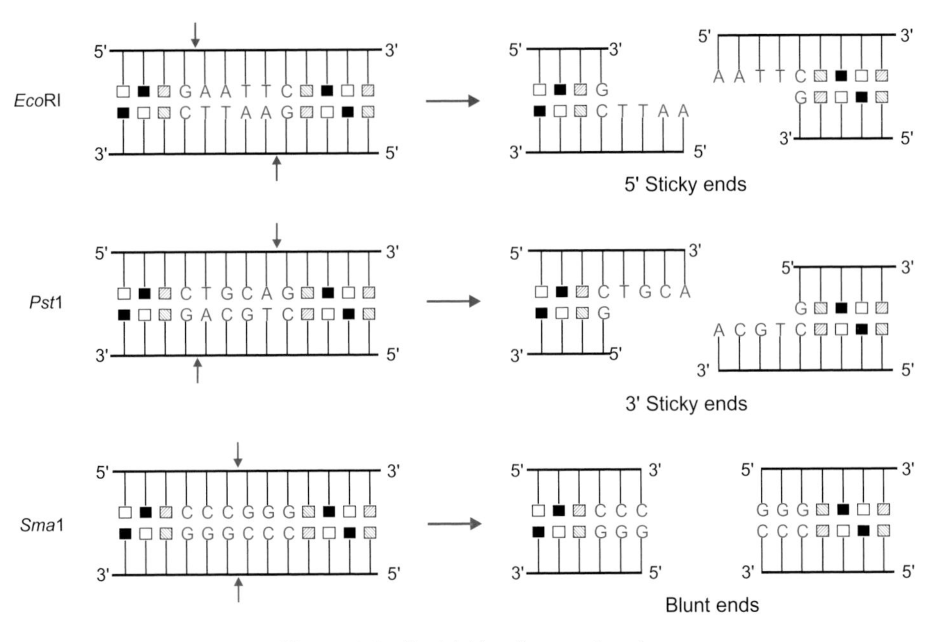

Figure 4.4 Restriction fragment ends

less efficiently ligated, have the advantage that they can be joined to any other blunt-ended fragment.

Some of the examples shown in Box 4.1 recognize the same restriction site. These are known as *isoschizomers* (Greek *iso*, equal; *skhizo*, to split). Some pairs of isoschizomers cut at different positions within the recognition site, which adds flexibility to our strategy. For example *Acc*65I and *Kpn*I both recognize the sequence GGTACC (Figure 4.5), but cut it at a different place, generating different sticky ends, while of another pair of isoschizomers (*Xma*I and *Sma*I), *Xma*I cuts asymmetrically and produces sticky ends that can be ligated to other *Xma*I fragments, while *Sma*I, as mentioned above, will generate blunt-ended fragments, allowing you to ligate the fragment to other blunt-ended DNA sequences.

4.2 Ligation

As outlined in Chapter 3, the next stage in cloning a gene is to join the DNA fragment to a vector molecule, such as a plasmid or bacteriophage, that can be replicated by the host cell after transformation. The joining, or ligation, of DNA fragments is carried out by an enzyme known as *DNA ligase*.

The natural role of DNA ligase is to repair single-strand breaks (nicks) in the sugar–phosphate backbone of a double-stranded DNA molecule, such as may occur through damage to DNA, as well as the joining of the short

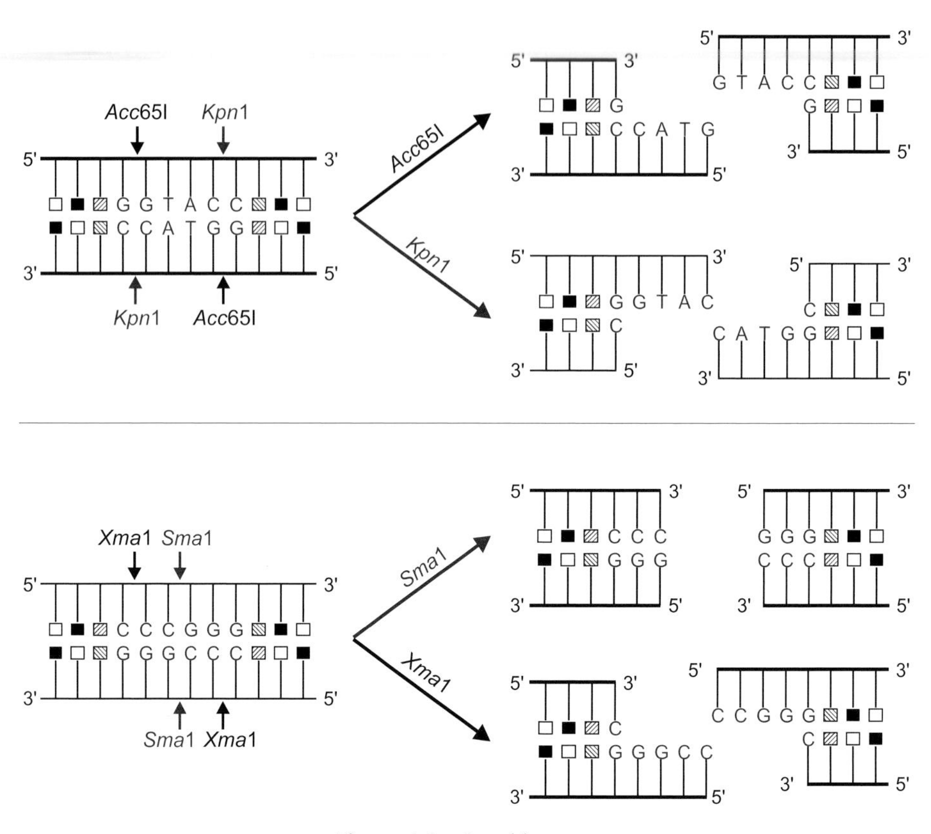

Figure 4.5 Isoschizomers

fragments produced as a consequence of replication of the 'lagging strand' during DNA replication. The action of the ligase requires that the nick should expose a 3′OH group and a 5′-phosphate (Figure 4.6). Digestion with restriction endonucleases cuts the DNA in this way, i.e. it leaves the phosphate on the 5′ position of the deoxyribose. The unstable pairing of two restriction fragments with compatible sticky ends can therefore be considered as a double-stranded DNA molecule with a nick in each strand, and is therefore a substrate for DNA ligase action.

Some DNA ligases, such as T4 DNA ligase (encoded by the bacteriophage T4), are also capable of ligating blunt-ended fragments, albeit much less efficiently, while others (notably the *E. coli* DNA ligase) are not; they require pairing of overlapping ends. To keep things simple, we will only consider the action of T4 DNA ligase, which is by far the most extensively used one. We will refer to this enzyme as T4 ligase (or just 'ligase') although there is also a (much less commonly used) T4 RNA ligase.

The T4 ligase requires ATP as a co-substrate. In the first step, the ligase reacts with ATP to form a covalent enzyme–AMP complex, which in turn

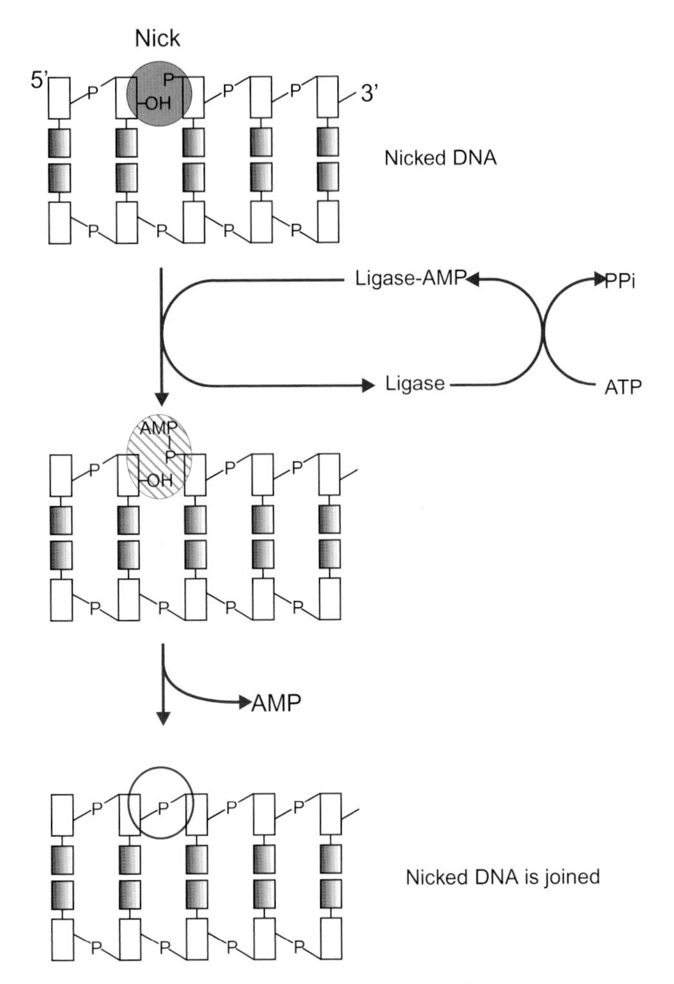

Figure 4.6 Action of T4 DNA ligase

reacts with the 5′-phosphate on one side of the nick, transferring the AMP to the phosphate group. The final stage is the attack by the 3′ OH group, forming a new covalent phosphodiester bond (thus restoring the integrity of the sugar–phosphate backbone) and releasing AMP. The absolute requirement for the 5′-phosphate is extremely important; by removing the 5′-phosphate we can prevent the occurrence of unwanted ligation (see below).

4.2.1 Optimizing ligation conditions

Ligation can be one of the most unpredictable (and often frustrating) steps in the cloning process. Factors that may compromise the success of ligation include the presence of inhibitory material contaminating our DNA

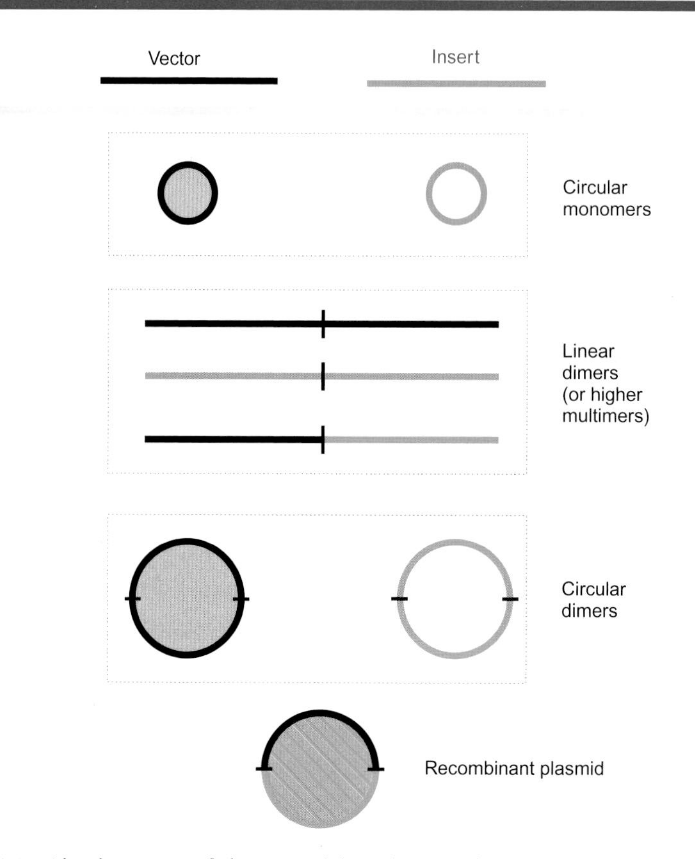

Figure 4.7 Ligation: some of the potential products. Only the shaded circles are expected to be replicated after transformation

preparations and degradation of the enzyme or the DNA (including loss of the 5′-phosphate). In addition, the conditions need to be adjusted correctly to achieve the optimum effect.

Since the reaction we want normally involves two different molecules of DNA (*inter*molecular ligation), we would expect it to be extremely sensitive to DNA concentration. It is therefore important to use high concentrations of DNA – but which DNA component? We have two components: the vector and the insert (ignoring for the moment the fact that the insert may itself be a heterogeneous mixture of fragments). There is therefore a variety of possible reactions that can occur (Figure 4.7), and by adjusting the relative amounts of the components, as well as the overall concentration of DNA, we can influence the likelihood of these different reactions.

At low concentrations of DNA, we are more likely to get the two ends of the same molecule joining (an *intra*molecular reaction), since the rate of a reaction involving one component will be linearly related to its concentration, whereas the rate with two different components will be proportional to the product of the two concentrations. (Or for a reaction involving two molecules of the same substrate, the rate will be proportional to the square of the concentration). Therefore, increasing the concentration will give a greater increase in the rate of the intermolecular reaction (between two molecules) than of the intramolecular reaction.

If we increase the concentration of the vector but not the insert, we will get a greater increase in the ligation of two vector molecules together (which we do not want) than in the production of the recombinant vector–insert product. Conversely, increasing insert concentration will give increased levels of insert–insert dimers (which we also do not want, although they are less of a problem as they will not give rise to transformants).

Therefore, we need not only to keep the overall DNA concentration high, but also to create an optimum vector: insert ratio. It is not easy to predict reliably what that ratio should be, but typically it would range from 3:1 to 1:3. Note that these are molar ratios and have to take account of the relative size of the vector and the insert. For example, if the vector is 5 kb and the insert is 500 bases, then a 1:1 molar ratio would involve 10 times as much vector, by weight, as insert (e.g. 500 ng of vector and 50 ng of insert). Conversely, for the same 5 kb vector but an insert size of 50 kb, if you used 500 ng of vector you would need 5 μg of insert to achieve a 1:1 molar ratio. In general, to convert the amount of DNA by weight into a value that can be used to calculate the molar ratio, divide the amount used (by weight) by the size of the DNA. Thus, if W_V and W_I are the weights used of vector and insert DNA respectively, and S_V and S_I are the sizes of vector and insert (e.g. in kilobases), then the vector: insert ratio is $(W_V/S_V):(W_I/S_I)$.

A further complication arises if we are working with a heterogeneous collection of potential insert fragments, as would be the case if we were making a gene library (see Chapter 6). Loading more insert DNA into the ligation mixture will increase the possibility of obtaining multiple inserts. In other words, we may produce recombinant plasmids that carry two or more completely different pieces of DNA. This is not a good idea. It can lead to seriously misleading results when we come to characterize the clones in the library and try to relate them to the structure of the genome of our starting organism.

Therefore, adjusting the relative amount of the vector and insert will not only influence the success of ligation, but will also affect the nature of the products. Fortunately we do not have to rely entirely on adjusting the levels of DNA in order to obtain the result that we want. In the next section, we will look at the use of alkaline phosphatase to remove 5′-phosphate groups (thus preventing some types of ligation), and in Chapter 5 we will consider designs of vector that allow us to distinguish between recombinant and non-recombinant transformants.

4.2.2 Preventing unwanted ligation: alkaline phosphatase and double digests

As described above, the ligation process depends absolutely on the presence of a 5′ phosphate at the nick site. If this phosphate group is removed, that site cannot be ligated. Removal of the 5′-phosphate groups is achieved with an enzyme known as alkaline phosphatase (because of its optimum pH) – one commonly used enzyme is calf intestinal phosphatase (CIP). Treatment of the vector molecule with CIP before ligation will remove the 5′-phosphates, and so make it impossible for self-ligation of the vector to occur. Ligation of vector to insert can occur, however, as the insert still has its 5′-phosphates. It is of course important to remove the CIP before ligation, and this is best done by phenol extraction and subsequent ethanol precipitation as heat inactivation is not sufficiently reliable. If you are using a standard vector, you can buy ready-made and tested dephosphorylated vector.

However, further inspection of the situation (Figure 4.8) will reveal that it is not quite that simple. At each junction between two DNA fragments there are *two* nicks that need repairing. At one of these, the 5′-phosphate is supplied by the insert, and that can be ligated normally. However, at the other nick, there should be a 5′-phosphate on the *vector*, and that has been removed. Therefore the second nick cannot be mended. How can we deal with this situation?

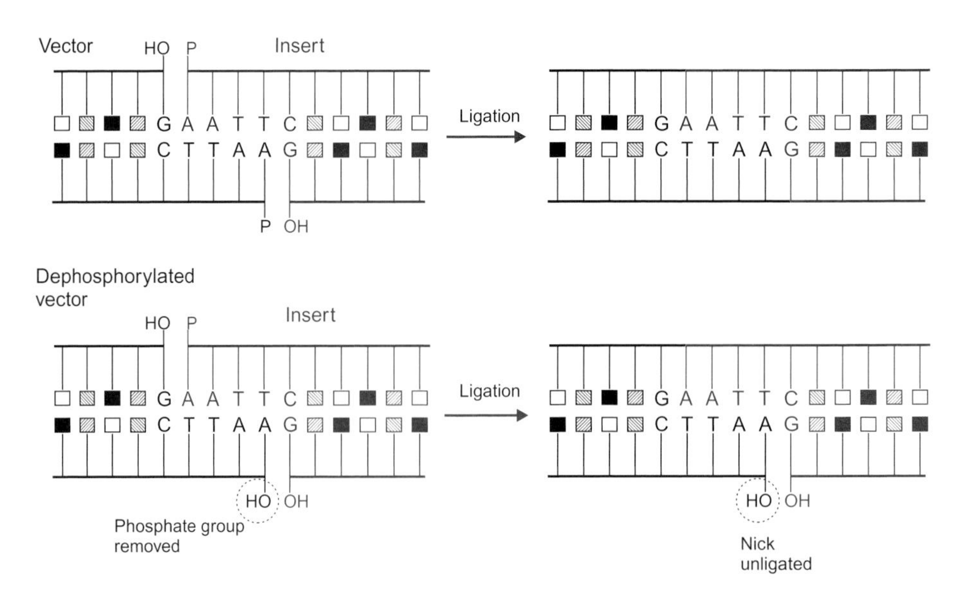

Figure 4.8 Ligation: effect of dephosphorylation of vector

Actually it is not a real problem. The recombinant plasmid will have unrepaired nicks in its (circular) DNA (one at each end of the inserted fragment), but it will hold together very stably by virtue of the base pairing all along the inserted DNA fragment. It will be maintained stably at 37°C, as a double-stranded molecule. When this nicked molecule is introduced into a bacterial cell, enzymes within the cell will rapidly repair the remaining nicks, adding the missing phosphates and ligating the broken ends. The plasmid will be replicated without any difficulty.

Some of the other problems still remain, potentially. We could still get insert dimers, or multiple inserts. However, since we no longer have to worry about self-ligation of the vector, we can increase vector concentration relative to the insert and thus drive the reaction in the direction we want.

Even so, if we are making a gene library, when the possibility of multiple inserts is at its most serious as a problem, we may want to turn this strategy on its head. Phosphatase treatment of the *insert* rather than the vector will prevent multiple inserts, and we can prevent the occurrence of non-recombinant transformants (carrying self-ligated vector rather than vector–insert recombinants) by using special vectors that are unable to produce clones unless they carry an inserted DNA fragment (see Chapter 5).

An alternative strategy is to cut both components (vector and insert) with two different restriction enzymes. Most modern vectors have multiple cloning sites (see Chapter 5) so you can cut the vector with, for example, *Eco*RI and *Bam*HI. Provided that both enzymes have cut efficiently (and that you have removed the small fragment between the two sites), then vector re-ligation will be impossible. If your insert fragment has been digested with the same two enzymes, then virtually all the colonies obtained will be recombinant, i.e. they will contain the insert fragment. This is also a useful strategy if you want to ensure that the insert fragment is in a specific orientation.

4.3 Modification of restriction fragment ends

Restriction fragments with sticky ends are useful as they can be readily ligated. However, there is a limitation on their usefulness, as they can only be ligated to another fragment with compatible ends. Therefore, an *Eco*RI fragment can be ligated to another *Eco*RI fragment, but not to one generated by *Bam*HI.

This is a potential nuisance, as the vector you are using will only have a limited number of possible sites into which you can insert DNA (and you may specifically want to put your insert in a particular position), and it may not be possible to generate a suitable insert with the same enzyme. The best strategy in this situation is to add short oligonucleotides (linkers or adaptors) to the ends of your insert fragments.

4.3.1 Linkers and adaptors

Linkers are short synthetic pieces of DNA that contain a restriction site. For example, the sequence CCGGATCCGG contains the *Bam*HI site (GGATCC). Furthermore it is self-complementary, so you only need to synthesize (or buy) one strand; two molecules of it will anneal to produce a double-stranded DNA fragment 10 bp long (see Figure 4.9). If this is joined to a blunt-ended potential insert fragment by blunt-end ligation, your fragment now will have a *Bam*HI site near each end. Cutting this with *Bam*HI will generate a fragment with *Bam*HI sticky ends, that can be ligated with a *Bam*HI-cut vector.

You may object that this still requires inefficient blunt-end ligation to join the linker to the potential insert. However, the efficiency of blunt-end ligation can be markedly improved by using high concentrations of at least one of the components. In this case, you can easily produce and use large amounts

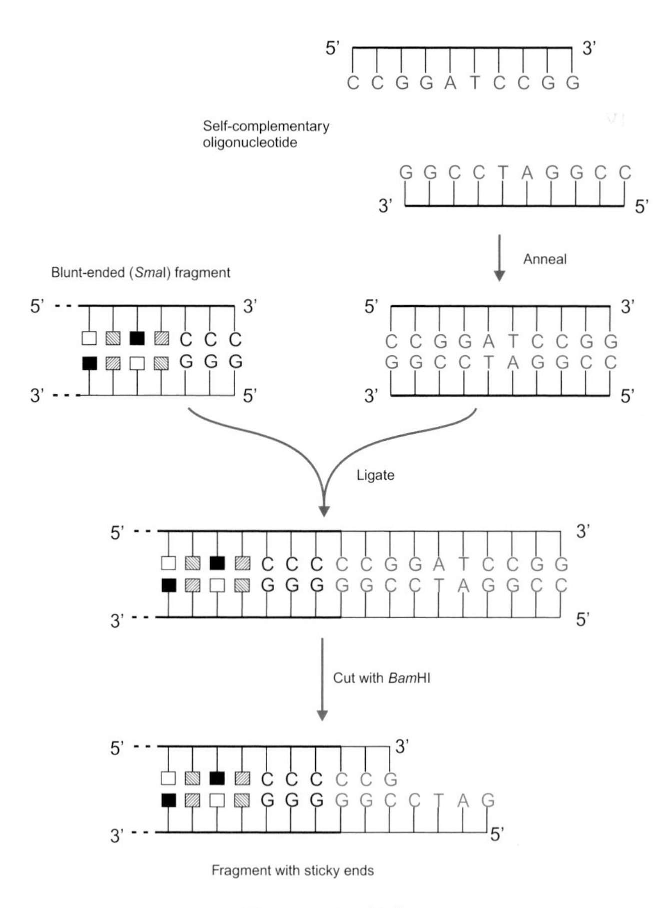

Figure 4.9 Linkers

of the linker. Furthermore, since the linker is very small (e.g. 10 bases), and it is the *molar* concentration that is important, even modest amounts of the linker by mass will represent an enormous excess of linker in molar terms. For example, if you use 100 ng of a 1 kb insert, then 10 ng of linker will represent a 10:1 linker:insert ratio. The high molar concentration of the linker will drive the reaction very effectively. Of course, this efficient ligation is likely to add multiple copies of the linker to the ends of your insert, but this is not a problem – the subsequent restriction digestion will remove them.

Further versatility can be obtained by the use of *adaptors*. These are pairs of short oligonucleotides that are designed to anneal together in such a way as to create a short double-stranded DNA fragment with different sticky ends (or with one sticky and one blunt end). For example the sequences 5′GATCCCCGGG and 5′AATTCCCGGG will anneal as shown in Figure 4.10 to produce a fragment with a *Bam*HI sticky end at one end and an *Eco*RI sticky end at the other, without needing to be cut by a restriction enzyme. Ligation of this adaptor to a restriction fragment generated by *Bam*HI digestion will produce a DNA fragment with *Eco*RI ends that can now be ligated with an *Eco*RI cut vector.

Alternatively, an adaptor with one sticky end and one blunt end can be used to convert blunt-ended DNA fragments, such as those generated by cDNA synthesis (see Chapter 6), into fragments with a sticky end, which

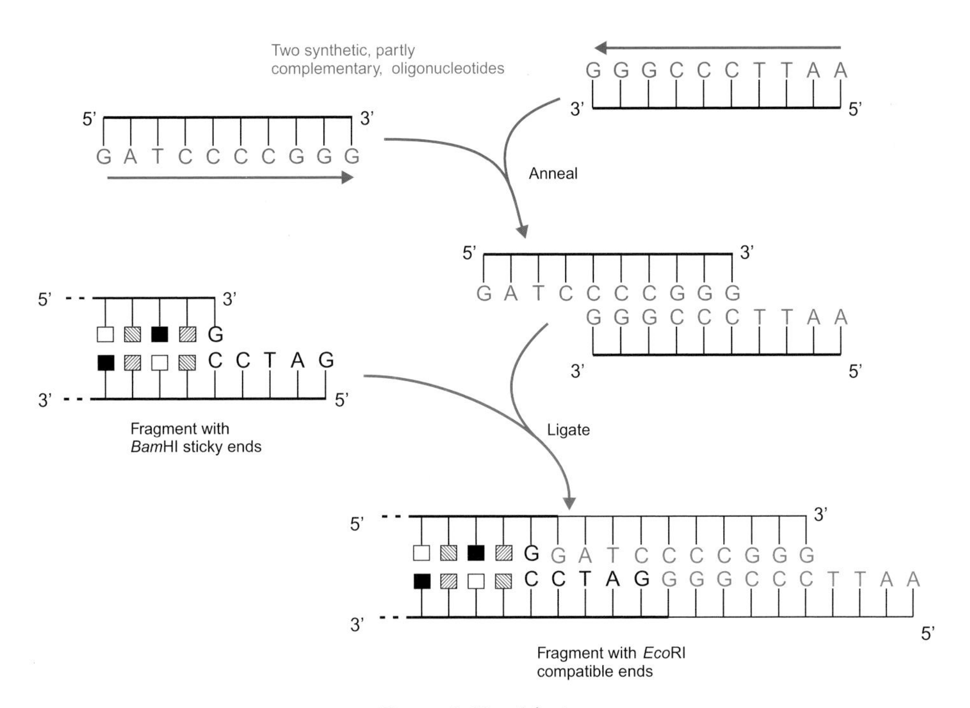

Figure 4.10 Adaptors

increases the cloning efficiency substantially. As with linkers, you can use high molar concentrations of adaptors to drive ligation very efficiently, and the use of adaptors with non-phosphorylated sticky ends ensures that you will not get multiple additions of the adaptor to the end of your DNA fragment.

Linkers and adaptors have applications much wider than adding restriction sites. For example, if you want to clone and sequence all the fragments from a sample of DNA, but only have a small amount of material, you can use the same methods to add, to the ends of all the fragments, short oligonucleotides that will act as recognition sites for a pair of PCR primers (see Chapter 8). PCR amplification will then give you an abundant supply of the full range of DNA fragments.

4.3.2 Homopolymer tailing

An alternative way of adding sticky ends to DNA molecules is to use the enzyme *terminal deoxynucleotide transferase* (or *terminal transferase* for short). When supplied with a single deoxynucleotide triphosphate (say dGTP), this enzyme will repetitively add nucleotides to the 3′OH end of a DNA molecule (Figure 4.11). This enzyme is different from DNA polymerases in that it does not require a template strand, so this reaction will produce a molecule with a single-stranded run of G residues at each 3′ end, hence the term homopolymer tailing. If the vector is treated in this way, and the insert fragment(s) are treated with terminal transferase and dCTP (generating a tail of C residues), the two tails are complementary and will tend to anneal to one another (see Figure 4.12).

It is not possible to ensure that the tails are all exactly the same length, so when the molecules anneal together there will be gaps in one of the sequences. This does not matter. If the tails are longer than about 20 nucleotides, then the pairing of the tails will be strong enough to be stable at room temperature, and the product can be used for transformation without repairing the gaps, or even without sealing the nicks with ligase. Both gaps and nicks will be repaired within the host cell after transformation.

One advantage of this strategy is that it is not possible for the vector to reform without an insert: the ends of the vector are not complementary to

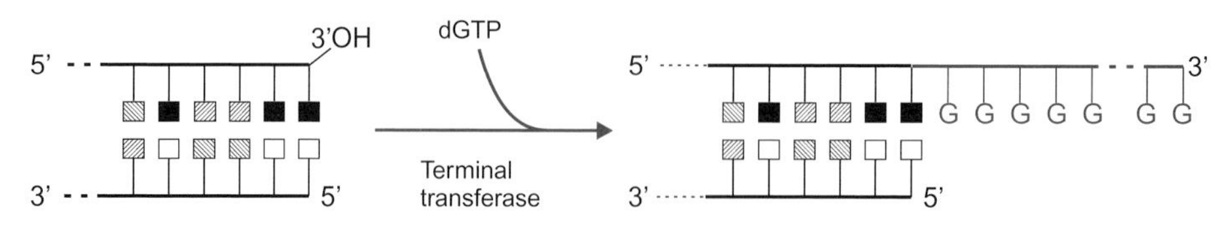

Figure 4.11 Tailing with terminal transferase

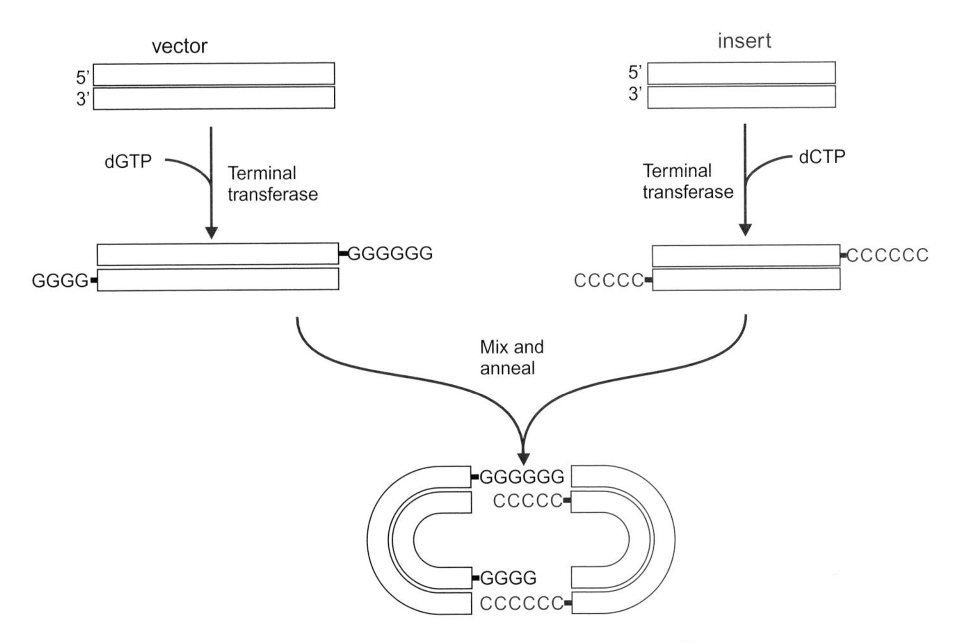

Figure 4.12 Cloning using homopolymer tailing

one another. The downside is that you have constructed a recombinant plasmid that contains a variable number of GC base pairs at either end of the insert, so it lacks the precision associated with the other methods described in this chapter. This is a disadvantage if you want to recover the insert from the recombinant vector, for example to reclone it in another vector. However, you can use restriction sites in the flanking region of the vector to release the insert. Or you can sequence the ends of the insert and use that information to design PCR primers (see Chapter 8) to allow you to specifically amplify the insert.

4.4 Other ways of joining DNA molecules

4.4.1 TA cloning of PCR products

Some of the polymerases used in PCR amplification (see Chapter 8) also have a (very limited) terminal transferase action: they add a single adenine residue to the 3′ ends of the synthesized strands. This can cause problems with cloning, as the ends are not truly blunt, so blunt-end cloning is likely to be unsuccessful. However, it provides an opportunity as well as a problem. There are commercially available 'TA vectors', which are supplied in linearized form with a single 3′T 'overhang' which is compatible with the unpaired

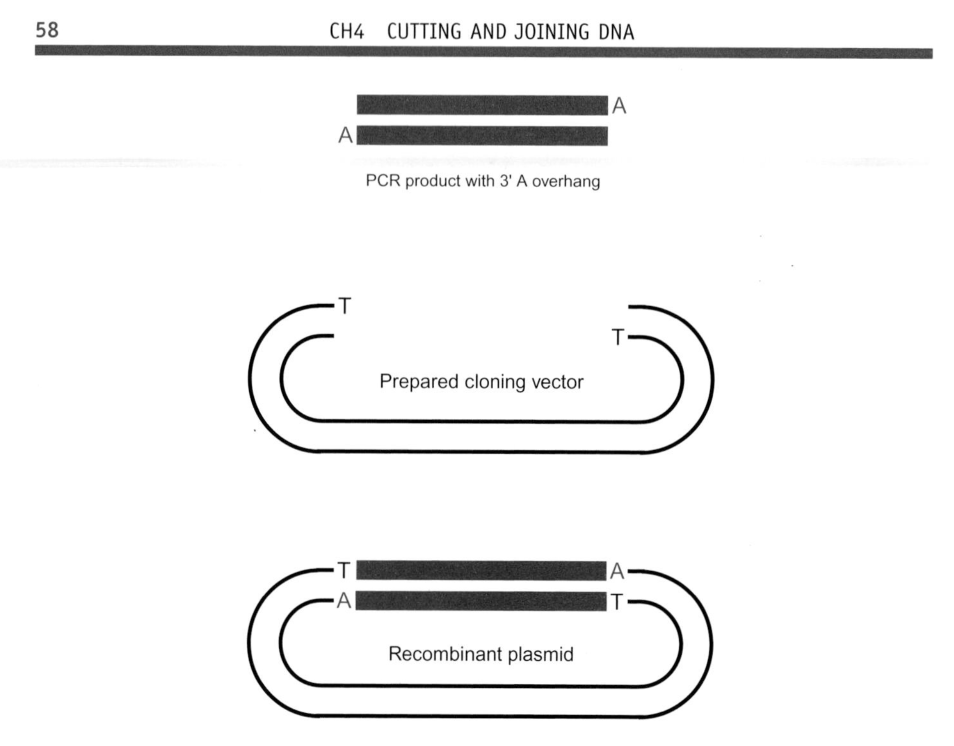

Figure 4.13 TA cloning

3′A on the PCR product (see Figure 4.13). This concept can be combined with topoisomerase cloning (see below) to form a highly efficient way of cloning such PCR products.

Alternative strategies include designing the PCR primers to incorporate a restriction site; the product can then be digested with the appropriate enzyme before ligation in the conventional manner. The strategies for cloning PCR products are considered further in Chapter 8.

4.4.2 DNA topoisomerase

Although DNA ligase is the most commonly used enzyme for joining two DNA molecules, it is not the only enzyme that can do this. DNA topoisomerase I is another example. Topoisomerases are responsible for controlling the degree of supercoiling of DNA (see Chapter 2); type I topoisomerases achieve this by cutting one DNA strand, which is then free to rotate (thus reducing the supercoiling); the enzyme subsequently rejoins the cut ends of the DNA strands. The enzyme remains covalently attached to the phosphate group at the end of the broken strand after cutting it, thus retaining the bond energy and being in place for the subsequent re-joining. In one commercially available system, a linearized vector is supplied with *Vaccinia* virus topoisomerase I covalently attached to phosphate groups at the 3′ ends

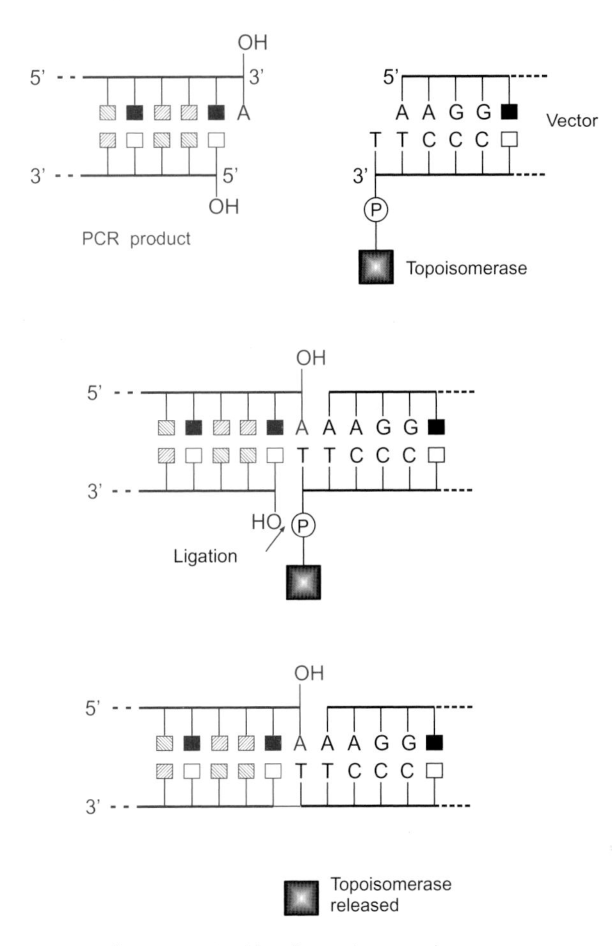

Figure 4.14 Ligation using topoisomerase

(Figure 4.14). When the prepared vector is mixed with the DNA fragment to be cloned, the enzyme transfers the phosphate linkages to the 5′ ends of the fragment, thus joining the insert to the vector. Since the topoisomerase is already attached to the vector, the reaction only requires two molecules (the vector and the insert) to come into contact, and is therefore quicker than conventional ligation which needs simultaneous contact between *three* molecules (vector, insert and ligase).

Note that, unlike the action of DNA ligase, the insert in topoisomerase ligation should not have a 5′-phosphate group. This system is mainly applied to the cloning of PCR products (Chapter 8), where the 5′ end (derived from the primers used in amplification) is not normally phosphorylated, unless you specifically order a 5′ phosphorylated primer.

Other enzyme systems are available for concerted breakage and rejoining of DNA strands, allowing for example a one-step transfer of a cloned gene from one plasmid to another. One example is the site-specific recombinase of bacteriophage lambda (the natural role of which is to insert lambda DNA into the bacterial chromosome, see Chapter 5). A second example is the Cre–*lox* system from the P1 bacteriophage. Cre is a highly site-specific recombinase, which will act only on the specific sequence known as the *lox* site. In both cases the enzyme will carry out recombination between two plasmids carrying the appropriate recognition sequences, leading to transfer of a cloned gene between two plasmids, or fusion of the plasmids.

In this chapter we have seen how restriction endonucleases can be used to cut DNA, how ligase can be used to join fragments together, and some of the ways in which these processes can be manipulated to improve the efficiency and versatility of the process. In the next chapter, we will look at the nature of the vectors into which the required fragment can be inserted, and how the resulting DNA can be introduced into a bacterial cell.

5 Vectors

In the previous chapter, we looked at the ways in which a potential insert can be ligated with a vector molecule, to enable that insert DNA to be replicated after insertion into a bacterial cell. We now need to consider further the nature of the vectors that can be used. Once again, we will initially be concentrating on the vectors that are used for cloning in bacteria (principally *E. coli*), and subsequently looking at the vectors needed for cloning in eukaryotes.

5.1 Plasmid vectors

Plasmids are by far the most widely used, versatile and easily manipulated vectors. They are the workhorse of the molecular biology laboratory. Plasmids occur widely in nature, and are found in most bacterial species, as extrachromosomal DNA molecules, usually circular, double-stranded and supercoiled. They vary considerably in size, from a few thousand base pairs up to several hundred kilobases, although the plasmids used as gene cloning vectors are usually small (typically 2–5 kb). Most of the commonly used ones are based on (or are closely related to) a naturally occurring *E. coli* plasmid called ColE1. Later in this chapter we will look at other types of vectors such as bacteriophages.

The most notorious property of plasmids lies in their ability to disseminate antibiotic resistance genes. They are responsible to a large extent for the spread of antibiotic resistance – although it should be noted that the plasmids used for gene cloning are nearly always unable to spread from one bacterium to another, and there are restrictions on experimental protocols to ensure that these experiments do not add new antibiotic resistance genes to clinically important pathogenic bacteria. Antibiotic resistance is not the limit of the

From Genes to Genomes, Second Edition, Jeremy W. Dale and Malcolm von Schantz

ability of plasmids, nor the reason for their existence. Interest tends to focus on antibiotic resistance because of its importance in medical microbiology, and because of the ease with which resistance genes can be isolated and studied. However, many naturally occurring plasmids code for other properties, or even for none at all, or at least none that we can discern. Plasmids exist because they can replicate within bacteria, and sometimes spread from one bacterium to another. That is all. They are a form of DNA parasite. Any advantage they confer on the host bacterium is a bonus that helps the plasmid to survive.

5.1.1 Plasmid replication

In genetic engineering, we make use of this ability of plasmids to be replicated, as it enables us to insert pieces of DNA which are then copied as part of the plasmid, and hence passed on to the progeny when the cell replicates. The most fundamental property of a plasmid, whether we are considering it as a cloning vector or as a natural phenomenon, is therefore the ability to replicate in the host bacterium. With the simplest plasmids, most or all of the enzymes and other products needed for this replication are already present in the host cell; the amount of information that the plasmid has to supply may be only a few hundred base pairs. This region of the plasmid that is necessary for replication is generally referred to as the origin of replication (*ori*), although literally the origin, or the site at which replication starts, is one specific base.

Plasmids that use the origin of replication from ColE1 or its relatives are multi-copy plasmids. Wild-type ColE1 is present at about 15 copies per cell, while most of the engineered vectors used today are present in numbers running into many hundreds of copies per cell. This is convenient in some ways as it makes it easier to purify large amounts of the plasmid, and if you want to express a cloned gene you also get a gene dosage effect. The presence of so many copies of the gene in the cell is reflected in higher levels of the product of that gene (see also Chapter 11). However, this can also be a disadvantage. Even without expression of the cloned gene, the large amount of plasmid DNA may make the cell grow more slowly. The effect may not be great, but it can lead to instability, as any cells that have lost the plasmid will outgrow those that retain it. This will be exacerbated if the cloned gene or its product is in any way harmful to the bacterium; in extreme cases, it can sometimes be very difficult to isolate the required clone. For some specific purposes therefore it is desirable to use alternative vectors that exist at low copy number (or to use different vectors altogether that do not require continued viability of the cell, such as some types of bacteriophage vector – see below).

Some plasmids are able to replicate in a wide variety of bacterial species (broad host-range plasmids), but most of those that are used for gene cloning are rather more restricted in their host range. In one way this is useful: if there is any question about potential health hazards or environmental consequences associated with cloning a specific fragment of DNA, then using a narrow host range plasmid makes it less likely that the gene will be transmitted to other organisms.

On the other hand, you may wish to carry out genetic manipulations in a bacterium other than *E. coli*, especially if your interest lies in studying the behaviour of specific bacteria rather than simply using them to clone pieces of DNA. It will then usually be necessary to isolate or construct new vector plasmids, based on a replication origin that is functional in your chosen species. The host range of your new vector will probably also be limited, and it may well be unable to replicate in *E. coli*. This is a disadvantage, because you are likely to want to use *E. coli* as an intermediate host for the initial cloning and for studying the structure and behaviour of the gene that you have cloned. However it is possible to insert two origins of replication into your plasmid, so that it will be replicated in *E. coli* using one origin, and in your chosen host using the alternative replication origin. Such a vector is known as a *shuttle plasmid*, because it can be transferred back and forth between two the species. We can also use shuttle vectors to transfer cloned genes between *E. coli* and a eukaryotic organism. We will be coming across various applications of shuttle vectors in subsequent chapters.

Therefore, the first essential characteristic of a plasmid cloning vector is the origin of replication, usually designated as *ori* in plasmid maps.

5.1.2 Cloning sites

The second characteristic that is necessary for a plasmid to be useful as a cloning vector is a *cloning site*, for insertion of the new DNA fragment. This must of course be located in a region of the plasmid that is not essential for replication or any other functions that we need. For a cloning site in a basic vector, we need a unique site at which a specific restriction enzyme will cut, i.e. the enzyme will cut the plasmid once only. If a circular molecule is broken at one position, it is converted into a linear molecule, and it is relatively simple to join the ends together to reform an intact circle. If an enzyme cuts more than once, the plasmid will be cut into two or more pieces, and joining them up again to make an intact plasmid will be much more inefficient.

A basic plasmid used as a cloning vector may naturally contain only one or two such unique restriction sites. With such a plasmid you are limited not only in your choice of restriction fragments that can be inserted and in the position of insertion, but also in the number of different fragments that can be inserted. This is because in most cases ligation of two restriction fragments, generated with the same enzyme, recreates the original restriction site. Thus when you insert say a *Bam*HI fragment into a site on the vector that

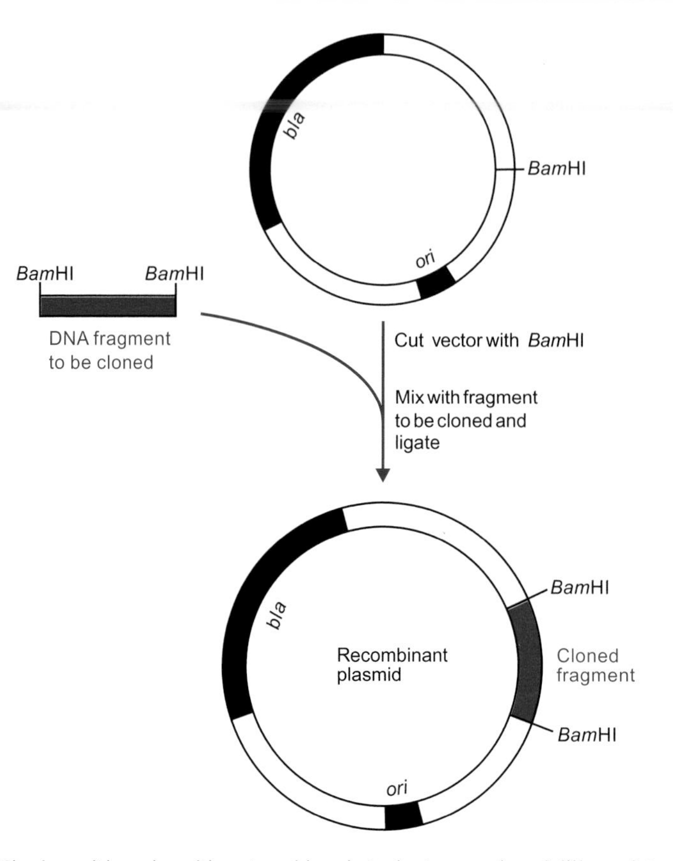

Figure 5.1 Cloning with a plasmid vector. *bla* = beta-lactamase (ampicillin resistance), selective marker; *ori* = origin of replication

has been cut with *Bam*HI, the resulting recombinant plasmid will have two *Bam*HI sites: one at each end of the inserted fragment (see Figure 5.1). Since the recombinant plasmid now has two *Bam*HI sites, it would be difficult to clone further *Bam*HI fragments into it.

The problem here is that in many cases we *do* want to insert several fragments into the same plasmid. We may want to combine the expression signals of one gene with the coding region of another, or we may want to insert additional markers that can be used to identify the presence of the plasmid. Or we may, as described above, want to insert the replication origin from another plasmid so as to create a shuttle vector, and still leave sites available for further inserts. The best way around this problem is to create a *multiple cloning site* (MCS), i.e. a short DNA region that contains recognition sites for a number of different enzymes. This is done by synthesizing a short piece of DNA with the required restriction sites (a *polylinker*), and inserting that into the plasmid in the usual way. Figure 5.2 shows the structure of pUC18,

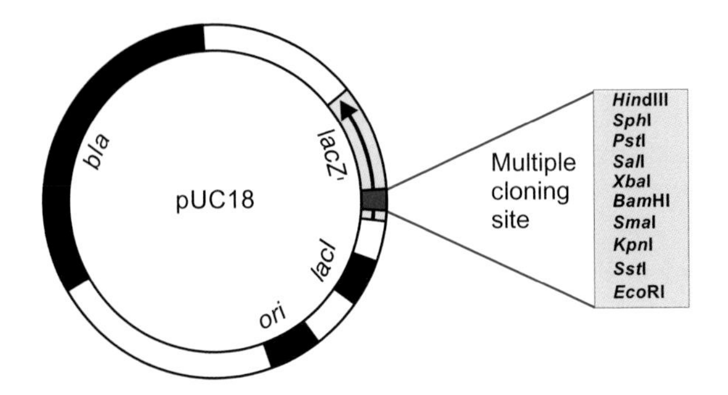

Figure 5.2 Structure of the plasmid cloning vector pUC18. *bla* = beta-lactamase (ampicillin resistance), selective marker; *ori* = origin of replication; *lacZ'* = beta-galactosidase (partial gene); *lacI* = repressor of *lac* promoter

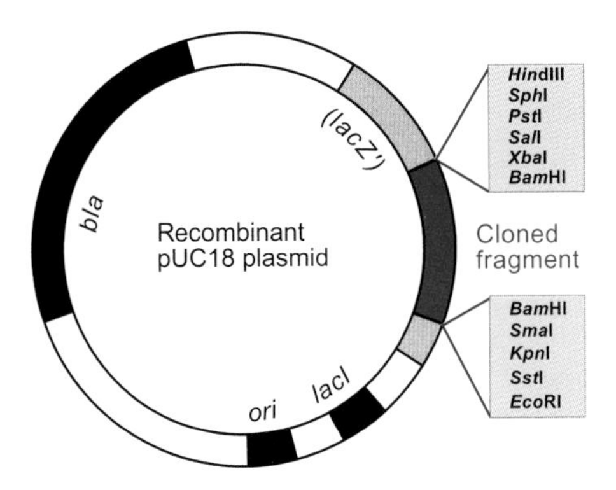

Figure 5.3 A recombinant plasmid formed using the cloning vector pUC18. The *lac* gene has been disrupted by insertion of a DNA fragment, resulting in white colonies on X-gal plates. *bla* = beta-lactamase (ampicillin resistance), selective marker; *ori* = origin of replication; *lacZ'* = beta-galactosidase (partial gene); *lacI* = repressor of *lac* promoter

one of a family of similar plasmids that are commonly used as cloning vectors, and you will see that pUC18 contains such a multiple cloning site. Insertion of a fragment into the *Bam*HI site, as in Figure 5.3, will still leave a selection of other sites available for further inserts.

5.1.3 Selectable markers

Therefore, we have a plasmid with a replication origin and one or more restriction sites suitable for cloning. One further feature is essential for a functionally useful vector, and that is a selectable marker. The need for this arises from the inefficiency both of ligation and of bacterial transformation. Even with the high efficiency systems that are now available for *E. coli* (see below), the best yield available, using native plasmid DNA, implies that only about 1% of the bacterial cells actually take up the DNA. In practice the yields are likely to be lower than this – and if you are using a host other than *E. coli*, many orders of magnitude lower. Therefore in order to be able to recover the transformed clones, it is necessary to be able to prevent the non-transformed cells (i.e. those cells that have not taken up the plasmid) from growing. The presence of an antibiotic resistance gene on the plasmid vector means that you can simply plate out the transformation mix on an agar plate containing the relevant antibiotic, and only the transformants will be able to grow. In Figure 5.2, you will see that pUC18 carries a beta-lactamase gene (*bla*), coding for an enzyme that hydrolyses beta-lactam (penicillin-like) antibiotics such as ampicillin (and hence often referred to as Amp^R, for ampicillin resistance).

5.1.4 Insertional inactivation

In Figure 5.2, you will see a further useful feature of pUC18. The multiple cloning site is located near to the 5′ end of a beta-galactosidase gene (*lacZ*). The synthetic oligonucleotide that creates the multiple cloning site was designed so that it did not affect the reading frame of the *lacZ* gene; it merely results in the production of beta-galactosidase with some additional amino acids near to the amino terminus of the protein. This does not affect the function of the enzyme; it is still able to hydrolyse lactose. (More accurately, we should say that pUC18 carries a *part* of the *lacZ* gene; we use *E. coli* strains that carry the remainder of the gene. The product of the host gene is unable to hydrolyse lactose by itself, and so the host strain without the plasmid is Lac^-, i.e. it does not ferment lactose. When pUC18 is inserted into the host, the plasmid-encoded polypeptide will associate with the host product to form a functional enzyme. We say that pUC18 is capable of *complementing* the host defect in *lacZ*.)

We can easily detect the activity of the beta-galactosidase by plating the organism onto agar containing the chromogenic substrate 5-bromo-4-chloro-3-indolyl-β-D-galactopyranoside (universally known as *X-gal*), together with the inducing agent isopropyl thiogalactoside (*IPTG*). The X-gal substrate is colourless but the action of beta-galactosidase releases the dye moiety, resulting in a deep blue colour. Colonies carrying pUC18 are therefore blue when grown on this medium. However, if we are successful in inserting a DNA fragment at the cloning site, the gene will (normally) be disrupted (Figure 5.3) and the resulting *E. coli* colonies are referred to as 'white'. The advantage of this *insertional inactivation* is that we can tell not only that the cells have been transformed with the plasmid (since they are able to grow in the presence of ampicillin), but also that the plasmid is a recombinant, and not merely the original pUC18 self-ligated. An insertional inactivation marker such as this is not an essential feature of a cloning vector but it does provide a useful way of monitoring the success of the ligation strategy and overcoming some of the problems referred to in the previous chapter.

One word of caution: if the insert is relatively small, and if it happens to consist of a multiple of three bases, transcription and translation of the *lacZ* gene may still occur and the enzyme may still have enough activity (despite the addition of still more amino acids at the N terminus) to produce a detectable blue colour. Conversely, a white colony is not a guarantee of cloning success as the deletion of even a single base at the cloning site, or the insertion of undesirable fragments (in other than multiples of three bases), will put the *lacZ* gene in the wrong reading frame and thus inactivate it.

The advantage of plasmid vectors, compared with the other vectors described subsequently in this chapter, is that they are small and easy to manipulate; also they are conceptually simple and universal. You can make and use plasmid vectors for a wide range of organisms without a detailed knowledge of the molecular biology of the host or the vector. On the other hand the basic plasmid vectors that we have been considering so far are limited in their cloning capacity, i.e. the size of the insert they can accommodate. Later in this chapter, we will look at other vectors that will accommodate larger inserts, but first we need to consider the ways in which we can introduce the recombinant plasmids into host bacterial cells.

5.1.5 Transformation

Bacterial transformation was discovered in 1928 with the demonstration by Fred Griffith that cultures of the pneumococcus (*Streptococcus pneumoniae*) that had lost virulence could have their pathogenicity restored by addition of an extract of a killed virulent strain. It was the identification, many years later (by Avery, MacLeod and McCarty, in 1944) of the 'transforming principle' as DNA that resolved the question of the chemical nature of the genetic material.

This experiment rests on the natural ability of the pneumococcus to take up 'naked' DNA from its surroundings. This ability is known as *competence*. Competence develops naturally in some bacterial species, but although the range of species that exhibit natural competence is much wider than was thought for many years to be the case, it is still too limited in scope (or too inefficient or too selective) to be of much use for genetic engineering. In particular, *E. coli* does not seem to exhibit natural competence. It was therefore necessary to develop alternative ways of introducing plasmid DNA into bacterial cells. Although these methods are radically different, they are all still referred to as transformation, which is defined as the uptake of naked DNA, to distinguish it from other methods of horizontal gene transfer in bacteria, namely *conjugation* (direct transfer by cell to cell contact) and *transduction* (which is mediated by bacteriophage infection).

The breakthrough came with the demonstration that competence in *E. coli* cells could be induced by washing them with ice-cold calcium chloride, followed by adding the plasmid DNA and subjecting the mixture to a brief, mild heat shock (e.g. 2 min at 42°C). It is then necessary to dilute the cells in growth medium and incubate for a while (30–60 min) to allow the bacteria to recover and to express the resistance marker introduced on the plasmid, before plating them onto a selective medium containing the appropriate antibiotic. Although this simple basic process represented a major, and essential, step forward, it was very inefficient, with yields of perhaps 10^4 transformants per microgram of pure, supercoiled plasmid DNA (and less with ligation mixtures or non-supercoiled DNA). Gradually, over the years, improvements have been made in the transformation process, both by modifying the preparation of competent cells (e.g. using salts other than calcium chloride) and also by selecting *E. coli* strains with mutations that make them easier to transform. With the best of these systems, it is now possible to obtain transformation frequencies in excess of 10^9 transformants per microgram of plasmid DNA. Where such high yields are required, it is cost-effective to purchase pre-prepared competent cells of a strain with high transformation efficiency. Note that, although transformation frequencies are generally quoted as the number of transformants per microgram of DNA, you would usually use much less DNA than that – so the real result might be 10^6 transformants with 1 ng of DNA. Transformation works best with low levels of DNA. If you increase the amount of DNA, the number of transformants does not increase in proportion, and may actually decrease.

Reference back to the discussion of ligation in Chapter 4 will disclose a quandary here. Ligation works best with high concentrations of DNA. The following step, transformation, works best with small amounts of DNA. The resolution is clear, although unpalatable: use only a small proportion of your ligation mix in the transformation step. If you really need very large numbers of transformants, scaling up the transformation step does not work very well – it is usually much better to carry out several separate small-scale transformations.

Transformation based on induced competence and heat shock can be used for bacterial species other than *E. coli*, but you immediately lose all the advantages that have been gained by optimization of transformation conditions for selected strains of *E. coli*. At best, therefore, transformation is likely to be very inefficient – and in most cases simply using an *E. coli* procedure will not work at all. Therefore laboratories that are interested in manipulating other bacterial species have had to develop alternative methods of transformation.

Electroporation is the most versatile transformation procedure. Bacterial cells, washed with water to remove electrolytes from the growth medium, are mixed with DNA and subjected to a brief pulse of high voltage electricity. This appears to induce temporary holes in the cell envelope through which the DNA can enter. The cells are then diluted into a recovery medium before plating on a selective medium in the same way as above. Although it is comparatively easy to obtain *some* transformants with a wide range of bacteria (or with most other cells), there are many parameters that need to be adjusted to obtain optimum performance, including the conditions under which the cells are grown, the temperature of the suspension, and the duration and voltage of the electric pulse.

Since the added DNA seems to simply diffuse through the holes created (briefly) by the electric pulse, the effect is not specific for DNA; other substances, notably RNA or proteins, can also be introduced into bacterial cells by electroporation. Nor is it directionally specific. Material within the cell can diffuse out as well, and the procedure has been used for isolating plasmid DNA from bacterial cells. It follows from this that, since the plasmid that comes out of one cell can enter another one, electroporation can be used to transfer plasmids from one strain to another, simply by applying it to a mixture of the two strains.

Other methods that are used more commonly with animal and plant cells, including microinjection, biolistics and protoplast transformation, are considered in Chapter 15.

5.2 Vectors based on the lambda bacteriophage

5.2.1 Lambda biology

Plasmid vectors are at their best when cloning relatively small fragments of DNA. Although there is probably no fixed limit to the size of a DNA fragment that can be inserted into a plasmid, the recombinant plasmid may become rather less stable with larger DNA inserts, the efficiency of transformation is reduced, and the plasmid will give a much smaller yield in cultures of *E coli.* Vectors based on bacteriophage lambda allow efficient cloning of larger fragments, which is important in constructing gene libraries. The larger the inserts, the fewer clones you have to screen to find the one you want (see Chapters 6 and 7). Lambda vectors have further advantages in gene library construction, as it is much easier to screen large libraries when using bacteriophage vectors; the results with bacteriophage plaques are much cleaner than those obtained with bacterial colonies.

In order to understand the nature and use of lambda cloning vectors, some knowledge of the basic biology of bacteriophage lambda is necessary. While we hope you are familiar with this, a recapitulation of the salient features (summarized in Figure 5.4) will be useful.

Lysogeny Lambda is a temperate bacteriophage, i.e. on infection of *E. coli* it may enter a more or less stable relationship with the host known as *lysogeny*, as an alternative to the usual lytic cycle, which would result in lysis of

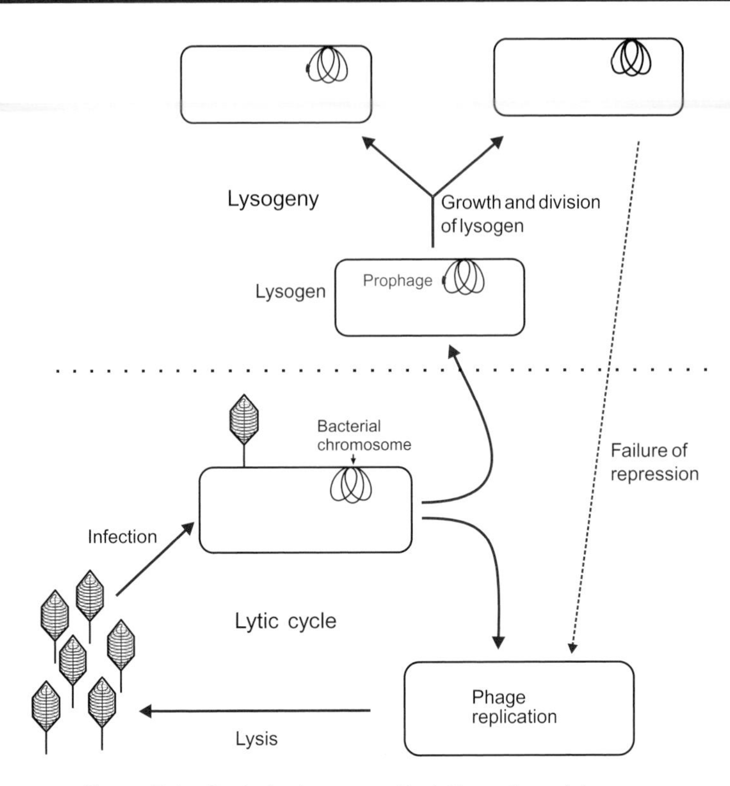

Figure 5.4 Bacteriophage growth: lytic cycle and lysogeny

the cell and liberation of a large number of phage particles. In the lysogenic state, expression of almost all of the phage genes is switched off by the action of a phage-encoded repressor protein, the product of the *cI* gene.

When you add a lambda phage preparation to an *E. coli* culture, some of the infected cells will become lysogenic, and some will enter the lytic cycle. The proportion of infected cells going down each route is influenced by environmental conditions, as well as by the genetic composition of the phage and the host. Some phage mutants will only produce lytic infection, and these give rise to clear plaques, while the wild-type phage produces turbid plaques due to the presence of lysogens which are resistant to further attack by lambda phage (known as *superinfection immunity*). On the other hand, some bacterial host strains carrying a mutation known as *hfl* (high frequency of lysogenization) produce a much higher proportion of lysogens when infected with wild-type lambda – which can be useful if we want a more stably altered host strain, for example if we are studying the expression of genes carried by the phage. Generally, when using lambda vectors we are more interested in the recombinant phage carrying the cloned genes, and the lytic cycle is the more relevant one in such cases.

Although the lysogenic state is relatively stable, that stability is not absolute. A culture of a bacterial lysogen will normally contain phage particles in the supernatant, due to a low level of spontaneous failure of the repression mechanism. This rate of breakdown of repression can be increased by treating the culture with agents that damage the DNA, such as UV irradiation; the DNA damage induces the production of repair enzymes which amongst other things destroy the *cI* repressor protein, allowing initiation of the lytic cycle. Some widely used lambda vectors carry a mutation in the *cI* gene which makes the protein more temperature-sensitive (*cI857* mutation). A bacterial strain carrying such a mutant phage can be grown as a lysogen at a reduced temperature and the lytic cycle can be induced by raising the temperature, due to inactivation of the repressor protein.

In the lysogenic state, lambda is normally integrated into the bacterial chromosome, and is therefore replicated as part of the bacterial DNA. However, this integration, although common amongst temperate phages, is not an essential feature of lysogeny. Lambda can continue to replicate in an extrachromosomal, plasmid-like state; with some bacteriophages (including P1, which we will encounter later in this chapter) this is the normal mode of replication in lysogeny.

Particles of wild-type bacteriophage lambda have a double-stranded linear DNA genome of 48,514 base pairs, in which the 12 bases at each end are unpaired but complementary (see Figure 5.5). These ends are therefore 'sticky' or 'cohesive', much like the ends of many restriction fragments – but the longer length of these sticky ends makes the pairing much more stable, even at 37°C. The ends can be separated by heating lambda DNA, and if it is then cooled rapidly you will get linear monomeric lambda DNA. At low temperatures, the ends of the molecule will move slowly, and therefore the re-annealing of the sticky ends will take a long time. Eventually however, it will resume a circular (although not covalently joined) structure. When lambda infects a bacterial cell and injects its DNA into the cell, it will form

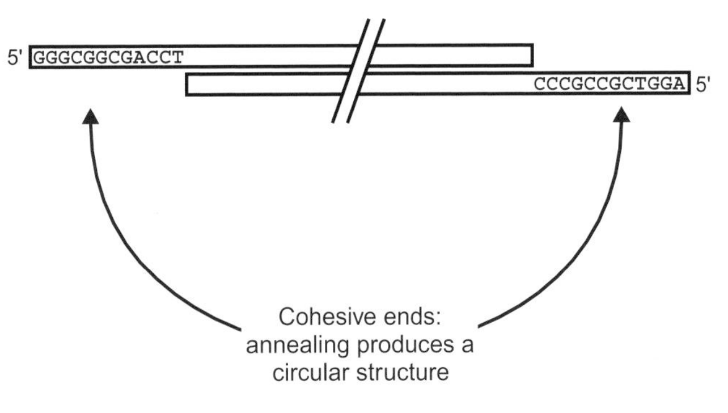

Figure 5.5 Cohesive ends of lambda DNA

a circular structure, with the nicks being repaired *in vivo* by bacterial DNA ligase. At about this time, a complex series of events occur that affect subsequent gene expression, determining whether the phage enters the lytic cycle or establishes lysogeny. We do not need to consider the details of the *lytic–lysogenic decision*, except to emphasize that it is essentially irreversible so that, once started on one or the other route, the phage is committed to that process. However, we do need to consider the events in the lytic cycle.

Lytic cycle In the lytic cycle, this circular DNA structure is initially replicated, in a plasmid-like manner (*theta* replication), to produce more circular DNA. Eventually, however, replication switches to an alternative mode (*rolling circle replication*) which generates a long linear DNA molecule containing a large number of copies of the lambda genome joined end-to-end in a continuous structure (see Figure 5.6). While all this is going on, the genes carried by the phage are being expressed to produce the components of the phage particle. These proteins are assembled first of all into two separate structures: the head (as an empty precursor structure into which the DNA will be inserted), and the tail (which will be joined to the head after the DNA has been packaged).

The packaging process involves enzymes recognizing specific sites on the multiple-length DNA molecule generated by rolling circle replication, and making asymmetric cuts in the DNA at these positions. These staggered breaks in the DNA give rise to the cohesive ends seen in the mature phage DNA; these sites are known as *cohesive end sites* (*cos* sites). Accompanying these cleavages, the region of DNA between two *cos* sites – representing a unit length of the lambda genome – is wound tightly into the phage head. This process is known as *packaging*. Following successful packaging of the DNA into the phage head, the tail is added to produce the mature phage particle, which is eventually released when the cell lyses.

Lysis of the bacterial cell is accomplished largely through the action of a phage-encoded protein, the product of gene *S*. Mutations in this gene can cause a delay or failure of lysis – which can be advantageous in increasing the yield of bacteriophages, as the replication of the phage will therefore continue for a longer time instead of being interrupted by lysis of the host cell. Many lambda vectors have such a mutation.

One of the most important features of this process from our point of view is that the length of DNA that will be packaged into the phage head is determined by the distance between two *cos* sites. If we insert a piece of DNA into our lambda vector, we will increase that distance, and so the amount of DNA to be packaged will be bigger. However, the head is a fixed size, and can only accommodate a certain amount of DNA. It can take somewhat more than is present in wild-type lambda (up to about 51 kb altogether, which is about 5%, or 2.5 kb, more than wild-type). As one of the reasons for using

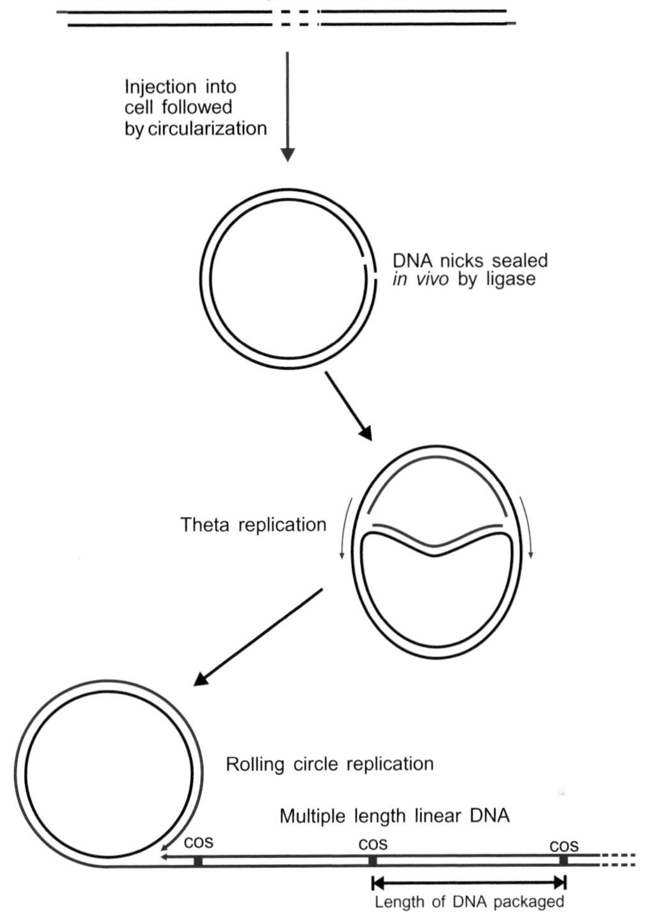

Figure 5.6 Replication of bacteriophage lambda DNA

lambda is to be able to clone large pieces of DNA, this would be a serious limitation. The way round it is to delete some of the DNA that is normally present. This is possible because the lambda genome contains a number of genes that are not absolutely necessary – especially if we only need lytic growth, when we can delete any genes that are solely required for the establishment of lysogeny. However, we cannot delete too much. The stability of the phage head requires a certain amount of DNA, so, even though there are more genes that are not required, we cannot delete all that DNA. To produce viable phage, there has to be a minimum of 37 kb of DNA (about 75% of wild-type) between the two *cos* sites that are cleaved.

The existence of these *packaging limits* is a very important feature of the design and application of lambda vectors, and also of cosmids, which we will discuss later.

5.2.2 *In vitro* packaging

Naked bacteriophage DNA can be introduced into a host bacterial cell by transformation (often referred to as *transfection* when talking about phage DNA), in much the same way as we described for a plasmid. The big difference is that, in this case, instead of plating on a selective agar and counting bacterial colonies, we would mix the transfection mix with a culture of a phage-sensitive indicator bacterium in molten soft agar and look for *plaques* (zones of clearing due to lysis of the bacteria) when overlaid onto an agar plate. Note that in this case we do not need an antibiotic resistance gene as a selective marker.

However, the large size of most bacteriophage DNA molecules, including that of lambda, makes transfection an inefficient process compared with plasmid transformation, and not suitable for the generation of gene libraries, which is the principal application of lambda vectors. However, there is a more efficient alternative. Some mutant lambda phages, in an appropriate bacterial host strain, will produce empty phage heads (as they lack a protein needed for packaging the DNA), while others are defective in the production of the head, but contain the proteins needed for packaging. The two extracts are thus complementary to one another. Use of the mixture allows productive packaging of added DNA, which occurs very effectively *in vitro* (including the addition of the tails). The resulting phage particles can then be assayed by addition of a sensitive bacterial culture and plating as an overlay, as above. Since *in vitro* packaging of lambda DNA is much more effective than transfection, it is the method that is almost always used.

One feature of this system that is markedly different from working with plasmid vectors is that the packaging reaction is most efficient with multiple length DNA. The enzyme involved in packaging the DNA normally cuts the DNA at two different *cos* sites on a multiple length molecule; monomeric circular molecules with a single *cos* site are packaged very poorly. Therefore, whereas with plasmid vectors the ideal ligation product is a monomeric circular plasmid consisting of one copy of the vector plus insert, for lambda vectors it is advantageous to adjust the ligation conditions so that we *do* get multiple end to end ligation of lambda molecules together with the insert fragments. The stickiness of the ends of the linear lambda DNA means this happens very readily.

5.2.3 Insertion vectors

The simplest form of lambda vector, known as an *insertion vector*, is similar in concept to a plasmid vector, in containing a single cloning site into which DNA can be inserted. However, wild-type lambda DNA contains many sites for most of the commonly used restriction enzymes; you cannot just cut it

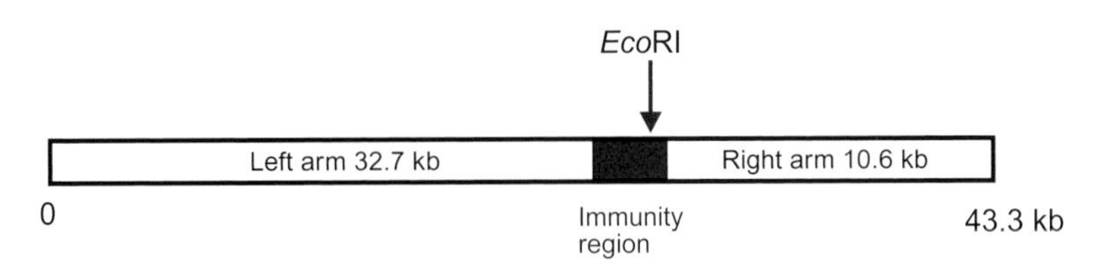

Figure 5.7 Lambda insertion vector gt10

with say *Hin*dIII and ligate it with your insert DNA. *Hin*dIII has seven sites in normal lambda DNA, and so will cut it into eight pieces. (Note the difference between a circular DNA molecule such as a plasmid, and a linear molecule like lambda: cut a circular DNA molecule once and you still have one fragment; cut linear DNA once and you have two fragments.) It would be almost impossible to join all these fragments (and your insert) together in the right order. To circumvent this, all lambda vectors have been genetically manipulated to remove unwanted restriction sites.

In Figure 5.7, we see one example of a lambda vector, known as lambda gt10. In this vector, there is only a single site at which *Eco*RI will cut the DNA. The manipulations that this phage has undergone have removed the unwanted sites, and have also reduced the overall size of the phage DNA to 43.3 kb (which is still large enough to produce viable phage particles, and still contains the genes that are needed for viability), and hence allows the insertion of foreign DNA up to a maximum of 7.6 kb.

Cutting this vector with *Eco*RI will produce two DNA fragments, referred to as the left and right arms. Although it therefore appears that the insert would have to be ligated to two different pieces of DNA, in practice this does not complicate the ligation as much as might be imagined. One end of each of the two arms is derived from the cohesive ends of the lambda DNA, and will therefore anneal quite stably at 37°C – so although not covalently joined they can be considered as a single DNA fragment.

Lambda gt10 also provides us with another example of how insertional inactivation can be used to distinguish the parental vector (which may form by religation of the arms without an insert) from the recombinants. The *Eco*RI site is found within the repressor (*c*I) gene, so the recombinant phage, which carry an insert in this position, are unable to make a functional repressor. As a consequence, they will be unable to establish lysogeny and will give rise to clear plaques, whereas the parental gt10 phage will give rise to turbid plaques. Therefore, you can achieve substantial enrichment of recombinant phage over the religated vector without having to use dephosphorylation.

Another example of a lambda insertion vector, lambda gt11, is used in a rather different way. Since it allows expression of the cloned fragment, it is considered later in this chapter, together with other expression vectors.

The packaging limits for lambda DNA are between 37 and 51 kb, as described above. In other words, we cannot make an insertion vector smaller than 37 kb, or we will be unable to grow it to produce the DNA that we need. In addition, we cannot insert a DNA fragment so large that it would make the product larger than 51 kb; the recombinant DNA would be impossible to package into the phage heads. It follows that the maximum cloning capacity for an insertion vector is (51 – 37) = 14 kb. This is larger than we would normally clone comfortably in a plasmid vector, but still smaller than we would like for some purposes. In order to increase the cloning capacity, we have to turn to a different type of lambda vector, known as a *replacement vector*.

5.2.4 Replacement vectors

The packaging limits that restrict the cloning capacity of insertion vectors are imposed by the physical requirements of the phage head rather than by the nature of the genes needed. There are further genes that are not essential for lytic growth and could be deleted, except that it would make the phage DNA too small to produce viable progeny. That provides a clue to an alternative design of lambda cloning vectors. Instead of merely inserting extra DNA, arrange the vector so that a piece of DNA can be removed and replaced by your insert – hence the term *replacement vector*.

Figure 5.8 shows an example of a lambda replacement vector, EMBL4. Instead of being cut just once by the restriction enzyme of choice (in this case *Bam*HI), there are two sites where the DNA will be cleaved. The vector

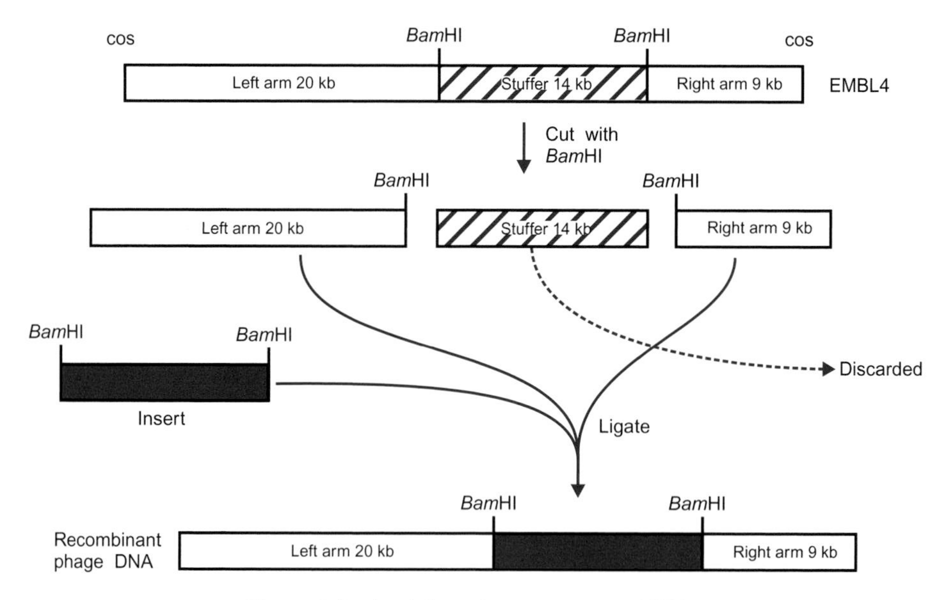

Figure 5.8 Lambda replacement vector EMBL4

DNA will therefore be cut into three fragments: the left and right arms (which will anneal by virtue of their cohesive ends) and a third fragment which is not needed (except to maintain the size of the DNA) and can be discarded. Since the only purpose of this fragment is to help to fill up the phage head it is known as a *stuffer fragment*.

In use, this vector would be cut with *Bam*HI and the fragments separated, e.g. by gel electrophoresis. The stuffer fragment would be thrown away, and the arms mixed with the restriction fragments to be cloned. These could be generated by *Bam*HI digestion, or by cleavage of the target with another enzyme that produces compatible ends, e.g. *Sau*3A (see Chapter 4). Ligation of the mixture produces recombinant phage DNA that can be packaged into phage heads by *in vitro* packaging. The cloning capacity of the vector is thus considerably increased; in this case the size of the arms combined comes to 29 kb and thus you can clone fragments up to (51 – 29) kb, or 22 kb.

There is a further advantage to such a vector. The combined size of the arms is only 29 kb, which is less than the minimum required for packaging. Any pairs of arms that are ligated without an insert will therefore be too small to produce viable phage particles. Such viable particles will only be produced if ligation results in an insert of at least (37 – 29) = 8 kb. The vector thus provides a positive selection for recombinants as opposed to parental phage, and furthermore for recombinants that contain an insert of at least 8 kb. The gene library will therefore be free of non-recombinant phage.

In Chapter 4 we discussed strategies for ensuring that we obtained recombinant plasmids rather than parental vector molecules, including alkaline phosphatase treatment of the vector to prevent recircularization. This is not necessary with replacement vectors. Since we do not have to treat the vector with phosphatase, we have another possibility – dephosphorylation of the *insert*. In the production of a gene library, the insertion of more than one fragment into the same vector molecule is a problem that can give rise to anomalies in characterizing the insert in relation to the genome it came from. Phosphatase treatment of the insert will prevent insert–insert ligation, and hence will ensure that all of the recombinants carry only a single insert fragment.

There is yet another useful feature that can be built into a replacement vector. Since the stuffer fragment is not necessary for phage production (apart from filling up the phage head), it does not have to be lambda DNA. It can be anything we want. So we could for example put in a fragment carrying a β-galactosidase gene. Then any plaques formed by phage that still carry the stuffer fragment would be blue (on a medium containing X-gal). Of course ideally there should be none. However, there may be some phage DNA molecules that have not been cut completely, or there may be some stuffer DNA contaminating the preparation of the vector arms, which could then be ligated back into the vector. Any plaques containing the stuffer will be blue – so you have an immediate check that everything has gone according to plan. Or not.

So we see that lambda vectors provide a highly versatile and efficient system for primary cloning of unknown fragments, especially in the construction of genomic and cDNA libraries (see Chapter 6). They extend the cloning capacity over that readily obtainable with plasmid vectors, and can easily generate the very large numbers of recombinants that are required for a gene library. However, some people do not like working with lambda systems, mainly because they require a different set of techniques for growing, assaying and maintaining phage preparations. There is nothing really difficult about it; it is just unfamiliar. The only real disadvantage to lambda cloning systems is the size of the vector DNA. With an insertion vector, your recombinant may contain 5 kb of insert and 45 kb of vector. This makes it more difficult to analyse or manipulate your insert – especially as the lambda vector will contain a substantial number of recognition sites for different restriction enzymes. The normal procedure therefore, having identified the recombinant clone of interest, would be to reclone the insert (or a part of it) into a plasmid vector for further analysis and manipulation.

Although lambda phages are the most widely used phage vectors, there are other phages, or vectors based on them, that are used for specific purposes, including P1 and M13. These are discussed later on in this chapter. However, first we need to look at a special class of vector that combines some of the features of lambda and plasmid vectors, and enables the cloning of even larger pieces of DNA. These are the *cosmids*.

5.3 Cosmids

The lambda packaging reaction has two fundamental requirements: the presence of a *cos* site, and the physical size of the DNA. Cosmids exploit this to provide cloning vectors with a capacity larger than can be achieved with lambda replacement vectors.

Basically, a cosmid is simply a plasmid which contains a lambda *cos* site. As with all plasmid vectors, it has an origin of replication, a selectable marker (usually an antibiotic resistance gene), and a cloning site. Digestion, ligation with the potential insert fragments and subsequent purification of recombinant clones, are carried out more or less as for a normal plasmid vector. However, instead of transforming bacterial cells with the ligation mix, as you would normally with a plasmid, you subject the ligation mixture to *in vitro* packaging as described above for lambda vectors. Since the cosmid carries a *cos* site, it can be a substrate for *in vitro* packaging – but only if it is big enough. The vector itself is quite small – in the example shown in Figure 5.9, it is 5.4 kb, which is much too small for successful packaging. The packaging reaction will only be successful if you have inserted a DNA fragment between about 32 and 45 kb in size. So you not only have an increased cloning

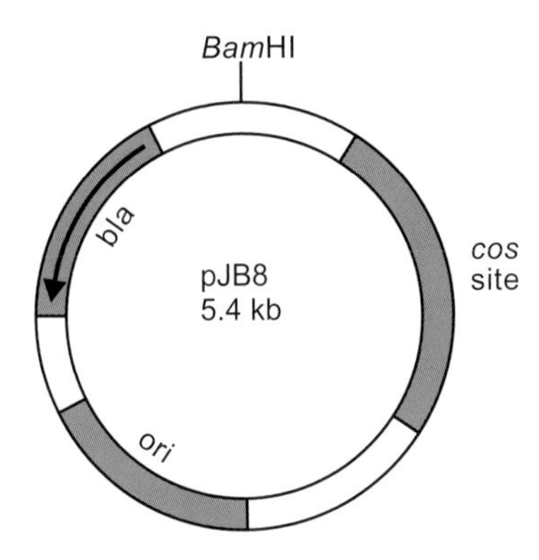

Figure 5.9 Structure of a cosmid. *ori* = origin of replication; *bla* = beta-lactamase (ampicillin resistance); *cos* site = region of lambda DNA required for packaging

capacity, but also a positive selection for an insert, and for an insert that is quite large.

Of course the products of the packaging reaction, although they are phage-like particles, will not give rise to more phages after infection of a host bacterium. They do not carry any of the genes that are needed for production of phage particles, nor for lysis of the cell, so you will not get phage plaques. However, the cosmid will replicate as a plasmid, and hence will give rise to more cosmid-containing cells, and these can be selected as colonies on agar containing an antibiotic (in this case ampicillin).

For very large genomes, such as mammalian ones, the cloning capacity of cosmids is still rather too small for convenience, and alternative vectors with even greater capacity are available (see below). However, for smaller genomes, cosmids can be extremely valuable. A complete bacterial genome can be covered by only a few hundred cosmids, which can be useful in genome mapping and sequencing projects.

Cosmids have some advantages over phage lambda, particularly in that they can be propagated and purified by conventional plasmid-oriented techniques, without having to become familiar with phage technology. Furthermore, with cosmids the vector is small compared with the insert, in contrast to lambda, where more than half of the DNA of a recombinant phage is derived from the vector. Subcloning of fragments of your insert, to obtain fragments carrying just the gene you need, is therefore rather easier with cosmids than with lambda vectors. On the other hand, a lambda library can be easily stored as a pool of phage particles in a single tube.

5.4 M13 vectors

Like lambda, M13 is a bacteriophage that infects *E. coli*, but is very different in many respects. M13 is a filamentous phage, which is 'sex-specific' or more

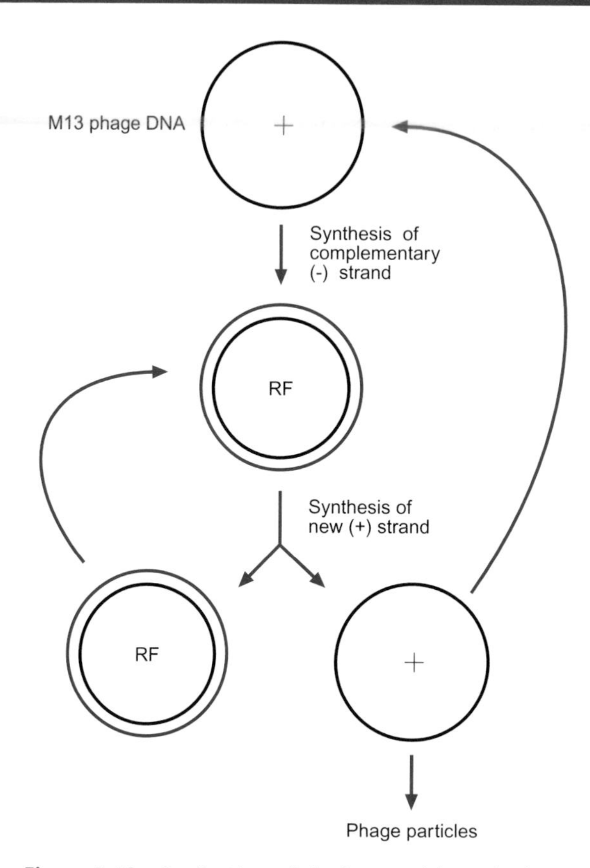

Figure 5.10 Replication of single-strand bacteriophages

accurately F-specific. It attaches to the tips of the pili that are produced on the surface of bacteria that carry an F-type plasmid, and therefore can only infect bacteria that carry such a plasmid. The phage particles contain a circular, single-stranded DNA molecule of about 6 kb. After this DNA enters the cell, it is converted to a double-stranded molecule (the *replicative form, RF*) by synthesis of the complementary strand. This molecule is replicated by producing a circular single-stranded copy of one strand of the RF. This single-stranded DNA is again converted to a double-stranded form (see Figure 5.10). This separation of the synthesis of the two strands is not unique to M13, but is found in some other bacteriophages and some plasmids. However, most of the phage DNA within the cell is double-stranded circular DNA (RF), and can be isolated by conventional plasmid purification methods.

Continued replication of the phage DNA leads to a build up of these plasmid-like DNA molecules within the cell. At the same time, expression of phage genes occurs, and the product of one of these genes binds to the single-stranded DNA, initiating production of phage particles. The production of phage particles occurs by extrusion of the single-stranded DNA through the cell membrane, during which process it becomes coated with phage proteins. The length of the filamentous phage particle is determined by the length of the DNA molecule, unlike lambda where the size of the particle is determined by the structure of the proteins of which it is composed. Hence, there are no absolute packaging limits for M13, although the phage does become increasingly fragile if large DNA fragments are inserted.

A significant, feature of M13 is that infection does not lead to bacterial lysis. Phage particles continue to be produced, and the cell remains viable, although it grows more slowly. Infection does result in the appearance of 'plaques' in a bacterial lawn, but these are zones of reduced growth rather than zones of lysis. As a consequence of the continuing viability of the host cell, very high titres of phage can be produced.

The main advantage of M13 is that it provides a very convenient way of obtaining single-stranded versions of a gene, which would be difficult to do in any other way. Single-stranded DNA obtained from M13 clones was originally essential for DNA sequencing (see Chapter 9), although nowadays double-stranded DNA templates are commonly used for sequencing. Another application where single-stranded DNA can be advantageous (although not essential) is site-directed mutagenesis (Chapter 11). M13 vectors also have another role, not connected with the production of single-stranded DNA: this is the technique known as *phage display* (see Chapter 14).

5.5 Expression vectors

The above discussion has assumed that all you want to do is to clone a piece of DNA. It does not consider the possibility that you might want to obtain expression of the gene encoded by that DNA. If you take a DNA fragment from another organism and clone it in *E. coli*, there are many reasons why it may not be expressed. At the simplest level, these relate to the signals necessary for initiating transcription (a promoter) and translation (a ribosome-binding site and start codon). The basic way of encouraging (although not ensuring) expression of the cloned gene is to incorporate these signals into the vector, adjacent to the cloning site. This is then known as an *expression vector*.

Expression vectors are of two main types (see Figures 5.11 and 5.12). If the vector just carries a promoter, and relies on the translation signals in the cloned DNA, it is referred to as a *transcriptional fusion* vector. On the other hand, if the vector supplies the translational signals as well (so you are inserting the cloned fragment into the coding region of a vector gene), then you have a *translational fusion*. Note that in this case the insert must be in frame with the start codon. The plasmid vector pUC18 that we looked at earlier is actually a translational fusion vector, although not often used as such. A better example

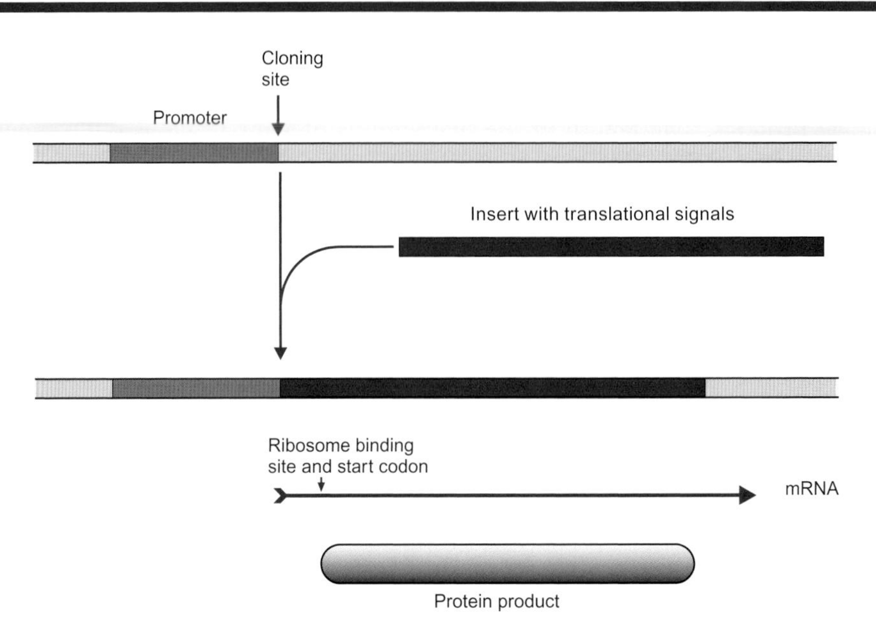

Figure 5.11 Expression vectors: transcriptional fusions

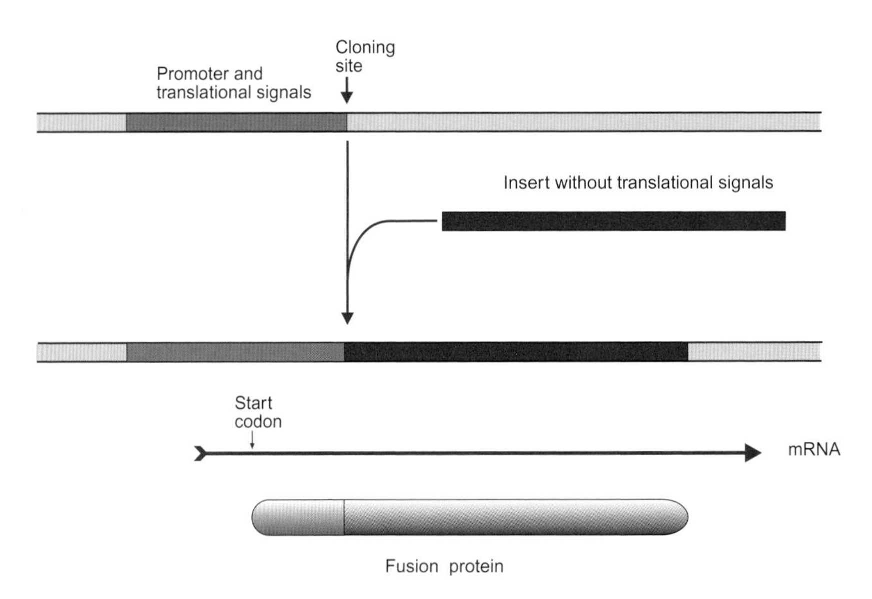

Figure 5.12 Expression vectors: translational fusions

is the lambda vector gt11 (see Figure 5.13). This is an insertional vector, 43.7 kb in length (making the maximum cloning capacity about 7 kb). It has been engineered to contain a beta-galactosidase gene, and has a single *Eco*RI restriction site within that gene – but in contrast to pUC18, the cloning site is towards the 3′ end of the beta-galactosidase gene. This confers two properties on the

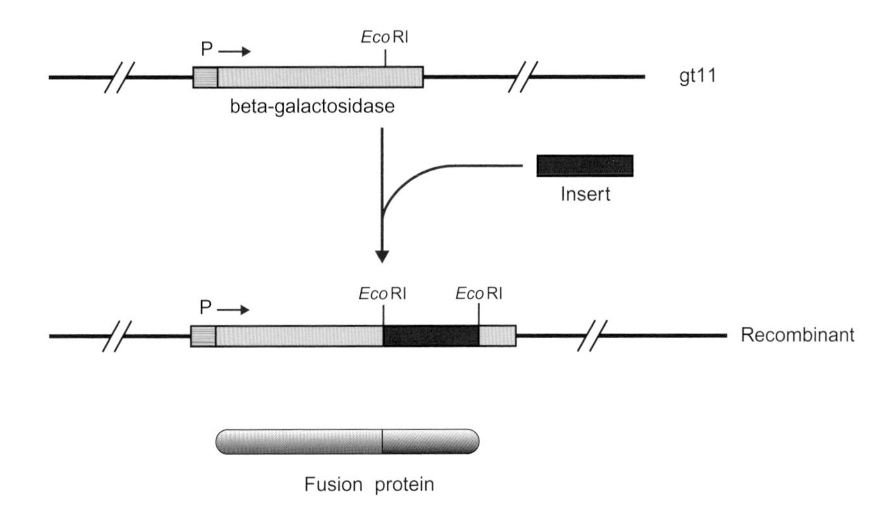

Figure 5.13 Use of lambda gt11 for generation of fusion proteins

vector. Firstly, insertion of DNA at the *Eco*RI site will inactivate the beta-galactosidase gene, so that recombinants will give 'white' (actually colourless) plaques on a medium containing X-gal. Secondly, the insert, if in the correct orientation and in frame, will give rise to a fusion protein containing the product encoded by the insert fused to the beta-galactosidase protein. This fusion protein is unlikely to have the biological functions associated with your cloned gene, but that is not the point. It *is* reasonably likely to react with some antibodies to the natural product, which makes it a useful way of detecting the clone of interest, as we will see in Chapter 7.

To make full use of an expression vector you need to be able to choose whether to have expression on or off. The regulation of most inducible bacterial promoters, such as that of the *lac* operon, is rather 'leaky' – that is, there is still some expression even in the uninduced or repressed state. Firmer control can be achieved by the use of promoters from bacteriophages, notably one from the bacteriophage T7. In T7, this promoter controls the expression of the 'late' genes, i.e. the genes that are only switched on at a late stage of infection. This promoter is not recognized by *E. coli* RNA polymerase, but requires the T7 RNA polymerase, a product of genes expressed earlier in the infection cycle. So if we clone our DNA fragment downstream from a T7 promoter, using an 'ordinary' *E. coli* host (lacking a T7 polymerase gene) we will get no expression at all. This can be useful, as the product might be deleterious to the cell. Once we are satisfied that we have made the right construct, we can isolate the plasmid and put it into another *E. coli* strain that has been engineered to contain a T7 polymerase gene, and hence will allow transcription of the cloned gene. If the expression of the T7 polymerase gene is itself regulated, for example by putting it under the control of a *lac*

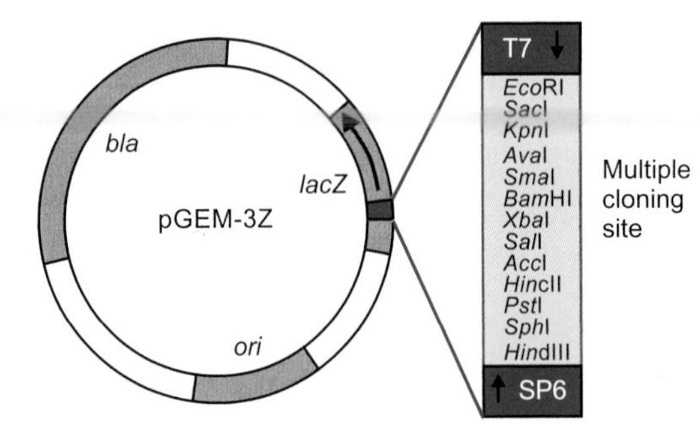

Figure 5.14 Structure of the expression vector pGEM-3Z. *ori* = origin of replication; *bla* = beta-lactamase (ampicillin resistance); SP6 and T7 are specific promoters recognized by the SP6 and T7 RNA polymerases respectively

promoter, then we can turn the expression of the T7 polymerase up (by adding IPTG) or down, so we still have control over the level of expression. Further devices can be included to inhibit the low level of T7 polymerase arising from the leakiness of the uninduced *lac* promoter.

The pGEM® series of vectors (see Figure 5.14) provide an example. In this case, there is a multiple cloning site adjacent to the T7 promoter, so any DNA inserted will be under the control of the T7 promoter. This is a transcriptional fusion vector, and is often more useful for generating substantial amounts of an RNA copy of your cloned fragment, which can then be used as a probe for hybridization. The vector actually has a second specific promoter, derived from another bacteriophage (SP6), at the other side of the multiple cloning site, so if you provide an SP6 polymerase you will get an RNA copy of the other strand, the antisense strand. The usefulness of this will be apparent when we consider applications of antisense RNA, such as the RNase protection assay (Chapter 10).

There are also a variety of translational fusion vectors with T7 promoters, which are designed for obtaining high, controllable, levels of protein expression. We will return to the concept of expression vectors, and other factors that have to be considered for the optimization of protein production, in Chapter 11.

5.6 Vectors for cloning and expression in eukaryotic cells

Most primary cloning (initial isolation of a gene or other DNA fragment from a target organism) is done with bacterial hosts (usually *E. coli*), because of the ease of manipulation and the range of powerful techniques available. Eukaryotic hosts are more commonly used for studying the behaviour of genes that have already been cloned (but now in an environment more closely related to their original source), for analysing their effect on the host cell and modifying it, or for obtaining a product which is not made in its natural state in a bacterial host. There is therefore more emphasis with eukaryotic vectors on obtaining gene expression rather than making gene libraries or primary cloning (with the notable exception of YAC vectors, see below). There is a bewildering variety of vectors available for cloning in different eukaryotic hosts, and a full review of them is beyond the scope of this book. In this chapter, we want to introduce some of the main concepts, many of which are similar in principle to those of bacterial cloning vectors, although there are significant differences; we will consider further the use of eukaryotic hosts for product formation in Chapter 11, and the genetic modification of animal and plant cells (or whole animals and plants) in Chapter 15.

5.6.1 Yeasts

Microbiologically, 'yeasts' are single-celled fungi, as opposed to filamentous fungi, but the term is quite imprecise. Not all 'yeasts' are related taxonomically, and indeed some filamentous fungi can also grow in a unicellular form that is referred to as a yeast form. Although in common usage the term 'yeast' would be taken to mean the brewer's/baker's yeast *Saccharomyces cerevisiae*, even molecular biologists are starting to have to recognize the existence of other yeasts (especially members of the genus *Pichia*, which we will encounter again in Chapter 11). However, for the moment we will limit ourselves to *S. cerevisiae*.

The vectors that will be most familiar, after reading about bacterial cloning vectors, are the yeast episomal plasmids (YEp). These are based (Figure 5.15) on a naturally occurring yeast plasmid known as the 2 μm plasmid, and they are able to replicate independently in yeast, at high copy number

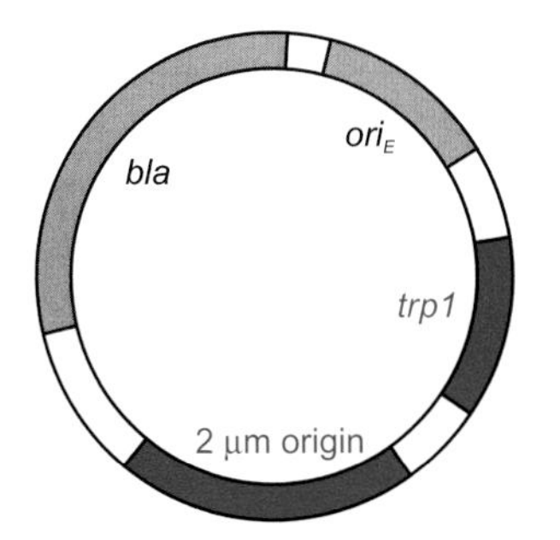

Figure 5.15 Structure of a yeast episomal vector. *ori* = origin of replication; *bla* = beta-lactamase (ampicillin resistance); *trp1* = selectable marker in *S. cerevisiae* auxotrophs; 2 μm origin = origin of replication in *S. cerevisiae*

(25–100 copies per cell). As is usually the case with vectors for eukaryotic cells, these plasmids also have an *E. coli* origin of replication, enabling them to be grown and manipulated in an *E. coli* host (i.e. they are *shuttle vectors*). There is one point of detail in which they differ from bacterial cloning vectors, and that is the nature of the selectable marker. For bacterial vectors, we can exploit the large number of antibacterial antibiotics (and the correspondingly large number of antibiotic resistance genes) to enable us to select our transformants. There are fewer antibiotics available to which yeasts are sensitive (although some fungicides can be used), and therefore selection more commonly makes use of complementation of auxotrophic mutations in the host strain. For example, a host strain of *S. cerevisiae* with a mutation in the *trp1* gene will be unable to grow on a medium lacking tryptophan. If the vector plasmid carries a functional *trp1* gene, then transformants can be selected on a tryptophan-deficient medium. Other markers that are commonly used for selection in a similar manner include *ura3* (uracil), *leu2* (leucine) and *his3* (histidine). These vectors would usually also carry an antibiotic resistance marker for selection in *E. coli*.

Vectors that replicate as plasmids in *S. cerevisiae* are often rather unstable, in that they tend to be lost from the culture as plasmid-free daughter cells accumulate. This is due to erratic partitioning of the plasmids during mitosis. Newer versions of YEp vectors, taking advantage of better understanding of the biology of the 2 μm plasmid, are more stable.

Autonomously replicating plasmids can also be constructed by inserting a specific sequence from a yeast chromosome; this sequence is known, unsurprisingly, as an *autonomously replicating sequence* or *ars*. The early versions of these plasmids were very unstable, but newer constructs that also include a centromere are more stable (Figure 5.16). In contrast to the YEp vectors, these yeast centromere plasmids (YCp), are normally maintained at low

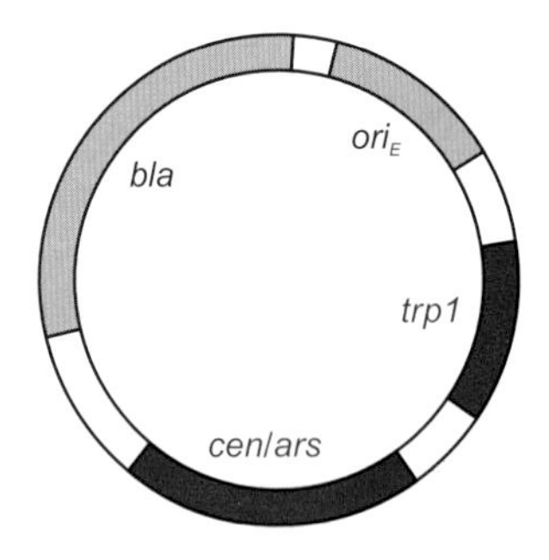

Figure 5.16 Structure of a yeast centromere vector. ori_E = origin of replication in *E. coli*; *bla* = beta-lactamase (ampicillin resistance, selectable marker in *E. coli*); *trp1* = selectable marker in *S. cerevisiae* auxotrophs; *cen/ars* = centromere and autonomously replicating sequence in *S. cerevisiae*

copy number (one or two copies per cell), which can be advantageous if your product is in any way harmful to the cell, or if you want to study its regulation.

Since one of the purposes of using yeast as a host is to express the cloned gene, these vectors are commonly designed as expression vectors. The principles involved are similar to those described above for bacterial expression vectors, except that of course the expression signals involved are those applicable to *S. cerevisiae* rather than *E. coli*. If required, this can include signals for secretion, or for targeting to the nucleus or other cellular compartments. Note that *S. cerevisiae*, although eukaryotic, has very few introns, and is not the host of choice if you want to ensure correct excision of introns.

The vectors described so far are maintained in yeast as circular DNA molecules, much like a bacterial plasmid. Two other classes of vectors deserve a mention. Firstly, there are the yeast integrating plasmids (YIp). These do not replicate independently but integrate into the chromosome by recombination (Figure 5.17). The frequency of transformation is very low, and it is difficult to recover the recombinant vector after transformation. The main advantage is that the transformants are much more stable than those obtained with the autonomously replicating plasmids.

Finally, there are the yeast artificial chromosomes (YACs), which carry telomeres that enable their maintenance in *S. cerevisiae* as linear structures resembling a chromosome. The use of these vectors, for cloning very large pieces of DNA, is quite distinct from the uses of the vectors described above. YACs are considered further, along with other vectors used for the same purpose, in a later section of this chapter.

5.6.2 Mammalian cells

In bacteria, the cloning vectors used replicate separately from the chromosome, as plasmids or bacteriophages. As we have seen above, the same is true of many types of vectors used in yeast. The situation with cloning in mammalian cells is somewhat different in that independent, plasmid-like, replication is often not sustained. Some vectors are capable of plasmid-like replication, especially those carrying the origin of replication from the virus SV40 (simian virus 40), which replicate episomally in some mammalian cells (such as COS cells). More stable clones are obtained by inserting the DNA into the chromosome, which happens readily in mammalian cells. In either case, the cloning vector enables you to organize your cloned gene in relation to a set of expression signals, many of which are derived from viruses such as SV40 or cytomegalovirus (CMV). The general features of such a vector are exemplified in Figure 5.18. A gene inserted at the multiple cloning site (MCS) enables high-level constitutive expression from the CMV promoter, while the presence of a polyadenylation signal increases mRNA stability. The

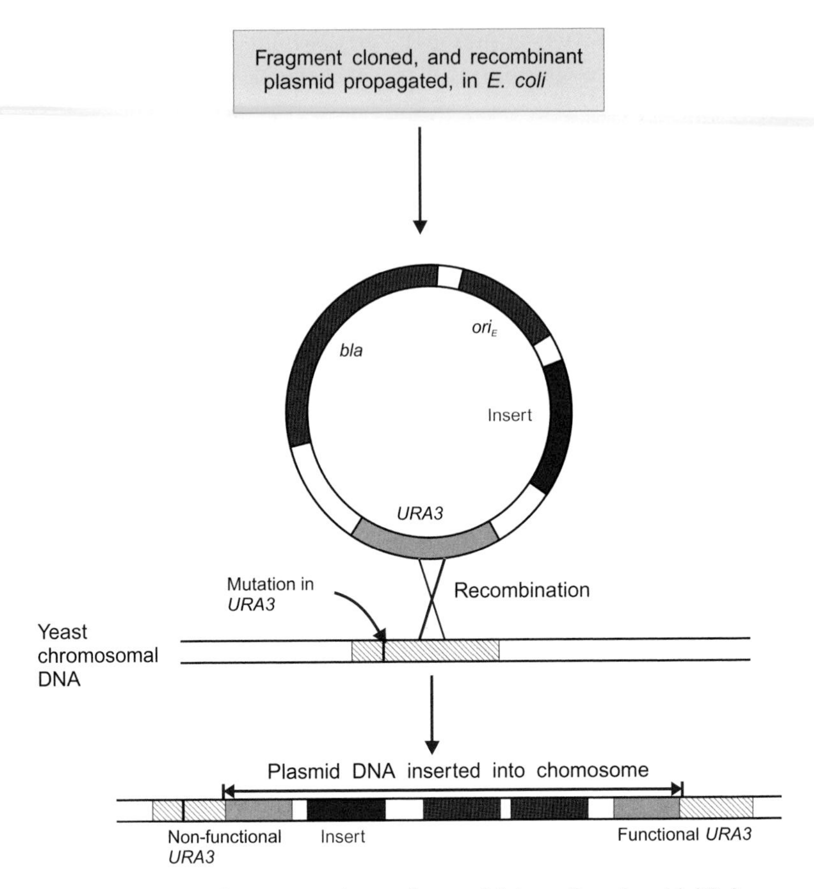

Figure 5.17 Structure and use of a yeast integrative plasmid (YIp)

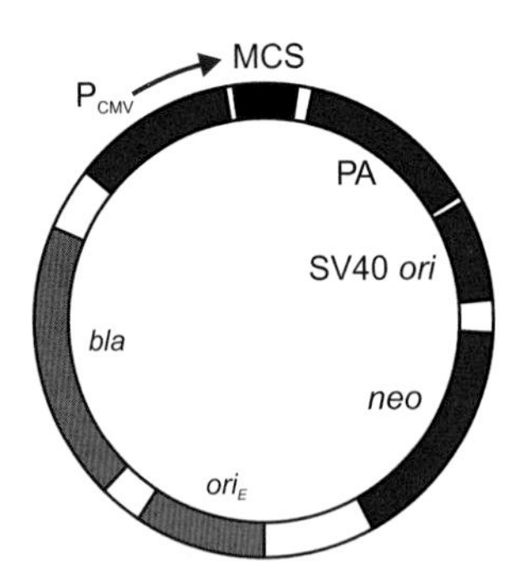

Figure 5.18 Structure of a basic episomal vector for gene expression in mammalian cells. $P_{CMV=CMV}$ promoter. High-level constitutive expression in mammalian cells; MCS = multiple cloning site; PA = polyadenylation signal; SV40 *ori* = origin of replication; episomal replication in COS cells; *neo* = neomycin phosphotransferase, resistance to G418, for selection in mammalian cells; ori_E = origin of replication in *E. coli*; *bla* = beta-lactamase (ampicillin resistance), for selection in *E. coli*

SV40 origin allows episomal replication in COS cells, and the neomycin phosphotransferase gene permits selection for resistance to the antibiotic G-418 (Geneticin®). Note that this is a shuttle vector, carrying an *E. coli* origin of replication and an ampicillin resistance gene (β-lactamase), so the construction can be carried out in *E. coli* before transferring the recombinant plasmid to a mammalian cell line. There is a wide variety of commercially available expression vectors for mammalian cells, with more sophisticated features than that shown. We will return to the topic of expression of cloned genes in mammalian cells in Chapter 11.

There are other types of vector available, based on various viruses, which can be used to transmit your cloned gene from one cell to another. Of these, the *retroviral* vectors deserve a special mention, and in order to understand these we need a brief account of retroviral biology. Retroviruses have an RNA genome. When a cell is infected, the RNA is copied into double-stranded DNA by the action of a viral protein, *reverse transcriptase*. This protein is present in the virion and enters the cell along with the RNA. (Reverse transcriptase is formally described as an RNA-directed DNA polymerase, and we will encounter it again in Chapter 6, where we consider its use in the production of cDNA from mRNA templates.) This DNA then circularizes and is integrated into the host cell DNA by the action of another virion protein known as *integrase*. The efficiency of integration of the DNA into the genome is one of the main attractions of this system for genetic manipulation of animal cells.

The integrated DNA is bounded by sequences known as *long terminal repeats* (LTR), which include a strong promoter for transcription of the integrated viral genes *gag*, *pol* and *env*. Full-length transcripts provide the viral RNA which is assembled into virus particles; one region of the virus, known as the *psi* site, is essential for this process. The packaged virus particles acquire envelope glycoproteins from the host cell membrane as they bud off from the cell, without lysis. These glycoproteins determine the type of receptor the virus uses to infect further cells.

Development of vectors based on retroviruses rests on the knowledge that most of these functions can be provided *in trans*, e.g. by genes from a defective helper virus already integrated into the genome of the host cell. The main features that are *cis*-acting, and therefore need to be located on the vector itself, are the LTR sequences and the *psi* site.

The basic features of the use of such a vector are outlined in Figure 5.19. The vector is a shuttle plasmid, so *E. coli* is used for construction of the recombinant plasmid by inserting the required gene at the multiple cloning site. This construct is then used to transfect a culture of a special cell line (*helper cells*) that contains the *gag*, *pol* and *env* genes required for virus production, integrated into the genome. The transfected cells will therefore be able to produce virus particles containing an RNA copy of your construct.

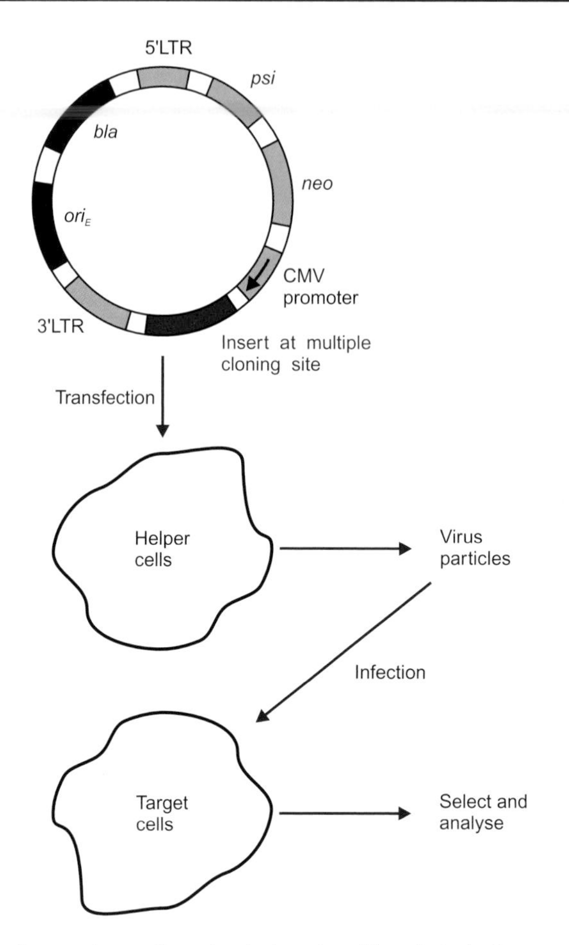

Figure 5.19 Structure and use of a retroviral vector; *bla* = beta-lactamase (ampicillin resistance), for selection in *E. coli*; ori_E = origin of replication in *E. coli*; *neo* = neomycin phosphotransferase, for selection in infected cells; CMV promoter is for transcription of cloned gene; LTR, *psi* = essential *cis*-acting retroviral sequences (see text)

These particles are able to infect other cells that do not contain the integrated essential genes; since the viral particles carry preformed reverse transcriptase and integrase, the RNA will be copied into DNA in such cells, and the DNA will be efficiently integrated into the genome. However, since these cells do not carry the essential genes, no further production of viral particles will occur. However, your gene is now stably integrated into the chromosome and can be expressed from the adjacent promoter derived from the vector.

The specificity of the viral particles for other cells will be determined by the envelope gene carried by the helper cells. Replacing that gene by other genes for envelope glycoproteins from other viruses, in particular the VSV-G gene from vesicular stomatitis virus, enables a wider range of target cells to be used, not just mammalian cells but extending to, for example chickens, oysters, toads, zebrafish and mosquitoes – in fact cells from virtually all non-mammalian (and mammalian) species can be infected.

The expression of genes in mammalian cells, and the incorporation of foreign genes into the genome of whole animals (*transgenesis*) is considered further in Chapters 11 and 15, respectively.

5.6.3 Plant cells

The most distinct way of getting foreign DNA into a plant cell involves the bacterium *Agrobacterium tumefaciens*. In nature, this causes a type of plant tumour known as a *crown gall*. The pathogenic ability of these bacteria is associated with the presence of a type of plasmid, known as a *tumour-inducing* (Ti) plasmid. The tumour is produced by the transfer of plasmid DNA (or more specifically a 23 kb fragment of the plasmid, known as T-DNA) from the bacterium to the plant cells, where it becomes integrated into the plant chromosomal DNA. This can be exploited by hooking foreign genes up to the plasmid so that the genes concerned are also transferred into the plant cells.

This is actually more difficult than it sounds. The natural Ti plasmids are very large (over 200 kb), and so it is not possible to clone genes directly into them. One strategy for overcoming this is illustrated in Figure 5.20. In this

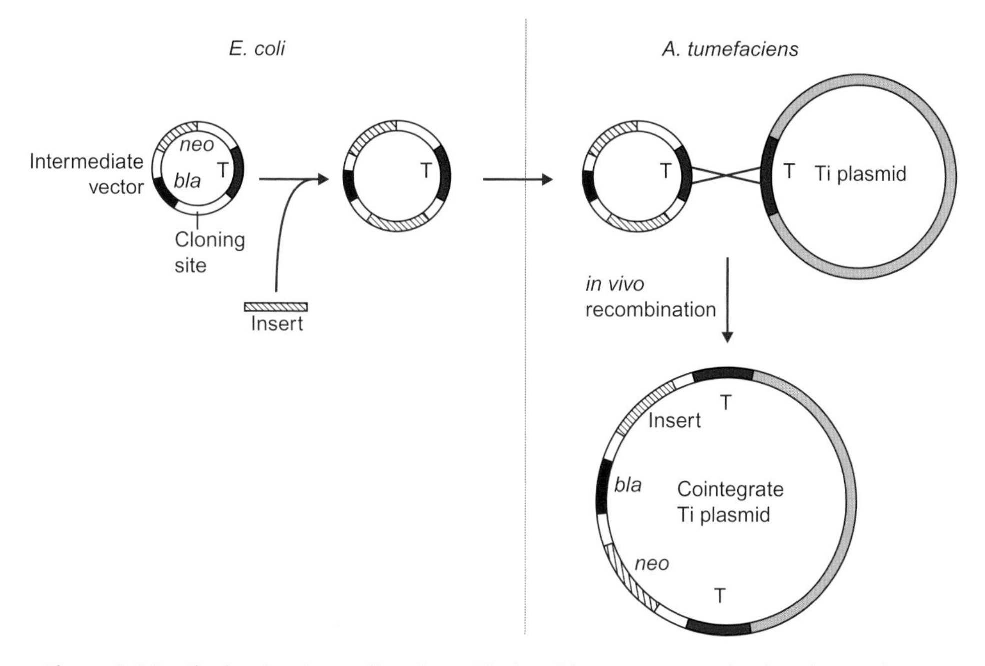

Figure 5.20 Cloning in plant cells using a Ti plasmid; *neo* = neomycin phosphotransferase, selction for resistance to G418 in plant cells. *bla* = beta-lactamase, selection for ampicillin resistance in *E. coli*; T = T-DNA fragment, transmitted to plant cells

procedure, the initial construct in *E. coli* is made in an *intermediate vector*, a small plasmid containing a part of the T-DNA fragment from a Ti plasmid. The recombinant intermediate vector can be transferred to *A. tumefaciens* by conjugation. Within the *A. tumefaciens* cell, a single recombination event between the homologous T-DNA sequences on the intermediate vector and the resident Ti plasmid will result in the incorporation of the entire intermediate vector plasmid into the T-DNA region of the Ti plasmid. The intermediate plasmid is not able to replicate in *A. tumefaciens*, so selection for antibiotic resistance will enable recovery of the bacterial cells carrying the cointegrate plasmid. Infection of a plant will now result in the transmission of the genes from the intermediate vector (including the cloned insert) into the plant cells along with the T-DNA.

Rather than infecting an intact plant, *A. tumefaciens* can be used to infect plant cells in culture and transformants can be isolated on a selective medium. All cells in a plant regenerated from a transformed cell will contain the cloned gene, and so the gene will be inherited in the subsequent progeny of that plant. However, plant cells infected with a wild-type Ti plasmid will not regenerate properly, because of the oncogenic effects of the T-DNA. However, only short sequences at the ends of the T-DNA region are necessary for transmission to the plant cells. The oncogenic genes between these sequences can be removed, resulting in *disarmed* Ti plasmids. Any DNA that is inserted between these two sequences, in place of the normal oncogenic genes, will be transmitted to the plant cells, and, in the absence of the oncogenes, the plant cells can be regenerated into mature plants carrying the foreign DNA. Further information on the genetic manipulation of plants and their applications will be found in Chapter 15.

5.7 Supervectors: YACs and BACs

Although cosmids were the first vectors that made the production and use of mammalian gene libraries feasible, their limited capacity would still not have sufficed for decoding the human genome. This was made possible by the development of novel supervectors that were able to carry 100 kb or more. The first of these were the YAC vectors (Figure 5.21). As with the shuttle plasmids we have referred to earlier, a YAC vector is propagated as a circular plasmid in *E. coli*. Restriction enzyme digestion removes the stuffer fragment between the two telomeres, and cuts the remaining vector molecule into two linear arms, each carrying a selectable marker. The insert is then ligated between these arms, as in the case of phage lambda, and transformed into a yeast cell, with selection for complementation of auxotrophic markers. This ensures that the recombinants contain both arms. Furthermore a successful recombinant must contain the *TEL* sequences at each end, so that the yeast transformant can use these sequences to build functional telomeres. The

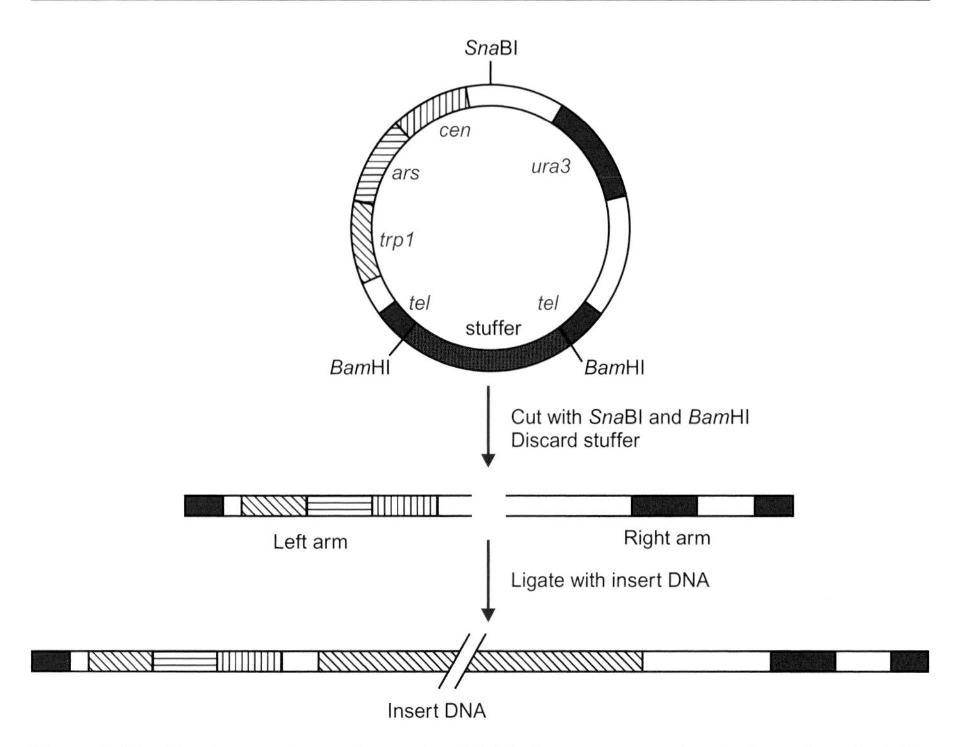

Figure 5.21 Structure and use of a yeast artificial chromosome vector. *trp1*, *ura3* = selectable markers; *cen/ars* = centromere and autonomously replicating sequence, enabling replication in *S. cerevisiae; tel* = telomere; for simplification, the *E.* coli origin of replication and selectable markers for *E. coli* are not shown

titans amongst vectors, YACs are routinely used to clone 600 kb fragments, and specialized versions are available which can accommodate inserts close to 2 Mb. As such, they will not only easily accommodate any eukaryotic gene in its entirety, but also the gene will be within its framework of three-dimensional structure and distant regulatory sequences. They have therefore been very useful in the production of transgenic organisms (see Chapter 15).

However, YACs have problems with the stability of the insert, especially with very large fragments, which can be subject to rearrangement by recombination. Furthermore, apart from the fact that many laboratories are not set up for the use of yeast vectors, the recombinant molecules are not easy to recover and purify. Thus, larger bacterial vectors are used more than YACs even though their capacity is lower. These include vectors based on bacteriophage P1, which are able to accommodate inserts in excess of 100 kb, and bacterial artificial chromosomes (BACs), which are based on the F plasmid and can accommodate 300 kb of insert. These vectors are more stable than

YACs, and play an important role in genome sequencing projects (see Chapter 12).

5.8 Summary

Earlier in this chapter we considered the advantage that can be gained in constructing gene libraries using vectors with higher cloning capacity such as lambda replacement vectors and cosmids. Even larger fragments can be cloned by employing other vector systems. For relatively small genomes, such as those of bacteria, there is not usually a need for such large inserts. If we simply find out that our gene is somewhere in a 300 kb insert, there is still a lot of work to do to find out exactly where it is. We might as well go straight to a library in a lambda replacement vector; with inserts of 15–20 kb, a library of a few thousand clones will provide adequate coverage of a bacterial genome. Screening a library of this size is quite straightforward. However, libraries of larger fragments are extremely useful for establishing physical maps of the chromosome and as an adjunct to genome sequencing projects (see Chapter 12).

However, for larger genomes, such as those of mammalian cells, a reasonably complete gene library in a lambda replacement vector would require hundreds of thousands of clones, and screening then becomes laborious. In these cases, the use of larger inserts is advantageous, and the use of additional systems such as YAC vectors becomes appropriate.

6 Genomic and cDNA Libraries

Gene libraries play a central role in gene cloning strategies, as will be apparent from the frequent references to them in previous chapters. Why are they so important? The answer to this question goes back to the basic concept of gene cloning, as outlined in Chapter 3. If you want to isolate a specific gene (or more generally a specific fragment of the genome), you can break the genome up into small enough pieces to be cloned using your chosen vector, and then screen that collection of clones to identify the one that contains the bit of DNA that you are looking for. (We will consider the ways of doing that in the next chapter.) You only want that one clone, so you could in principle throw all the rest away. However, if at some point in the future you want to clone another piece of DNA from the same source, you would have to go through the whole process again. Therefore, instead of discarding all the clones you do not want, you can store them away somewhere, pooled together, so that next time you can just screen the same collection to identify the clones that carry this other gene. You can share the collection with colleagues, or ask to share theirs for another cloning need. You have made a gene library – a resource that can be used to retrieve any of the genes from your starting material. Even if you do not intend to keep it and re-screen it, it is still a gene library.

A gene library, therefore, is a collection of clones which between them represent the entire genome of an organism. More specifically, we should refer to such a library as a *genomic* library, to distinguish it from a different sort of gene library which is constructed from DNA copies of the mRNA present in the originating cells at the time of isolation. These DNA copies of mRNA are referred to as *copy* or *complementary DNA* (*cDNA*), and hence such a library is referred to as a *cDNA library*. In this chapter we will look at the construction of these two types of libraries; ways of screening them are considered in Chapter 7.

From Genes to Genomes, Second Edition, Jeremy W. Dale and Malcolm von Schantz

6.1 Genomic libraries

The first step in producing a genomic library is to fragment the genomic DNA into pieces of a suitable size for cloning in an appropriate vector. It might be considered that the simplest way to do this is to digest the DNA to completion with a restriction endonuclease such as *Eco*RI. However, therein lies a problem. In Chapter 4, we saw that the average fragment size generated by *Eco*RI is about 4 kb (given certain assumptions about DNA composition). However, this is only an average. Even if restriction sites are randomly distributed we would expect some fragments to be very much bigger, and some would be very small. Ligation tends to work best with smaller fragments, so these would be over-represented in the library, while some of the largest fragments may be too big to be cloned efficiently, if at all.

That is only part of the problem. Identifying a clone carrying a specific DNA fragment is often not the end of the story. For many purposes, notably for genome sequencing, we are likely to want to be able to isolate the adjacent DNA as well. This would enable us to piece together all the small bits of DNA represented by individual clones so as to build up a bigger picture. If we make a library of, say, *Eco*RI fragments, then we have no way of knowing how they fit together. There is no information in the library that connects one clone with another. To provide that information, we need a library of overlapping fragments. Figure 6.1 shows how these overlapping fragments enable the identification of clones on either side of the one that

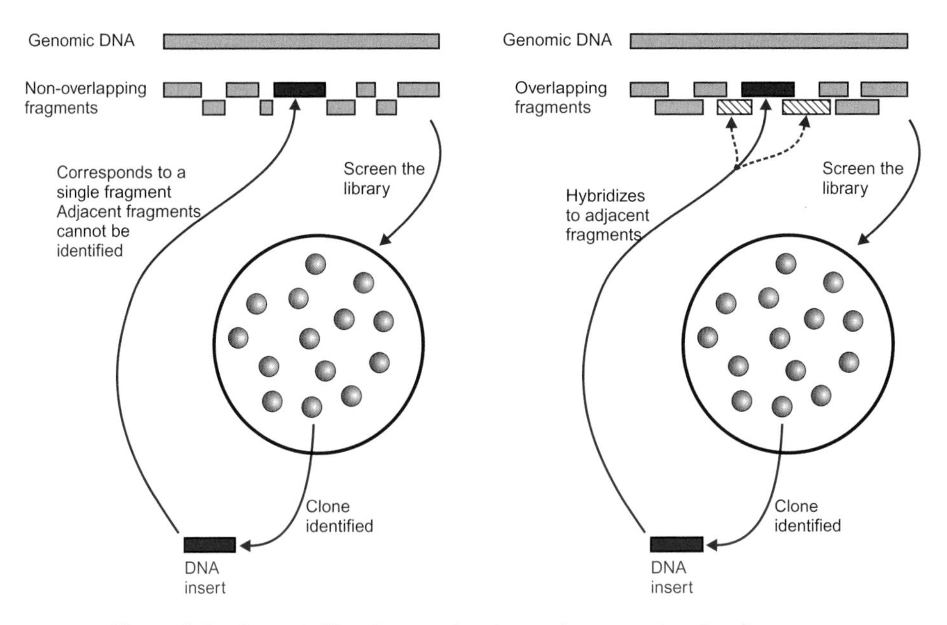

Figure 6.1 Genomic libraries: overlapping and non-overlapping fragments

we originally selected. These clones could then be used to identify further overlapping clones, and so we can move along the chromosome in either direction. This is the basis of a technique known as *chromosome walking*, which we will come back to in Chapter 12.

6.1.1 Partial digests

One way to construct a library of overlapping fragments is to use partial digestion. This means using conditions, such as short digestion times, that result in only a small proportion of the available sites being cut. A similar effect can be obtained by using very small amounts of enzyme, or by incubating the digest at a reduced temperature, or a combination of these. The digested material may then be fractionated by electrophoresis to obtain fragments of the required size range before cloning in an appropriate vector. If there is an equal probability of cutting at any site, the result will be a series of overlapping fragments (Figure 6.2), which would overcome the difficulties referred to above. If we do this with an enzyme, such as *Eco*RI, that recognizes a six-base sequence, the average size of fragments in a partial digest will be too large for cloning in typical plasmid or lambda vectors, but vectors that can accommodate large inserts (such as BAC or YAC vectors; see Chapter 5) can be used for generating a genomic library of such fragments from a large genome. For smaller genomes, such as those of bacteria, we would tend to use partial digests with an enzyme that cuts more frequently, e.g. one such as *Sau*3A that has a four-base recognition site (a 'four-base cutter'), and clone the products using a lambda or cosmid vector.

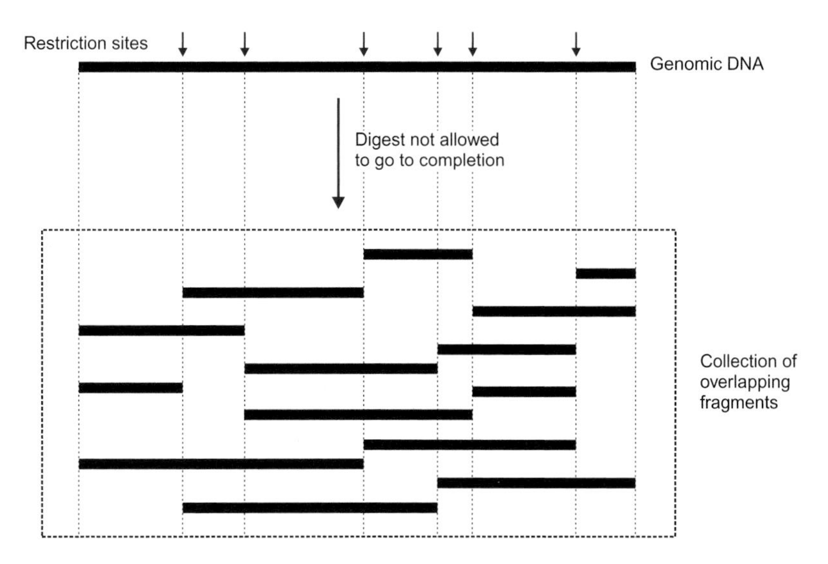

Figure 6.2 Using a partial digest to produce a collection of overlapping fragments

A four-base cutter such as *Sau*3A will statistically produce fragments of 256 bp on average (assuming an even distribution of sites). The fact that the fragments suitable for a lambda replacement vector are 15–20 kb in length implies that we are cutting only a few per cent of available sites. The reason for requiring such a low degree of digestion is not only to obtain overlapping fragments, but also because the distribution of restriction sites is not even. In some places there will be a number of sites very close together, and if we digest too much we will get only very small fragments from this region. These will be eliminated by size fractionation, or will not be suitable for cloning in vectors such as cosmids or lambda replacement vectors, where only recombinants carrying large inserts are viable. This would result in lack of representation of that region in the library. Thus, in order to minimize this problem, it is important to ensure that you get your partial digestion just right. The converse situation – lack of restriction sites in certain regions – is less likely to be an overwhelming problem in this case. There are unlikely to be any regions where neighbouring *Sau*3A sites are more than 20 kb apart, so there should be *some Sau*3A fragments in the appropriate size range (although the representation of different parts of the genome may be far from equal).

There is a further potential problem. We made the assumption that, in a partial digest, there is an equal probability of the enzyme cutting at any site. However, this is often not true. In a partial digest, some sites may be cut more efficiently than others. The nature of the adjacent sequences, and the formation of secondary structures in the DNA, may cause some sites to be cut more rapidly and/or efficiently than others. If this occurs, then our library of partial digest fragments will not be completely overlapping. When we come to try to fit the clones together, we will find discontinuities in the map so that we will be unable to identify the adjacent gene (or the clone carrying it). One way of overcoming this is to use separate partial digests with different enzymes.

Even with a library made in this way, we have not completely guaranteed that all our problems are solved. There is still the possibility that some regions of the genome will be over-represented and other regions will occur less frequently in the library. The only way of avoiding this, and ensuring that all parts of the chromosome are equally represented, is to abandon the strategy of using restriction fragments, and instead to use a truly random way of fragmenting the genome. This may be done by mechanical shearing, for example by passing the solution rapidly and repeatedly through a small syringe needle. This is an effective way of generating relatively large fragments, and has the advantage of requiring no special equipment, but the vulnerability of the DNA to shearing diminishes as the average fragment size decreases. A more elegant strategy is to use ultrasonication.

Although the complete randomness of mechanical shearing makes it more attractive in principle than partial digests, most workers continue to use restriction digests. Not only is it easier to control the extent of degradation, but also the restriction digests can be directly ligated with the vector. In contrast, the fragments generated by mechanical shearing have either blunt or 'ragged' ends – that is they may contain variable lengths of single stranded regions at the 5′ or 3′ ends. These are not suitable for ligation, and have to be converted to blunt ends by trimming back the unpaired ends with a single-strand specific exonuclease. In addition, since blunt end ligation is a relatively inefficient process, addition of linkers or adaptors (see Chapter 4) will be necessary in order to generate enough clones to constitute a representative library. However, mechanical shearing does have a role to play in generating the sets of overlapping fragments that are needed for *shotgun sequencing* (see Chapter 9).

6.1.2 Choice of vectors

Of the various types of vectors described in Chapter 5, which should we choose for constructing our genomic library? This choice is influenced by three interlocking parameters: the size of the insert that these vectors can accommodate; the size of the library that is necessary to obtain a reasonably complete representation of the entire genome; and the total size of the genome of the target organism.

Superficially, you might think that if we start with a genome of 4 Mb (4×10^6 bases; most bacterial genomes are of this order of magnitude) and produce a library of fragments which are 4 kb (4×10^3 bases) long, then you should be able to cover the entire genome with 1000 clones, since $(4 \times 10^6)/(4 \times 10^3) = 10^3$. However, this would only be possible if (a) all the clones are different, and (b) the clones are non-overlapping. Yet we are considering a library of random fragments, so the first condition is not met, and, as discussed above, we want the library to contain overlapping fragments, so the second condition does not hold either.

In a random collection of clones, the more clones we look at, the more likely it is that some of them will be completely or partially identical. In other words, the larger and more complete the library is, the more redundancy there will be. As we increase the size of the library, it becomes less and less likely that each additional clone adds any new information. Ultimately, it is not possible to produce a library that is *guaranteed* to carry all of the genetic information from the original genome. We have to use probabilities. We could require a 90 per cent probability ($p = 0.9$) of having the gene that we want, or a 99 per cent probability ($p = 0.99$). The level at which we set this probability will affect the required size of the library.

More specifically, the number of independent clones needed can be calculated from the formula

$$N = \frac{\ln(1-P)}{\ln(1-f)}$$

where N = required number of clones, P = the probability of the library containing the desired piece of DNA and f = the fraction of the genome represented by an average clone, which is calculated by dividing the average insert size by the total genome size.

Note that this calculation refers to the number of *independent* clones, i.e. the number of cells that were transformed originally (or the number of phage particles arising from a packaging reaction). This is usually determined from the number of colonies or phage plaques produced. Once you have plated out the library, and resuspended the colonies or plaques, you have *amplified* the library, and each clone is represented by thousands of individual cells or phage in the tube containing your library. You cannot increase the size (or complexity) of the library by plating out larger volumes. If your original library contains 1000 clones, plating it out to produce 10 000 plaques will simply mean each clone is present (on average) 10 times. You are *not* screening 10 000 independent clones.

Box 6.1 shows that, with the example of a bacterial genome of, say, 4 million bases and a plasmid vector carrying inserts with an average size of 4 kb, we will need a library of nearly 5000 clones to have a 99 per cent chance of recovering any specific sequence. (This assumes that all pieces of DNA are equally likely to turn up in the library, which is not entirely true; some fragments may be lethal, or may be difficult to clone for other reasons.)

Box 6.1 Estimates of the required size of genomic libraries

Organism	Genome size	Vector type	Average insert size	P	Library size
Bacterium	4×10^6 bases	Plasmid	4 kb	0.99	4.6×10^3
		Lambda replacement	18 kb	0.99	1.0×10^3
		Cosmid	40 kb	0.99	458
		BAC	300 kb	0.99	59
Mammal	3×10^9 bases	Plasmid	4 kb	0.99	3.5×10^6
		Lambda replacement	18 kb	0.99	7.7×10^5
		Cosmid	40 kb	0.99	3.5×10^5
		BAC	300 kb	0.99	4.6×10^4

The values shown for the genome sizes of bacteria and mammals are examples of the purpose of this calculation. The actual genome sizes vary quite widely from one organism to another. The insert sizes for specific vectors wil also vary.

Screening a library of 5000 plasmid clones is possible, but we can shorten the procedure by using vectors with a greater capacity. We can obtain a representative bacterial genomic library with about 1000 clones, using a lambda replacement vector. If we use a cosmid vector, the required size of the library is smaller – and we can reduce it even more by using vectors with a higher cloning capacity. However, there is a trade-off. One of the main purposes of producing a gene library is to be able to identify clones carrying a specific gene (or more generally a specific fragment of DNA, which could include regulatory sequences or other features), so that we can isolate and characterize that gene. The larger the insert size, the more work we have to do subsequently to find out which bit of that insert carries the gene we are interested in. (This can be illustrated by extending the argument to the absurd limit: the smallest gene library would be represented by a single clone carrying the entire genome, which would get us no nearer to identifying the gene that we want!) Therefore, for bacterial genomic libraries, lambda replacement vectors are usually the best compromise (although other vectors such as cosmids are sometimes used).

Note, however, that genomic libraries can be used for other purposes as well, especially for mapping and sequencing genomes. For these purposes larger fragments, which are capable of bridging any gaps that might be present, can be invaluable, even with bacterial genomic libraries. Vectors with a larger cloning capacity, such as cosmids and BACs, are often used for this purpose (see Chapter 12).

With larger genomes, such as those of mammals, the situation is rather different. As can be seen from Box 6.1, a library created in a lambda replacement vector would have to consist of nearly a million clones to be reasonably representative and would thus be more laborious to screen. On the plus side, a lambda library is reasonably easy to construct in a small laboratory and without special knowledge or equipment. However, wherever feasible, the use of vectors with larger cloning capacity can reduce the required size of the library to more manageable proportions.

A further factor that operates in favour of the use of larger inserts with mammalian (and other eukaryotic) genomes is that genes commonly contain introns. The overall size of the gene may therefore be too large to be contained within even a lambda replacement vector. (Although some bacterial genes do carry introns, they are relatively uncommon.) Therefore, if we need to obtain a clone carrying the entire gene, we will have to use a vector which can accommodate a large enough DNA fragment.

In the discussion so far, we have implicitly assumed that the library will be screened by replicating the clones to a filter, followed by hybridization with a labelled nucleic acid probe (see Chapter 7). Lambda vectors have a technical advantage over the other vectors in this process in that it is easier to screen phage plaques at high density than bacterial colonies. Not only do bacterial colonies tend to grow into one another if plated at high density, but also they tend to smudge when blotted, which can make it difficult to identify the required colony.

6.1.3 Construction and evaluation of a genomic library

The basis of the construction of a genomic library has been covered partly in earlier chapters and partly by this chapter so far. By way of recapitulation, the genomic DNA is fragmented, as randomly as possible, into suitably sized pieces for insertion into your chosen vector. The vector is prepared by digestion with the appropriate enzyme, and (for a lambda replacement vector) removing the stuffer fragment. The vector is then ligated with the complete mixture of genomic fragments. If you have a chosen a lambda vector, or a cosmid, you will need to mix your ligation products with packaging extracts for the assembly of infectious phage particles (see Chapter 5). If you are using a plasmid vector, you will introduce the mixture of ligated DNA into a bacterial cell by transformation or electroporation. The library will then be obtained as bacterial colonies (if using plasmid or cosmid vectors, or BACs) or phage plaques on a bacterial lawn (with lambda vectors). A library in a yeast artificial chromosome vector (YAC) would similarly be electroporated into yeast cells. For storage of the library, you would make a pooled suspension of the bacterial colonies, or of the phage harvested from the plate(s).

You then want to know how good your library is. Firstly you count the number of colonies or plaques (at an appropriate dilution so that you get countable colonies/plaques). If you do this using the original plates, it tells you the size (or complexity) of your library. If you determine the titre (the number of colonies or phage plaques) of the stored library recovered from the original plate, you will get a falsely elevated estimate of the size of the library, since this library has already been amplified. The amplification of a gene library is illustrated (Figure 6.3) with a plasmid vector. The concept with a phage vector is the same, except that plaques are produced and the amplified library will consist of phage particles. The library would consist of thousands or millions of clones, rather than the few shown.

Secondly, you will want to determine the quality of the library, i.e. what proportion of the clones actually contain an insert, and how large are those inserts? Both of these questions can be answered by picking a number of clones, growing them up individually, extracting the plasmid or phage DNA and subjecting it to restriction digestion followed by agarose gel electrophoresis. The details will vary according to the vector and cloning strategy, but at the simplest level (insertion of restriction fragments into a plasmid) you will see that each clone has one fragment of constant size corresponding to the vector, and fragments of various sizes which are your insert fragments (Figure 6.4). By determining the size of these fragments, you can estimate

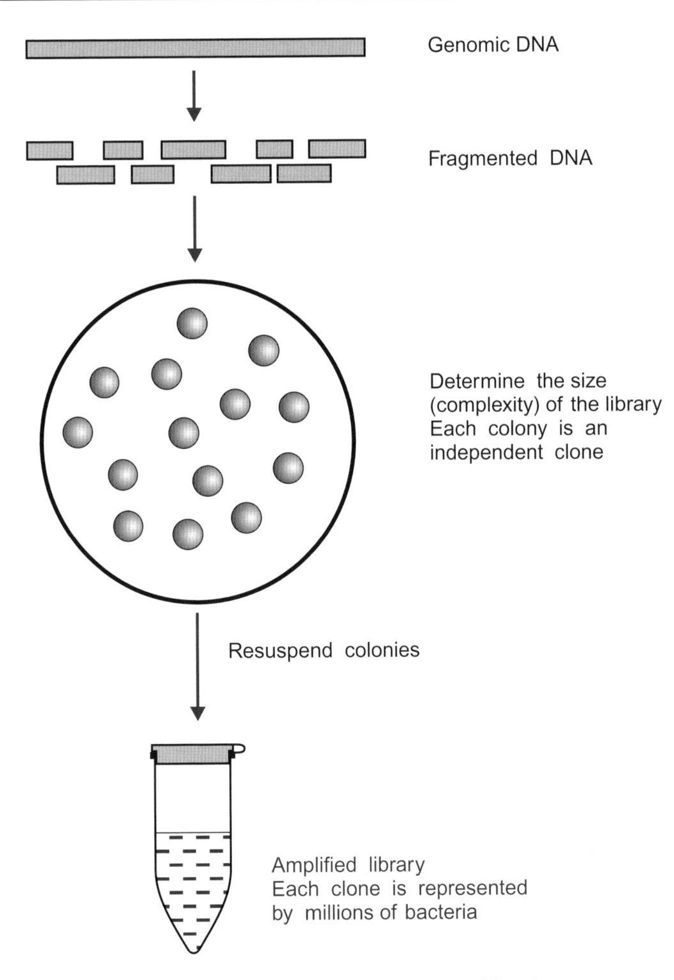

Figure 6.3 Amplification of gene libraries

the average insert size in your library (assuming you have picked a representative sample). Those without an insert band (see track 5) are probably religated vector, so you can estimate what proportion of clones do not have an insert at all. Clearly you do not want too many of these, as it will reduce the efficiency of your screening.

In Chapter 5, we described how some vectors, such as lambda gt11 or plasmid vectors of the pUC family (such as pUC18), provide you with a more direct estimate of the proportion of clones lacking an insert. With these examples, the non-recombinant vectors (without an insert) will produce blue plaques/colonies on an X-gal/IPTG containing plate (due to the β-galactosidase gene in the vector being uninterrupted by a cloned insert). If you have successfully introduced an insert fragment into the vector, the insert will (usually) disrupt the β-galactosidase gene, giving 'white' clones. Any

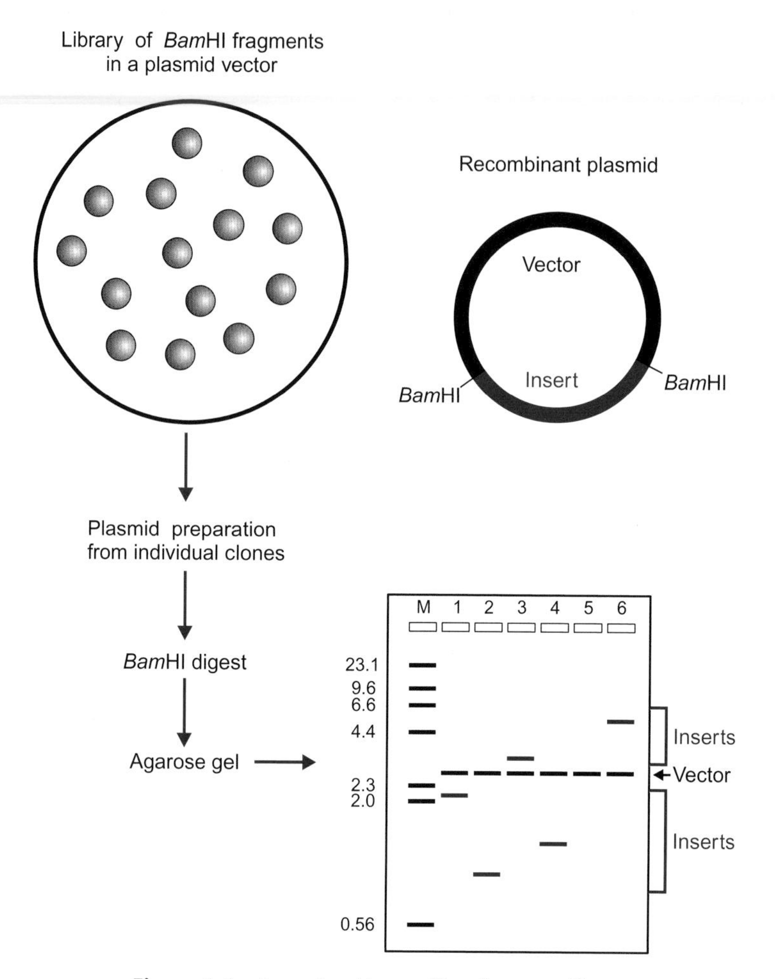

Figure 6.4 Assessing the quality of a gene library

religated or intact vector will give blue plaques or colonies. Lambda replacement vectors should not produce any clones lacking an insert (as they would be too small to be packaged), but intact vector may persist, or the stuffer fragment may not have been completely removed. Many replacement vectors contain additional markers to enable further differentiation or selection of recombinant clones. Cosmids also provide positive selection for the presence of inserts, again because the religated empty vector is too small for the packaging reaction.

Ideally, each clone should contain a single insert fragment. Multiple inserts can cause problems later on, as they will provide misleading information about the relationship between different parts of the genome. With those vectors that provide positive selection for inserts (lambda replacement vectors and cosmids) it is possible to dephosphorylate the insert (using alkaline phosphatase) – rather than the vector – thus virtually eliminating the possibility of multiple inserts. If you do this with other vectors, you run the risk of obtaining an unacceptably high frequency of vector religation. It may be worth running this risk, especially if you are using a vector such as pUC18 or lambda gt11 when you can tell straight away what your insertion frequency is. If you get too many blue colonies/plaques, then you throw the library away and try again.

6.2 Growing and storing libraries

Once a library has been made, it represents a potentially useful resource for subsequent experiments, as well as for the initial purpose for which it was produced. You will therefore want to store it safely for future use. A random library will consist of a tube containing a suspension of pooled colonies from a plate (if you used a plasmid or cosmid vector), or pooled bacteriophage particles (for a phage vector). This would normally be kept at –80°C; bacterial cells in a plasmid library are protected from the adverse effects of freezing by glycerol, while phage libraries are cryoprotected by dimethyl sulfoxide (DMSO). When the library is to be screened, a small lump of the frozen stock is removed and thawed. Plasmid libraries in cells are simply spread out on agar plates containing the appropriate antibiotic, which ensures that the cells do not shed their plasmids. Phage libraries first need to be mixed with, and infect prepared bacterial cells before being plated out. It is preferable to divide your library into aliquots before freezing it, so that each aliquot is only thawed once. This avoids the loss of clones that will accompany repeated freezing and thawing.

The first task is to determine the titre of the library, which almost inevitably will have dropped since the original stock was frozen. A dilution series is produced and each dilution is spread out on an agar plate (on a lawn of bacteria, in the case of phage libraries) and grown at 37°C overnight. This allows you to calculate the titre of the library, and to determine how much is to be used for each plate in the actual screening. The number of clones required for the screen can be calculated using the formula described above. Thus, the number of clones that need to be screened is dependent on (1) the size of the fragments in the library and (2) the size of the genome. However, note the earlier comments about library amplification. If your original library contained 1000 clones, you cannot screen 10 000 clones by growing up more of the library and plating out a larger amount. You are merely screening the same 1000 clones 10 times.

In the next section, we describe the construction of a different sort of library, using cDNA produced by reverse transcription of mRNA. To determine the number of clones in a cDNA library that you need to screen, you have to take account of the abundance of the relevant mRNA. If you are searching for a cDNA clone for a reasonably abundant mRNA, you might have to screen 20 000 clones (10 plates), while for a very rare transcript, you might need 200 000 clones (100 plates).

6.3 cDNA libraries

A genomic library represents all the DNA in the genome, whether it is expressed or not. However, very often it is really the genes that are being expressed that are our main target. This is likely to be quite a small proportion of the total DNA, especially in mammalian or plant cells. Therefore, if we base our library on the mRNA extracted from the target cells, rather than on their DNA, we will be able to focus more closely on the real target, and make the identification of the required clones much more efficient.

The advantages of cDNA libraries, and the contrast with genomic libraries, extend further than that. First of all, since the introns are removed by processing the mRNA, the cDNA clones will reflect only the coding regions of the gene (exons) rather than the much longer sequence contained in the genome. This is especially relevant if we want to try to express the gene in a bacterial host, which will be unable to splice out the introns.

Secondly, in any organism, some or most of the DNA does not appear to code for anything, or to have any other identifiable purpose. This is often referred to as *junk DNA* – but we need to be careful in this interpretation, as some of these sequences may have a function that we have not yet ascertained (see Chapter 12). The amount of apparent junk in the genome is to some extent related to genome size. Bacterial genomes carry mobile elements and phages, and some other repetitive elements, but in most cases relatively few obvious pseudogenes, and less non-coding DNA than mammalian genomes.

A library based on mRNA, rather than a genomic library, will reflect only those genes that are actually expressed in a particular cell or tissue sample at a particular time. Any non-transcribed regions will be excluded. However, it goes further than that. A cell will use only a part of its genetic capability at any one time. A bacterial cell will switch genes on or off, depending on its environment and its stage of growth, and in a multicellular organism, differentiation of cells into tissues and organs will be reflected in more or less permanent changes in the nature of the genes that are expressed. Therefore a cDNA library from, say, liver will be different from a kidney cDNA library from the same individual. Furthermore, some genes will be active only at certain times, such as during specific developmental stages, or different times of the day. You may also choose to make a library from a cancerous sample, or from an individual who suffers from a genetic disease. A library of this sort will reflect the nature of the cells from which the mRNA was obtained. As we will see later on, this not only reduces very substantially the number of clones needed for a representative library, but it also provides us with a variety of ways in which we can focus attention on the differences between various cells or tissues, and thus identify genes that are selectively expressed in different environments or in different tissues.

However, we cannot clone the mRNA directly. We have to produce a complementary DNA (cDNA) copy; hence the designation of such a library as a *cDNA library*. The synthesis of the cDNA is carried out using an enzyme known as *reverse transcriptase*. (Since transcription refers to the production of RNA from a DNA template, the opposite process – RNA-directed DNA synthesis – is known as reverse transcription, RT.) Although this is not a normal process in most cells, some types of viruses, such as leukaemia viruses and HIV, replicate in this fashion; the viral particle contains RNA, which is copied into DNA after infection, using a virus-encoded enzyme. Some cellular DNA polymerases also have reverse transcription capability.

6.3.1 Isolation of mRNA

Most of the RNA in a cell is not messenger RNA. The initial RNA preparation will contain substantial amounts of ribosomal RNA and transfer RNA. For construction of a cDNA library it is highly desirable to purify the mRNA. With eukaryotic cells, we can take advantage of the fact that mRNA carries a tail at the 3′ end – a string of A residues that is added post-transcriptionally. The polyadenylated mRNA will anneal to synthetic oligo(dT) sequences (i.e. short polymers of deoxythymidine, or in other words short stretches of synthetic DNA containing just T residues). Other RNA species and non-RNA components will not anneal and can be washed off (Figure 6.5). Although some mRNA in bacteria does have polyA tails, these are much shorter and only a small proportion of mRNA is polyadenylated. Therefore, this does

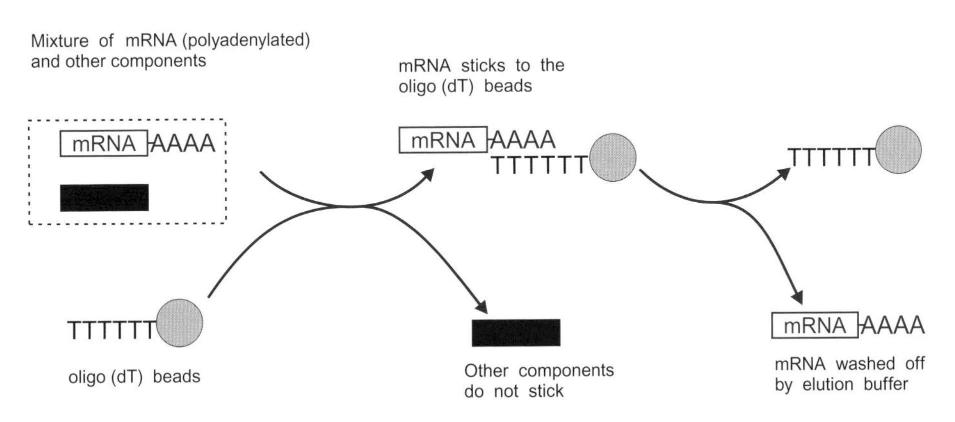

Figure 6.5 Principle of oligo-dT purification of mRNA

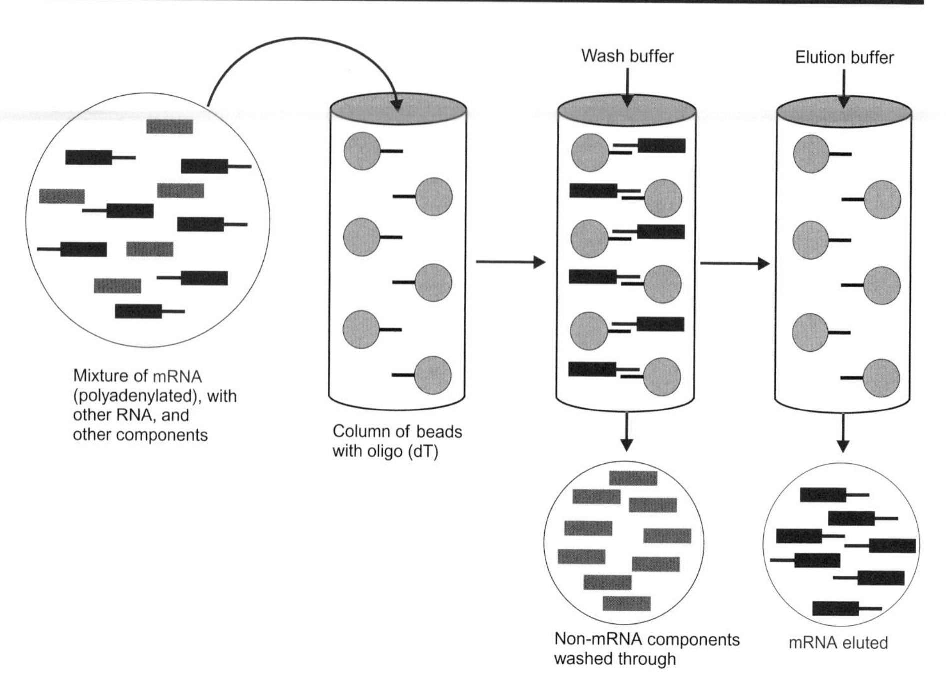

Figure 6.6 Purification of mRNA through an oligo (dT) column

not provide a reliable way of isolating bacterial mRNA. The question of the production of bacterial cDNA is discussed below.

When the RNA preparation is passed through a column of a polymer coated with synthetic oligo(dT) fragments, the polyA tail will anneal to the oligo(dT) residues and will be retained on the column while other RNA species will pass through. This is in effect a hybridization process, and as such the hybrids can be made unstable by lowering the salt concentration and raising the temperature, enabling the elution of purified mRNA from the column (Figure 6.6). The eluate will be a complex mixture of all the mRNA species present in the cell at the time of extraction. The relative amounts of the different transcripts will vary substantially, which has major implications for the ease of obtaining certain cDNA clones. This is a further clear distinction from a genomic library. Although the following description is presented in terms of a single mRNA, bear in mind that we would in reality be dealing with a complex mixture.

6.3.2 cDNA synthesis

The presence of the polyA tail is also useful in the reverse transcription step (Figure 6.7). Reverse transcriptase, like DNA-directed DNA polymerase, requires a primer for initiation. An oligo(dT) primer will anneal to the polyA

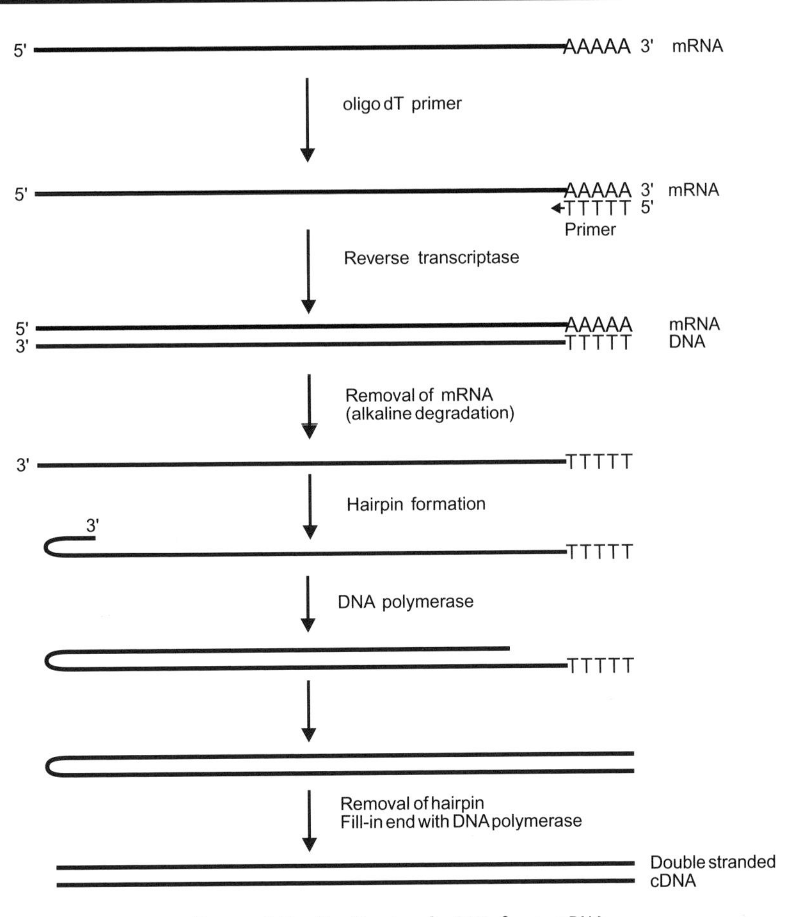

Figure 6.7 Synthesis of cDNA from mRNA

tail; reverse transcriptase will then extend this primer, using the mRNA as the template, and will produce a single-stranded cDNA copy.

We now have a double-stranded heteroduplex molecule that is a hybrid between DNA and RNA. In order to obtain a molecule that is stable and can be cloned into a vector, we need to replace the RNA with a DNA strand of the same sequence. One way of doing this is firstly to degrade the RNA strand by alkali treatment. This leaves the cDNA strand largely in single-stranded form. Single-stranded nucleic acid molecules tend to form secondary structures, looping back on themselves, because of the hydrophobicity of the bases. The single-stranded cDNA will therefore tend to form a hairpin loop at the 3′ end. This hairpin loop is used by DNA polymerase I to prime synthesis of the second strand. The product is a double-stranded DNA molecule, with a hairpin loop at one end. That loop is then removed by treatment with S1 nuclease (which will cut single-stranded DNA, including exposed loops).

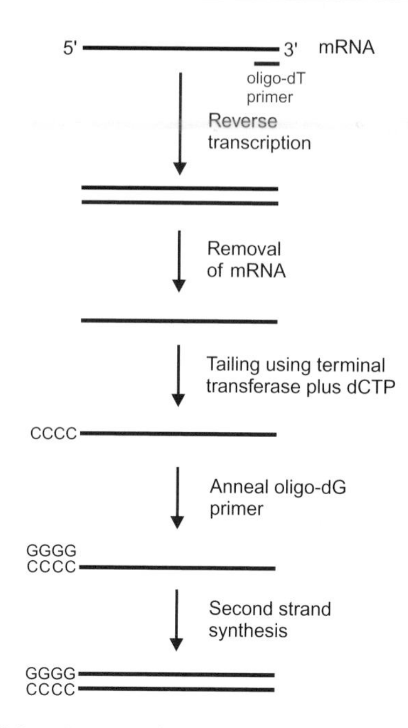

Figure 6.8 cDNA synthesis: using homopolymer tailing

This method suffers from several disadvantages. In particular, the requirement for hairpin formation to prime second-strand synthesis, and the need to cleave the hairpin with S1 nuclease, can cause the loss of sequence at the 5′ ends of the mRNA. It has therefore been largely superseded by alternative strategies, such as using homopolymer tailing to add a tail to the 3′ end of the first cDNA strand (see Figure 6.8). As described in Chapter 4, terminal transferase, if provided with, say dCTP, will add a string of C residues to the 3′ ends of DNA molecules. This enables the use of an oligo(dG) primer to initiate second strand synthesis, without requiring hairpin formation and cleavage.

Alternatively, degradation of the RNA strand by RNaseH (rather than alkali treatment) will leave small RNA fragments, which act as primers for second-strand synthesis. This also avoids the need for cutting the hairpin with S1 nuclease.

For cloning the cDNA, adaptors (see Chapter 4) are added, by blunt-end ligation, to make the cDNA molecules compatible with the chosen vector (Figure 6.9). After size-fractionation, eliminating excess adaptors and small, abortive cDNA fragments, the library is inserted into the vector in a second ligation. The size of cDNA molecules makes it unnecessary to consider vectors with a large cloning capacity. The choice is essentially between a plasmid vector, or a phage lambda insertion vector such as gt10 or gt11 (see Chapter 5).

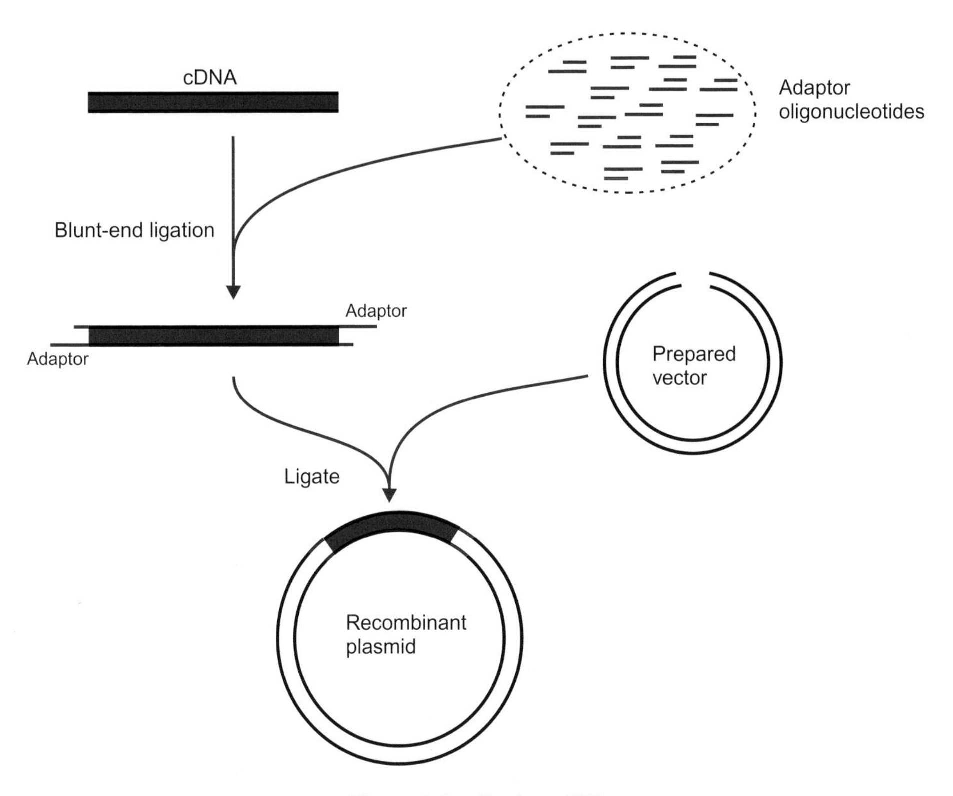

Figure 6.9 Cloning cDNA

One limitation of the basic procedures described above is that you may not get full-length cDNA. Constraints such as elements of secondary structure in the mRNA may interfere with reverse transcription, so that the enzyme rarely, if ever, reaches the end of the mRNA. As a result, the regions at the 5′ end of the mRNA may be under-represented in the cDNA library. This can be partially addressed by using random primers rather than oligo-dT primers. These random primers will initiate first-strand cDNA synthesis at intermediate points, and hence the enzyme will be more likely to reach the 5′ end of the mRNA (see Figure 6.10). You are unlikely to get any clones containing full-length cDNA, but these clones containing the 5′ end can be compared to other clones carrying the missing 3′ portion, making it possible to devise strategies for obtaining full-length molecules. The use of random primers rather than oligo(dT) primers also overcomes the problem that some RNA molecules (e.g. bacterial mRNA and genomic RNA from viruses) are not polyadenylated.

Many variations of these strategies have been devised to obtain full-length cDNA. The most powerful strategy, known as *rapid amplification of cDNA ends* (*RACE*), exploits the amplification power of the polymerase chain reaction; this is discussed in Chapter 8.

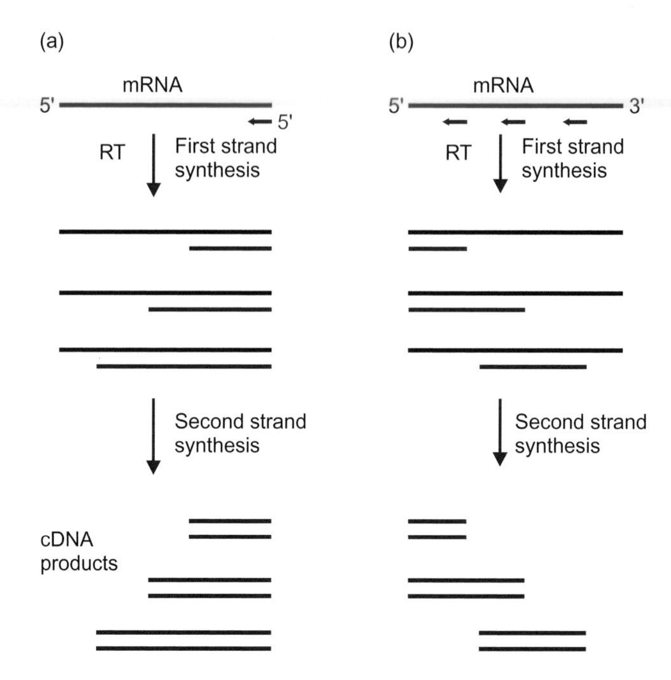

Figure 6.10 cDNA synthesis: enhancing representation of 5′ mRNA ends. (a) Oligo (dT) primer, incomplete products; (b) random primers – enrich for 5′ ends

Alternatively, if you can predict the sequence of the ends of the mRNA, for example from genome sequence data, you can simply use a pair of specific primers for RT-PCR amplification, which can generate your specific, full-length, cDNA directly from even small amounts of mRNA in your starting material. This bypasses completely the need for production and screening of a cDNA library.

6.3.3 Bacterial cDNA

The arguments in favour of cDNA, rather than genomic, libraries carry much less force with bacterial targets. The smaller size of bacterial genomes, and the (virtual) absence of introns, means that a genomic library is usually quite adequate – and a lot easier to construct. There are also technical difficulties in producing cDNA with bacteria. Not only is the mRNA not consistently polyadenylated, but also it is remarkably unstable: many bacterial mRNA species have a half-life (*in vivo*) of only a minute or two. Furthermore, the organization of bacterial genes into polycistronic operons (groups of genes that are transcribed into a single long mRNA) means that a bacterial mRNA can be as much as 10–20 kb in length. Not only is it difficult to isolate this mRNA intact, but it would be very difficult to produce a full-length cDNA copy from it.

As a consequence, bacterial cDNA libraries are rarely produced. However, for some purposes, such as the analysis of gene expression (see Chapter 10), cloning of bacterial cDNA can play an important role, for example in identifying those transcripts that are relatively abundant in the bacterial cells under selected conditions.

6.4 Random, arrayed and ordered libraries

So far we have envisaged a gene library as a single tube containing a mixture of a large number of clones. When you want to screen the library, you plate it out to obtain a large number of bacterial colonies or phage plaques. For many purposes, this is perfectly adequate. However, there are circumstances when you do not want to treat the library as a random collection of a large number of clones. For example, if you have a very complex screening procedure, you may be able to test only a portion of the library at a time. This might mean taking a small aliquot of the library and testing say 100 clones. If you then take another aliquot, and again test 100 clones, then some of those clones may be the same ones that you have already tested. (Remember that, in an amplified library, each independent clone is present in many copies.)

One way around this is to produce an *arrayed* or *gridded* library. If we go back to the original transformation step (or infection with the packaged phage particles), then instead of simply pooling all the clones and storing them in one tube, we can pick them individually and store them separately (Figure 6.11). This can be done using individual wells in microtitre trays, or sometimes on filters. If you are doing this manually, then you are limited in the size of library that you can handle (depending on how patient you are) –

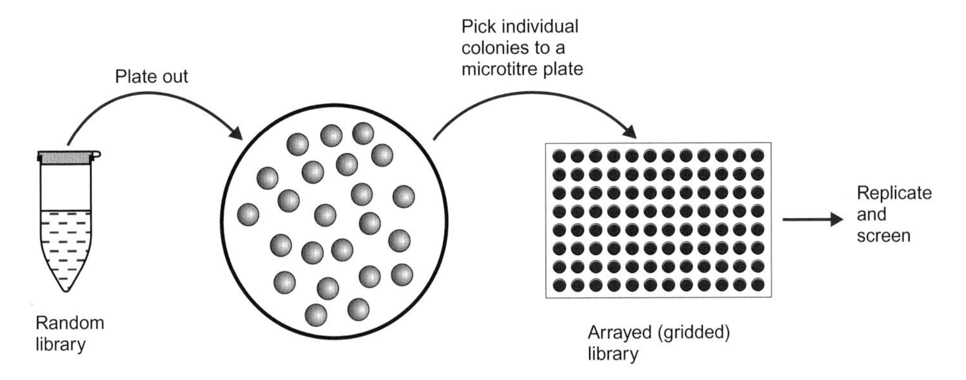

Figure 6.11 Production of an arrayed, or gridded, library

but it is quite possible for a bacterial library of a few thousand clones. However, there are machines available that will identify colonies on a plate, pick them individually and transfer them to individual wells in a microtitre tray, enabling the production of much larger gridded libraries. Once you have an arrayed library, it is then easy to subculture the clones to trays with fresh culture medium, creating multiple and identical copies of your library. You can then work your way through the library, testing each clone individually, knowing that every well contains a different clone. Alternatively, the library can be replicated to agar plates, or directly to membranes, which can then be screened to identify specific clones, as described in Chapter 7.

Replicating such a library onto a membrane is one way of producing another form of array, consisting of spots of DNA rather than viable clones. For a larger library, for example a human gene library, you would want to screen the library at a much higher density. That is, you would want to put the spots on the filter much closer together than would be obtained by merely replicating from the microtitre tray. This requires very precise positioning of the spots on the filter, which is achieved by using another robot.

The availability of genome sequence data makes it possible to produce DNA arrays without constructing a gene library, using PCR products or synthetic oligonucleotides. These either use nylon membranes (*macroarrays*) or glass slides (*microarrays*), and are especially useful for analysing variations in genome content and genome-wide analyses of transcription. The use of such arrays is considered in Chapters 13 and 14.

It is not necessary to be able to produce your own arrays, or to own your own robots. Arrays representing genomic and cDNA libraries from a considerable range of organisms are readily available from public and commercial resource centres, as well as facilities for producing such arrays from your own libraries. The most comprehensive collection is housed by imaGenes (http://www.imagenes-bio.de/). After obtaining the array and screening it, the investigator can order the positive clones from the resource centre.

An arrayed library as shown in Figure 6.11 still consists of a random set of clones. Without screening it, we have no information as to the nature of the insert in each clone, or the relationship between clones. A further development of the concept is to establish which clones overlap so that we can produce a set of clones that can be arranged in order, so as to cover the whole genome. This can be done by hybridization between sets of clones. For example, if we pick one clone at random (clone 1) and hybridize it to the library (Figure 6.12), we will find other clones that overlap. We can pick one of these (clone 2) and repeat the hybridization, identifying the next adjacent, overlapping clone (clone 3), and so on, until we have covered the whole genome. Obviously it is not quite as simple as that in practice. Note that this exemplifies how important it is to have overlapping clones in the library.

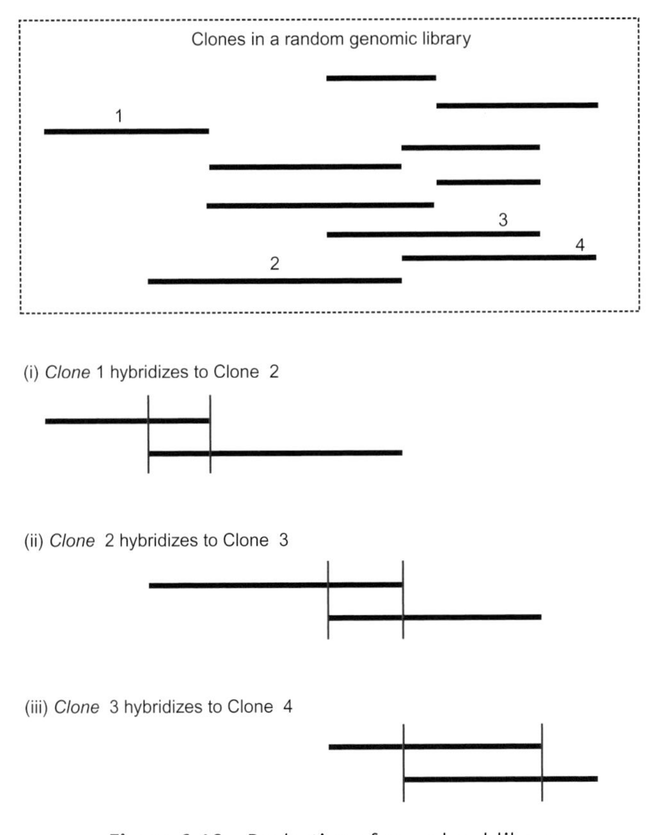

Figure 6.12 Production of an ordered library

Because of the inherent redundancy in a random library, the number of clones required in an *ordered* library is much less; some of the clones in the random library will not be needed. The actual number of clones needed will depend on the degree of overlap between the chosen clones, as well as the insert size.

An ordered library can be a valuable resource. Ordered libraries have contributed substantially to some of the genome sequencing projects – notably the public human genome sequencing consortium used an ordered library of BAC clones. However, the work required to establish an ordered library is substantial, even for a relatively small genome such as that of a bacterium. More recent developments in sequencing technology, and in the computer techniques for assembling sequence data from large numbers of small fragments (see Chapter 12), provide alternatives to the production of ordered libraries as strategies for genome sequencing.

7 Finding the Right Clone

7.1 Screening libraries with gene probes

In the previous chapter, we described how to create a gene library. Whether it is a genomic library or a cDNA library, it is a large collection of random clones, and we still have to find a way of identifying which clone(s) carry the gene that we are interested in. (We can ignore the more specialized topic of ordered libraries). This means that we need ways of rapidly *screening* very large numbers of clones. This is most commonly done using a nucleic acid *probe* (DNA or RNA), which will *hybridize* to the DNA sequence you are looking for in a specific clone. The principle involved is that the library (in the form of bacterial colonies or phage plaques) is replicated onto a filter, which is then treated to release the DNA and bind it to the filter. The filter then carries a pattern of DNA spots, replicating the position of the colonies or plaques on the original, and can be hybridized with the probe, which has first been labelled so that it can be easily detected. This allows you to detect which DNA spots hybridize to the probe, and recover the corresponding clones from the original plate.

In order to appreciate the power of this technique it is necessary to consider more closely the question of hybridization. Later in this chapter we will consider an alternative strategy, which involves using antibodies to screen an expression library.

7.1.1 Hybridization

Hybridization is based on the difference in stability between the covalent bonds in the nucleic acid backbone of each strand, and the much weaker hydrogen bonds that bind the two strands in the double helix together by base pairing. Thanks to this arrangement, the two strands can be safely

From Genes to Genomes, Second Edition, Jeremy W. Dale and Malcolm von Schantz

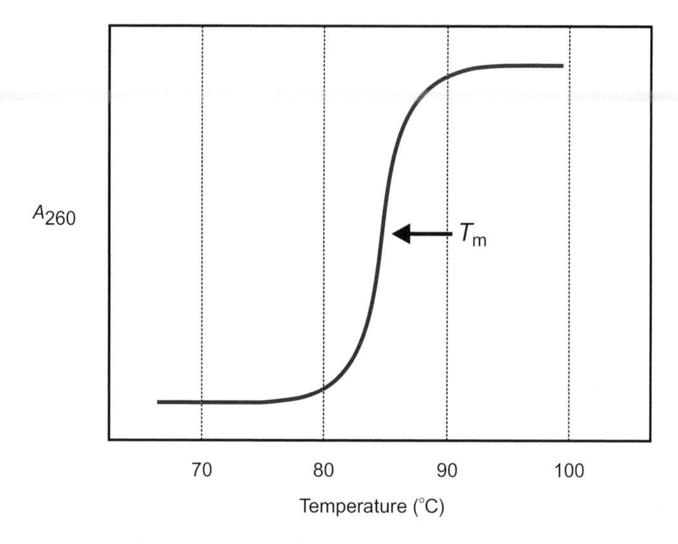

Figure 7.1 Melting (denaturation) of DNA

separated under conditions that are much too mild to pose any threat to the covalent bonds in the backbone. This is referred to as *denaturation* of DNA, and unlike the denaturation of most proteins it is readily reversible. Because of the complementarity of the base pairs, the strands will easily join together again and *renature*. In the test tube, DNA is readily denatured by heating, and the denaturation process is therefore often referred to as 'melting' even when it is accomplished enzymatically (e.g. by DNA polymerase) or chemically (e.g. by NaOH). The separation of the strands during renaturation causes a radical change in the physical properties of DNA, such as optical density (Figure 7.1). During melting of DNA, the optical density changes dramatically during a short temperature interval, and then stabilizes after the strands have separated entirely. The midpoint of this temperature interval is denoted the *melting temperature* (T_m). Under physiological conditions, the T_m is usually between 85 and 95°C (depending on the base composition of the DNA). In the laboratory, we can adjust other factors, such as the salt concentration (see below) to bring the melting temperature down to a more convenient range.

The reason that the T_m varies according to the base composition of the DNA is that guanine – cytosine base pairs are joined together by three hydrogen bonds, whereas adenine – thymine base pairs have only two. If we take reasonably large DNA molecules, such as might be obtained by isolating total DNA from an organism, we can make an estimate of its base composition by measuring the T_m. Alternatively, if we know the base composition, we can calculate the T_m. For shorter sequences, such as the 20–30 base synthetic oligonucleotides that are commonly used as *primers*, other factors have to be

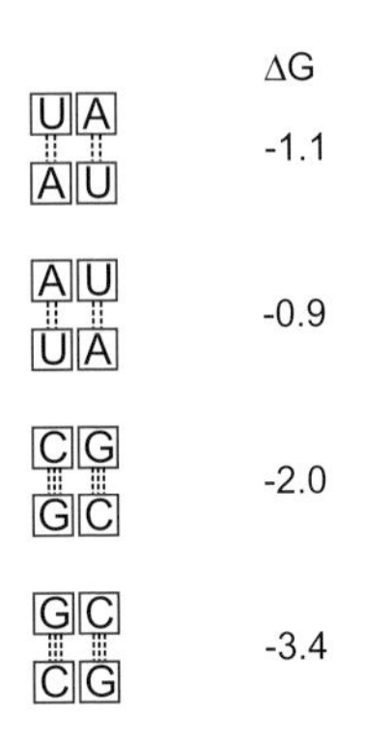

Figure 7.2 Free energy of base pairing

taken into account. The strength of the association between two bases (expressed as ΔG, the energy released on formation of a base pair) also depends on the adjacent bases, because hydrophobic interactions between adjacent bases (*stacking*) also affects the stability of the pairing. Some examples are shown in Figure 7.2, the negative values indicating that energy is released on formation of a base-paired structure. More energy released means greater stability. It can be seen that the free energy of base pairing (kcal/mol) for the CG/GC doublet is not the same as for the GC/CG doublet. For short oligonucleotides, calculation of the T_m therefore has to take account of the context of each base in the sequence (computer programs are readily available to do this for you). This is not likely to be such an important factor in determining the T_m of a longer stretch of DNA because there will be a tendency for these differences to average out.

Although the normal base pairs (A – T and G – C) are the only forms that are fully compatible with the canonical Watson – Crick double helix, pairing of other bases can occur, especially in situations where a regular double helix is less important (such as the folding of single-stranded nucleic acids into secondary structures; see below).

In addition to the hydrogen bonds, the double-stranded DNA structure is maintained by hydrophobic interactions between the bases on opposite strands. (This is in addition to the hydrophobic stacking of adjacent bases on the same strand, as referred to above.) The hydrophobic nature of the bases means that a single-stranded structure, in which the bases are exposed to the aqueous environment, is unstable; pairing of the bases enables them to be removed from interaction with the surrounding water. In contrast to the hydrogen bonding, hydrophobic interactions are relatively non-specific, i.e. nucleic acid strands will tend to stick together even in the absence of specific base-pairing, although the specific interactions enable a stronger association. The specificity of the interaction can therefore be increased by the use of chemicals (such as formamide) that reduce the hydrophobic interactions.

What happens if there is only a single strand, as is the case normally with RNA, and sometimes with DNA? In that case, removal of bases from the surrounding water is accomplished by the formation of secondary structures in which the nucleic acid folds up on itself to form localized double-stranded regions, including structures referred to as hairpins or stem – loop structures. At room temperature, in the absence of denaturing agents, a single-stranded nucleic acid will normally consist of a complex set of such localized secondary structure elements.

A further factor to be taken into account is the negative charge on the phosphate groups in the nucleic acid backbone. This works in the opposite direction to the hydrogen bonds and hydrophobic interactions; the strong negative charge on the DNA strands causes electrostatic repulsion that tends to repel the two strands. In the presence of salt, this effect is counteracted by the presence of a cloud of counterions surrounding the molecule, neutralizing the negative charge on the phosphate groups. However, if you reduce the salt concentration, any weak interactions between the strands will be disrupted by electrostatic repulsion – hence the use of low salt conditions to increase the specificity of hybridization (see below).

If two similar, but different, double-stranded DNA fragments are mixed, melted and then left to renature, some of them will anneal with their perfectly complementary halves, but others will have formed hybrid, imperfectly matched, pairs (Figure 7.3). This is known as a *heteroduplex*. This would happen, for example, if you were to mix the cDNA molecules encoding the human red- and green-sensitive photopigments, or the actin genes from mouse and rat. Any base pairs that do not match will cause imperfections in the resulting heteroduplex. A higher number of mismatches will lead to a

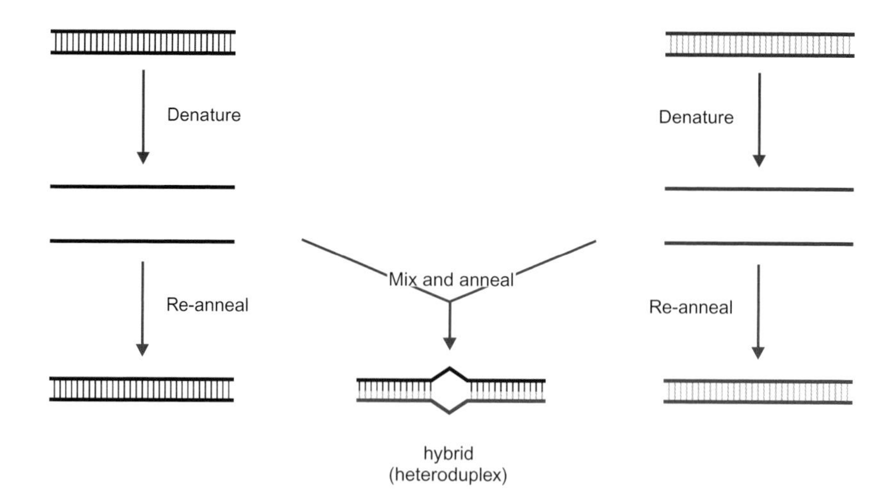

Figure 7.3 Formation of hybrid DNA between similar but non-identical DNA molecules

less stable hybrid – in other words one that would have a lower melting temperature.

This is very important when single strands of nucleic acids are hybridized in the laboratory. The investigator can choose conditions that would be more or less forgiving of partial mismatches, depending on whether they are wanted or not. This is referred to as varying the *stringency* of the hybridization. The most important and obvious way to increase the stringency is to raise the temperature. Secondly, hybrid DNAs are more stable at higher salt concentrations. At low salt concentrations, the negative charges of the phosphate groups in the backbones cause an electrostatic repulsion between the two strands. At higher salt concentrations, the presence of positive counterions will relieve this repulsion. In the laboratory, salt concentration is usually regulated as multiples of SSC (standard saline citrate; 1 × SSC is defined as 0.15 M sodium chloride and 0.015 M sodium citrate). Sometimes, formamide is added to the hybridization solution. Formamide lowers the melting temperature and is therefore used in situations where hybridization temperatures need to be kept low, such as when carrying out *in situ* hybridization (see Chapter 10).

The basic types of hybridization that are used in the laboratory are filter hybridization, solution hybridization, and *in situ* hybridization. In general, and in this chapter in particular, a specific nucleic acid fragment (DNA or RNA) is labelled and used as a *probe* to detect a corresponding DNA strand, the *target*, in a complex mixture such as a gene library. However, we will also come across situations where *reverse hybridization* is used. For example, in the use of arrays (see Chapters 13 and 14), a series of specific DNA fragments is arranged on a filter or a slide, which is hybridized with a labelled complex mixture such as fragments of total genomic DNA. In this situation, we will refer to the labelled complex mixture as the probe, rather than the specific fragments, but the terminology is not always consistent.

7.1.2 Labelling probes

A fundamental feature of nucleic acid hybridization is that the probe is labelled in a way which will make it possible to detect it, and thereby the target it has bound to, after the hybridization. The classic labelling method is to incorporate a radioactive isotope into the probe molecule. Isotopes that are used for this include ^{32}P, ^{33}P, ^{35}S and ^{3}H. A more energetic isotope, such as ^{32}P, gives a stronger signal, but less good resolution. This makes it useful for experiments where sub-millimetre resolution is irrelevant, such as Southern blot hybridization (see below). At the other extreme, ^{3}H gives a weaker signal and thus requires a much longer exposure time to be detectable. On the other hand, it provides excellent resolution, and can be used in experiments such as *in situ* hybridization (Chapter 10), where it is important to be able to assign the probe labelling not only to specific cells, but even to specific regions of chromosomes.

The classical way to detect radiolabelled probes is to place the probe–target hybrid on an X-ray film. Radioactive particles will expose the region of the film with which they are in contact just like X-rays or visible light would. A more modern method for detecting binding of radioisotopes is phosphoimaging. This method requires specialized and expensive apparatus, but is on the other hand faster, and the phosphoimaging plates can be reused, in contrast to X-ray films.

Many investigators have abandoned radioactive labelling methods in favour of non-radioactive ones. These have advantages in terms of worker safety, detection speed and cost. They are also unaffected by the continuous decay of radioisotopes, and are therefore more stable, so your labelled probe can be kept and re-used. There are a variety of non-radioactive labels that can be used. Nucleotides substituted with biotin or digoxigenin can be incorporated into the probe, and then detected with specific antibodies (or, in the case of biotin, using avidin, which binds very strongly and specifically to biotin). The antibody (or avidin) that is used is itself labelled with an enzyme, such as horseradish peroxidase (HRP) or alkaline phosphatase. These enzymes can be detected using a chromogenic substrate (i.e. a substrate that yields a coloured product when reacted with the enzyme) or a chemiluminescent substrate (where the initial reaction product is unstable and light is emitted as it breaks down). In the latter case, the emitted light will darken X-ray film just like a radioisotope would. As with radioactive methods, specialized equipment can be used instead of X-ray film.

Alternatively, the probe can be labelled directly with HRP, or a fluorescent label can be incorporated, which allows direct detection of the labelled probe. Such probes are especially useful for *fluorescent in situ hybridization* or *FISH* (Chapter 13).

7.1.3 Steps in a hybridization experiment

Nucleic acid probes have a tendency to bind non-specifically to other materials on the filter, or even to the filter itself. To minimize this non-specific probe binding, the hybridization solution contains various blocking agents, which may include detergents, bovine serum albumin, and non-homologous DNA. It is often advantageous to pretreat the filter with the hybridization solution without added probe (*prehybridization*).

If the probe is double-stranded DNA, it will need to be heat-denatured by boiling prior to hybridization in order to make it accessible to the target. After hybridization, which typically takes place in a controlled-temperature chamber overnight, non-specifically bound probe is removed by washing. This is the step where you can most conveniently decide how tolerant the

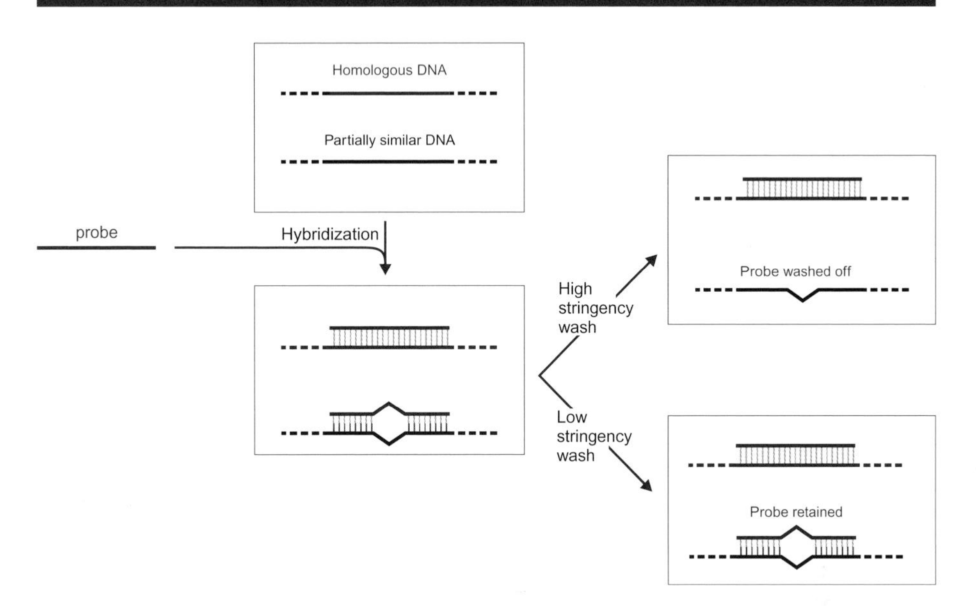

Figure 7.4 High and Low stringency washing

experiment should be of partially mismatched probe–target hybrids, by choosing an appropriate combination of temperature and salt concentration. If, for example, the intention is to hybridize a cDNA probe with genomic DNA from the same species, then high stringency conditions (i.e. high temperature and low salt concentration) should be chosen (Figure 7.4). This will ensure that the probe will remain bound to the membrane only where it has annealed to the correct complementary sequences. Where the probe has annealed to DNA that is partially similar (for example members of the same gene family), the annealing will be disrupted by these conditions, and the probe will be washed off the filter. If, on the other hand, you want to detect, say, a chicken gene with a probe made from human DNA (a heterologous probe), then low stringency (i.e. low temperature and high salt concentration) should be chosen in order to protect the expected partially mismatched hybrids.

The hybridization of the filter with an appropriate probe is the first example of *filter hybridization* that we will come across in this book. In filter hybridization, as the name implies, the target is immobilized on a filter and the probe is free in the hybridization solution until it binds to the target. Other important application of filter hybridization include *Southern blots* (see later in this chapter) and *northern blots* (Chapter 10). In both cases, the target nucleic acids are first size-separated in an electrophoresis gel before being transferred onto the membrane. The difference between them is that in Southern blots the target nucleic acids are fragments of DNA while northern blots are used for the identification of RNA. The subsequent hybridization of these filters follows the same principles as outlined above.

7.1.4 Screening procedure

We can now return to the question of screening a gene library, and how you use hybridization with a gene probe to identify a specific clone. Remember that at this stage your gene library is in the form of a number of colonies (or phage plaques) on an agar plate (or a number of such plates). The first step is to produce a replica filter of each plate, using a procedure known as a *colony lift* (or a *plaque lift*). A nitrocellulose or nylon membrane is placed on top of the plate, for a minute or so. During this time, part of each bacterial colony, or some of the phages from each plaque, will bind to the membrane. The membrane is then soaked in a sodium hydroxide solution, which releases the DNA from the cells or phage, and denatures it. After neutralization with a buffer solution, the single-stranded DNA molecules are fixed to the membrane by heat or UV irradiation. The membrane filter is then hybridized with the appropriate probe. Following hybridization, the filter is washed, under the chosen stringency conditions, before detection of the probe, using procedures appropriate for the nature of the label.

If you are successful, you will now have an X-ray film with one or more black spots on it that marks the position of the clones that you want on the original plate. You then return to that plate, align it with the X-ray film image, pick that colony and subculture it. This is the clone that you want. This procedure is summarized in Figure 7.5.

In practice, especially if you are screening a library at high density, you are unlikely to be able to pick an individual clone without also collecting some of the neighbouring ones. You will then have to use this mixture of clones for rescreening at a lower density. Furthermore, some of the clones will be false positives (for various reasons), and so it will be necessary to submit them to further testing to verify their identity.

7.1.5 Probe selection and generation

Screening a gene library with a nucleic acid probe implies that you have access to a suitable DNA (or RNA) fragment. Yet the whole purpose of screening a gene library is to isolate a clone carrying a novel piece of DNA. There are a variety of ways around this circular argument. Firstly, there are cases where you actually *are* screening with a probe emanating from the same gene and the same species; this is called *homologous probing*. In this chapter we will describe the use of homologous and heterologous probes, and probes generated by *back translation* (as well as the use of antibodies to screen *expression libraries*). Chapter 8 will describe another approach, which is the

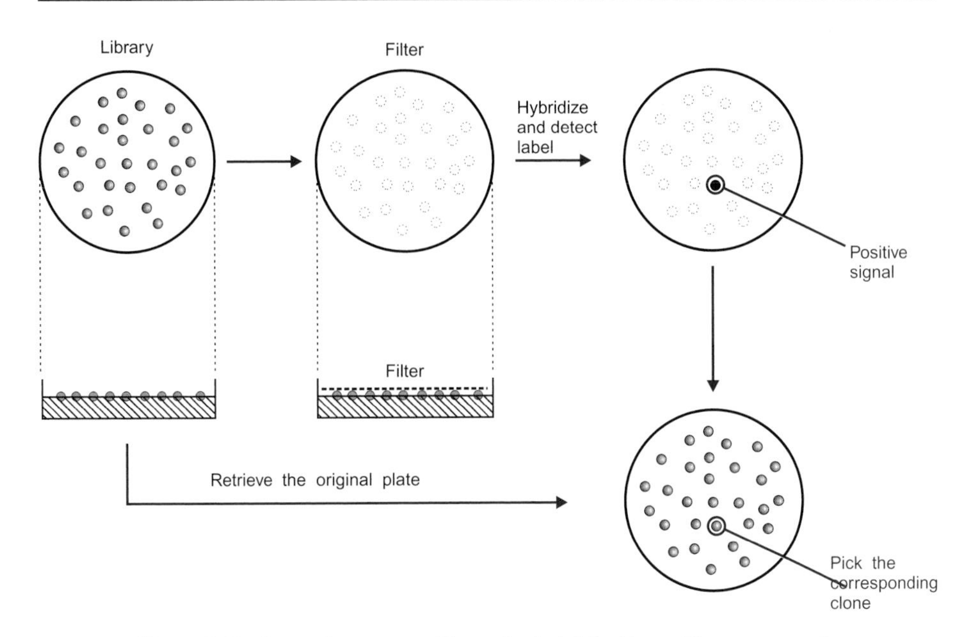

Figure 7.5 Screening a gene library by hybridization with a gene probe

use of PCR-based techniques for generating suitable probes. Additional methods based on differential and subtractive hybridization are described in Chapter 14.

Homologous probes Why would you use a probe coming from the same gene in the same species? This is not as strange an idea as it may seem. Firstly, you may already have a clone, but an incomplete one. You may not have the complete cDNA sequence, or you may want to identify adjacent sequences, such as promoter elements. You may also have access to a cDNA clone, but desire to know the genomic sequence that encodes it. Finally, your business may be to study genetic variation between individuals or strains, whether polymorphisms or mutations causing disease. However, for at least some of these purposes it would be much easier to use PCR (see Chapter 8) to amplify that bit of DNA from the genome than to screen a gene library with the homologous probe.

Heterologous probes If the gene we are trying to clone is not completely unknown, but a related one has already been cloned and characterized from another source, then we can use that clone as a probe. So in isolating the human insulin gene for example, it was possible to use the previously characterized rat insulin gene to probe a human gene library. The availability of genome sequences (Chapter 12) has made this approach much easier. You can now browse through these sequences to find genes that may be related to the one you are attempting to clone, and use the sequence data to design appropriate probes.

This takes advantage of the fact that genes with the same function are often very similar in different organisms. (We will see examples of this in Chapter 9.) As we would expect, the similarity tends to be greatest for the most closely related species (you might not like to think it but we are closely related to rats in this sense!). The extent of the relationship varies from one gene to another; some genes are very highly conserved while others are more variable. Generally, the closer the relationship between the two species, the more reliable the screening is likely to be.

If we move in the other direction, towards less closely related species, then the similarity is likely to decline to a level at which we can no longer detect hybridization using high stringency conditions. This does not mean that the probe is useless. It is quite common to screen gene libraries with heterologous probes that require lower stringency (e.g. lower temperature or higher ionic strength) in order to hybridize to the target clone. However there is a price to be paid. As you lower the stringency in order to allow your heterologous probe to hybridize, so you also allow the probe to hybridize to other genes which are similar to the gene you are trying to clone. In other words, you are more likely to get false positive signals. A certain proportion of false positives can be tolerated (indeed you are likely to encounter some false positives, however good your probe is); these have to be eliminated by further screening and checking (see below). However, if the ratio of false to genuine positives becomes too high, then the re-screening becomes prohibitively time-consuming. (The precise point at which this ratio becomes unacceptable will depend on how badly you want the clone and whether you can think of any other way of getting it.)

Back translation An alternative strategy becomes possible if you are able to obtain in pure form the protein that is encoded by the gene of interest. It is then possible to determine part of the amino acid sequence of the protein. Technological advances, especially using mass spectrometry (see Chapter 14), have made this a much more attractive option, since partial amino acid sequence can be obtained from a protein spot or band on a gel. This information can then be used to infer the likely sequence of the gene itself. Since this process is the reverse of normal translation it is referred to as *back translation* or *reverse translation*. Once you have done that, you can then synthesize an appropriate nucleic acid probe.

However, you have to remember the redundancy of the genetic code. If you know the DNA or RNA sequence you can predict accurately the amino acid sequence of the protein (subject to a few peripheral assumptions). The fact that several different codons code for the same amino acid does not affect the inference in the real translation direction, but in the reverse direction it is a different matter. If you know there is say a leucine residue in the protein at a specific position, then you have considerable uncertainty over what the DNA sequence actually is. It could be any one of six codons (see Box 7.1).

Box 7.1 Genetic code: possible codons for each amino acid

Amino acid	Possible codons						Number of codons
Alanine	GCU	GCC	GCA	GCG			4
Arginine	CGU	CGC	CGA	CGG	AGA	AGG	6
Asparagine	AAU	AAC					2
Aspartate	GAU	GAC					2
Cysteine	UGU	UGC					2
Glutamate	GAA	GAG					2
Glutamine	CAA	CAG					2
Glycine	GGU	GGC	GGA	GGG			4
Histidine	CAU	CAC					2
Isoleucine	AUU	AUC	AUA				3
Leucine	UUA	UUG	CUU	CUC	CUA	CUG	6
Lysine	AAA	AAG					2
Methionine	AUG						1
Phenylalanine	UUU	UUC					2
Proline	CCU	CCC	CCA	CCG			4
Serine	UCU	UCC	UCA	UCG	AGU	AGC	6
Threonine	ACU	ACC	ACA	ACG			4
Tryptophan	UGG						1
Tyrosine	UAU	UAC					2
Valine	GUU	GUC	GUA	GUG			4
Stop	UAA	UAG	UGA				3

You can easily accommodate ambiguity in the sequence by programming the DNA synthesizer to include a mixture of bases at that position. However, every such ambiguity reduces the specificity of the probe. If you allow too much ambiguity you will have a probe that will react with an unacceptable number of non-specific clones in the library. One way of reducing the ambiguity of the probe is to take account of the codon usage of the organism. If this shows that there is a very marked preference for one codon over another synonymous codon, then you can use that information in selecting the codon to be included in your probe. This clearly involves taking a chance, and in many organisms this preference is not marked enough to be very helpful. Instead, you can reduce the required ambiguity by careful selection of the region of the amino acid sequence to be included, so as to avoid amino acids such as leucine, arginine or serine (with six possible codons) in favour of those amino acids with unique codons (methionine, tryptophan) or with only two possible codons (e.g. tyrosine, histidine). Oligonucleotides (oligomers) based on reverse-translation deduction are often called *guessmers*.

To decrease the number of false positives, you can make duplicate lifts from your library and produce two guessmer probes from different regions . The chances of a clone giving a false positive signal in both of these is much smaller than with one. This approach can also be used for the design of PCR primers for the identification of unknown genes (see Chapter 8).

7.2 Screening expression libraries with antibodies

Libraries made with an expression vector such as lambda gt11 (see Chapter 5) allow an alternative method of screening, using antibodies (see Box 7.2). For this, the library is cultured at conditions which permit the expression of fusion proteins. This means growing at 42°C, to inactivate the temperature-sensitive phage repressor, and in the presence of isopropyl galactoside (IPTG), in order to induce expression from the *lacZ* promoter. The plates are then overlaid with a nitrocellulose or nylon filter in the same way as with nucleic acid probe screening, except the filters are not treated with sodium hydroxide. Then, instead of a gene probe, they are incubated with a diluted antibody to the protein in question. This antibody can be produced by purifying the protein in question and injecting it into an animal such as a rabbit. Alternatively, a synthetic peptide can be produced corresponding to a known or assumed part of the protein, and used similarly for immunization.

It is also possible to use lambda gt11 libraries to identify antigens that have not been characterized, for example by using antisera from experimentally infected animals, or from human subjects recovering from an infectious

Box 7.2 Antibodies

Antibodies are made by animals in response to antigens – proteins or peptides, for our purposes. An antiserum contains a range of antibodies, which recognize different parts of the protein, or *epitopes*. Some bind to short amino acid sequences (*linear* epitopes), irrespective of conformation. If you immunize with a short peptide, you will *only* get antibodies that recognize linear epitopes. Other antibodies bind to *conformational* epitopes, formed by folding the amino acid chain. Such antibodies only react with native, not denatured, protein. Since many procedures use denatured protein, it is important that the antibody can react with a linear epitope. Furthermore, if the immunizing protein is glycosylated (or has other post-translational modifications), some of the antibodies will recognize modified epitopes, and will be no use for screening gene libraries. However, a typical antiserum will contain a mixture of antibodies, and some of these will recognize linear, non-modified epitopes.

Antibodies vary in specificity – some will react with proteins other than your target. The antiserum may also contain antibodies to other antigens to which the animal has been exposed. In particular rabbits often have antibodies to *E. coli* antigens, which is a problem if you are testing expression in *E. coli*. You have to test the specificity of the antiserum – and sometimes purify it by absorbing out cross-reacting antibodies.

Alternatively, isolation and culture of individual antibody-producing cells will yield a single antibody species rather than the mixture you get in an antiserum. Antibody-producing cells cannot be maintained in culture, but they can be fused with other cells to form a *hybridoma,* which can be grown and used to produce a single antibody. This is a *monoclonal antibody*, made by a single clone of antibody-producing cells. Although monoclonal antibodies are important where specificity is needed (e.g. in diagnosis), they are not *inherently* specific. Their specificity arises through selecting clones that make antibodies that *are* specific. For us, this specificity can be a disadvantage. If the monoclonal antibody recognizes a glycosylated or conformational epitope, it will be useless for many of our applications.

The other advantage of monoclonal antibodies is their constancy. With conventional antisera, once you have used up your supply you have to immunize another animal. The antiserum will differ in both titre and specificity. (Larger animals, such as goats, can supply antibodies for years – but they are expensive.) However, a monoclonal antibody is produced from a permanent cell line, so you can obtain continuous supplies of identical antibody.

The antibody needs to be labelled, for example with an enzyme, such as horseradish peroxidase, which can be detected using chromogenic or chemiluminescent substrates. It is often more convenient to use a second antibody. For example, mouse anti-rabbit IgG is an antibody raised in a mouse that reacts with rabbit antibodies. A wide range of labelled second antibodies can be bought off the shelf.

disease. This has proved to be a very powerful way of identifying those antigens that are especially important in the natural course of infection, or for protection against infection.

After incubating the filter with the antibody, excess antibody is washed off, and the filter incubated with a labelled *secondary antibody* that will bind to the first one, such as for example anti-rabbit immunoglobulin, allowing detection of the clones that reacted with the primary antibody (see Box 7.2).

Note the limitations of using antibody screening. Screening a gene library with antibodies obviously needs expression of the cloned gene – hence the common use of an expression vector such as lambda gt11. Furthermore, the protein that is expressed in *E. coli* may not fold into its correct, natural conformation – especially if you are using a vector like lambda gt11 that generates a fusion protein. You therefore need an antibody that binds to a *linear epitope* (see Box 7.2). In addition, it must recognize a non-glycosylated epitope, since the protein you are trying to detect will not be correctly glycosylated in *E. coli.*

An antiserum derived from immunizing an animal with a purified protein will contain a variety of different antibodies, some of which will recognize linear, non-modified epitopes. The problem with conformational or modified (e.g. glycosylated) epitopes is most likely to occur with monoclonal antibodies (see Box 7.2), where it is necessary to confirm the nature of the recognized epitope before using it for screening libraries. Similar considerations arise when using antibodies for western blotting (see Chapter 10).

Whichever method you use for screening your library, some clones will be false positives, and they are also likely to be contaminated by surrounding clones. The clones you recover will therefore need to be subjected to further screening, at lower density, until all clones on the plate produce a positive signal.

7.3 Subcloning

In subcloning, the recombinant insert is transferred to another vector, almost invariably a plasmid, which is more amenable for growth, purification and analysis. Clearly, if the original vector is a plasmid, subcloning may be unnecessary. For recombinant clones with larger inserts, such as lambda or cosmid clones, it may be possible to divide the insert up into a set of defined restriction fragments and clone each of those fragments, or to identify which fragment carries the DNA that hybridizes to your probe and just clone that fragment into a plasmid vector.

If the insert is even larger than that, such as those found in yeast and bacterial artificial chromosomes (YACs and BACs), this approach is likely to be too time-consuming. Instead you would probably want to fragment the insert from the selected clone into a number of smaller fragments (*shotgun subcloning*) and produce a mini-library of those fragments which would then be re-screened in the same way as the original library (although much more easily).

7.4 Characterization of plasmid clones

Whether your clone is produced directly or by subcloning from a larger fragment from a gene library, at some stage you are certain to want to characterize the plasmid, and the fragment that it contains. The basic techniques for doing this were described in Chapter 3, so we only need a recapitulation of the major features here. The most basic procedure is to run total DNA extracts or crude plasmid preparations on an agarose gel and look for a change in the size of the plasmid compared with the original vector. If the insert is a reasonable size compared with the vector, you will be able to distinguish recombinant plasmids (carrying the insert) from the parental vector. However you need to be aware that it is not easy to reliably determine the size of the insert in this way. Conventional size markers are linear DNA fragments, while the intact plasmid will be a supercoiled circular structure, and will therefore run differently on the gel.

7.4.1 Restriction digests and agarose gel electrophoresis

More accurate and more reliable estimates of the size of the insert can be obtained by digesting the plasmid with an appropriate restriction enzyme to generate linear fragments that can be accurately sized on an agarose gel. A common procedure is to use the same enzyme that was involved in the cloning step. For example, if you have inserted an *Eco*RI fragment into a unique *Eco*RI site on the cloning vector, then digestion with *Eco*RI will yield two fragments: one corresponding to the linearized vector, and the other the inserted fragment. Using a standard marker containing a set of linear DNA fragments of known size enables the gel to be calibrated so that the size of the insert can be determined (within the limits of accuracy of the system).

Further restriction digestion can be used to verify the construct. If you know the location of other restriction sites within your cloned fragment, you can confirm the nature of the insert by additional digests. The use of a site that is asymmetrically placed within the insert will enable you to check the orientation of the insert. Figure 7.6 shows an example. In this case, we have

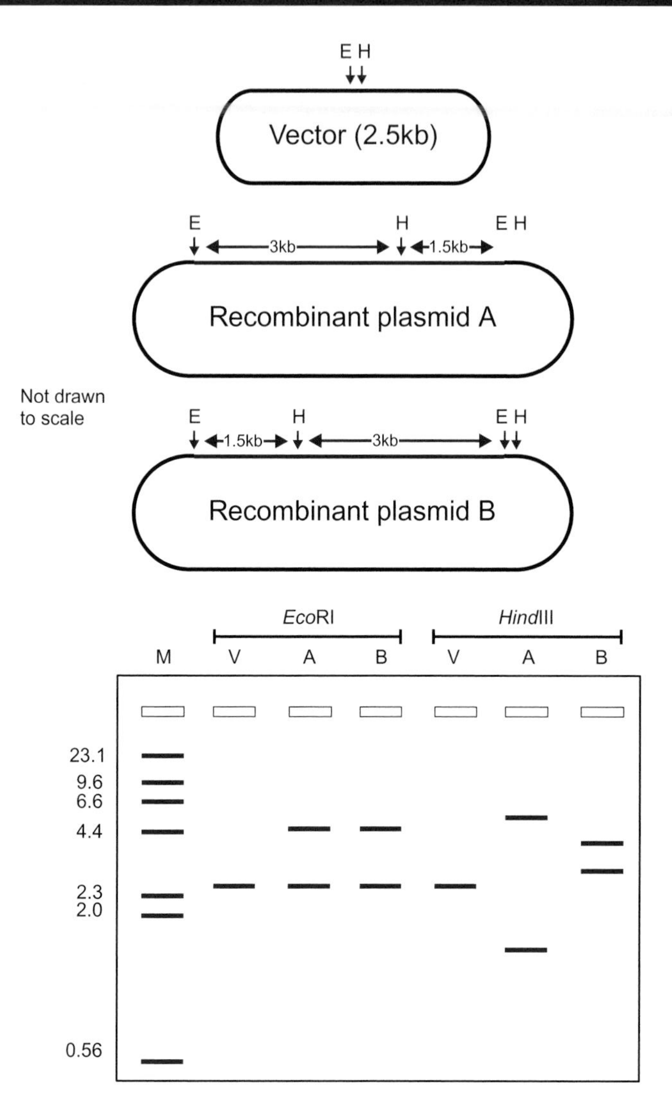

Figure 7.6 Determining the orientation of an insert fragment

inserted a 4.5 kb *Eco*RI fragment into the *Eco*RI site (E) of a 2.5 kb vector plasmid. The *Eco*RI digest confirms that both plasmids A and B have a 4.5 kb insert. However, the insert could be either way round. Fortunately, there is a *Hin*dIII site asymmetrically placed within the insert, and another in the vector, adjacent to the *Eco*RI site. Therefore, a *Hin*dIII digest will settle the matter. The gel shows that plasmid A yields two *Hin*dIII fragments, of 5.5 and 1.5 kb, while plasmid B has the insert in the other orientation, yielding fragments of 3 and 4 kb. Since this is simple to do, it is possible to investigate a number of colonies to find one in which the insert is the right way round.

However, the best way of characterizing your insert, including its orientation, is to determine the sequence of that part of the recombinant plasmid. (See Chapter 9 for the method of doing this.) The procedure above is now only useful if you need to rapidly screen a large number of clones to obtain the right one.

7.4.2 Southern blots

Further confirmation of the nature of your insert comes, once again, from hybridization. You cannot hybridize a probe to DNA fragments while they are in an agarose gel. You need to transfer them to a filter first. The technique for doing this is one that we have already alluded to: Southern blotting, named after E.M. Southern, who developed it. The basis of the method is illustrated in Figure 7.7. A membrane (nitrocellulose or, now more commonly, nylon-based) is placed on the gel and a stack of dry paper towels on top of the filter. Buffer is drawn up through the gel and filter by capillary action, and carries the DNA fragments with it. They are trapped on the membrane, which thus acquires a pattern of DNA bands that corresponds to the position of those fragments in the agarose gel. The arrangement shown in the figure looks very crude, and there are much more elegant pieces of equipment available; nevertheless, many molecular biologists prefer to use their own home-made apparatus. Following transfer of the DNA to the filter,

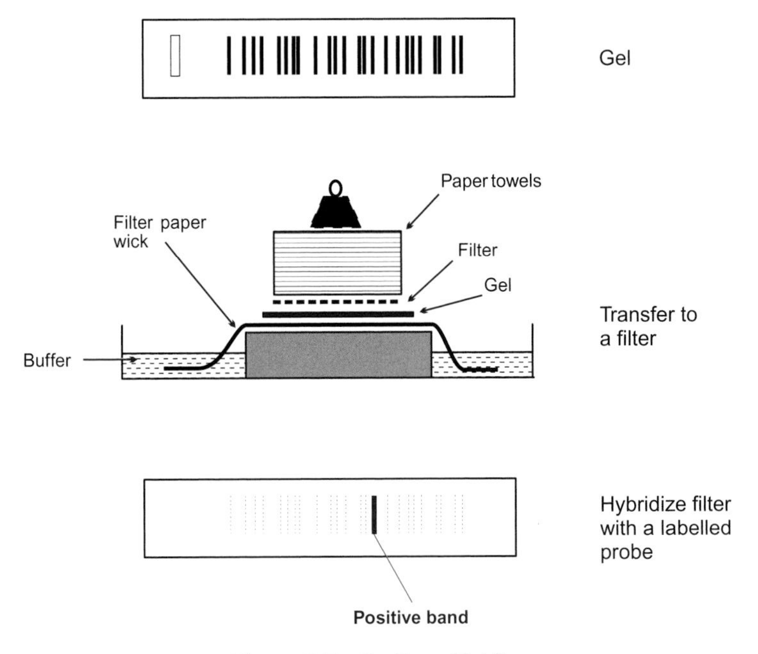

Figure 7.7 Southern blotting

the relevant DNA fragment can be identified by hybridization in the ways already described.

Southern blotting is a very useful way of verifying that the band you can see on a gel is indeed the insert that you have been seeking. However, that is only the beginning of the applications of Southern blotting. We will come across it again later on, particularly in Chapter 13, where it is invaluable for detecting specific gene fragments in a digest of total chromosomal DNA (which just looks like a smear if stained with ethidium bromide), and for comparing the banding patterns obtained with specific probes in DNA from different strains or different individuals.

7.4.3 PCR and sequence analysis

Usually the sequence of the vector is known. It is therefore simple to design a pair of primers that will hybridize to a region at either side of the cloning site, and use these primers to PCR amplify a fragment containing the inserted DNA (see Chapter 8). In this case there is no need to purify the plasmid DNA from each recombinant clone. You can simply resuspend a portion of the colony in water and boil it. PCR amplification will then produce a product of a characteristic size, if you have cloned the right piece of DNA. You can screen a substantial number of potential recombinants in this way. Once you have found one that does produce the correct size of product in the PCR, you can determine the sequence of the PCR product to confirm the nature of the insert.

If you are more confident about the nature of your recombinant product – for example if you are simply re-cloning a known fragment rather than recovering a recombinant from a library, you can skip the PCR step and simply carry out sequencing on purified plasmid DNA from the chosen colony.

8 Polymerase Chain Reaction

On 10 December 1993, Kary B. Mullis received the Nobel Prize in Chemistry from King Carl XVI Gustaf of Sweden for the invention of the polymerase chain reaction (PCR) method. Mullis had published and patented this invention only eight years earlier, in 1985. The same year, the Physiology and Medicine prize was awarded to Richard J. Roberts and Phillip A. Sharp for their independent discovery that eukaryotic genes are split into introns and exons. These laureates had waited twice as long, 16 years, since their fundamental discovery made in 1977. This will give you some idea of the immense impact PCR has had. Although it is (as we shall see) quite a simple method with obvious limitations, the applications of this method have revolutionized both basic and applied biology. This has been particularly dramatically illustrated in forensic science, where numerous old open cases have been solved, and in clinical applications, where it has suddenly become possible to make diagnoses in hours that previously took weeks.

The key factor in transforming the initial method into one that could have such an impact was the introduction of a thermostable DNA polymerase called *Taq* polymerase. It was originally isolated from *Thermophilus aquaticus*, a thermophilic archaebacterium that thrives in hot springs at temperatures close to the boiling point of water. As a result, all enzymes in this organism have evolved to withstand high temperatures where most proteins from most other organisms would denature immediately and irreversibly. Apart from this distinct feature, *Taq* polymerase (not to be confused with *Taq*I, a restriction enzyme from the same species) is a normal DNA polymerase. It will synthesize a new DNA strand complementary to a single-stranded DNA template. Like all other DNA polymerases, it requires a *primer*, a more or less short strand of complementary DNA, to start its synthesis from. In fact, as we will see, it is not a particularly outstanding DNA polymerase. Although it has high processivity (which means it has a

From Genes to Genomes, Second Edition, Jeremy W. Dale and Malcolm von Schantz

propensity for remaining bound to the DNA and continuing to add successive nucleotides without dissociating from the DNA), it lacks proofreading activity, so it is unable to correct erroneously incorporated nucleotide bases. This is significant for some applications, as the product may not be a completely accurate copy of the original sequence, and other thermostable polymerases are available which have proofreading ability.

PCR uses *Taq* polymerase (or other thermostable DNA polymerases) for the exponential amplification of a DNA fragment from a longer initial template, which could be as long as a whole chromosome. The amplified fragment is defined by two short synthetic oligonucleotides, *primers*, that are complementary to the opposing DNA strands of the template that is being amplified. This introduces another limitation to the method. You must know the sequence for at least part of the DNA molecule you wish to amplify – or you must at least be able to make an educated guess.

So how do we go from a small amount of, say, total human genomic DNA to a large amount of one short region that has been exponentially amplified to the extent that it entirely dominates the reaction mixture? This will be clear if you go through the first few cycles in a PCR amplification.

8.1 The PCR reaction

In this example, we will be starting with genomic DNA purified from human buccal cells, collected by swabbing the inside of the mouth with a cotton bud. The preparation will contain sheared chromosome fragments. Only a small amount is required, even from such a complex template – in fact, as little as a single cell has been used as starting material.

We will also need two primers. These can be synthesized to your specifications from a specialist supplier at a very low cost within a couple of days. In this example, we will replicate the work of Mullis and his colleagues and use primers flanking a region for the human beta-globin gene that is mutated in sickle cell anaemia. A great excess of primer molecules is added to the reaction. (This of course refers to an excess in molar, or molecular, terms, as the primer is very much smaller than the template – see the discussion in Chapter 4.)

The binding, *annealing*, of the primer to the template is a typical DNA: DNA hybridization reaction, and follows similar principles to the hybridization of probes as described in Chapter 7. First, the double-stranded template needs to be *denatured*. The temperature used for this in PCR, 94°C, hardly does any damage to the *Taq* polymerase molecule during the minute or so that the PCR reaction is heated to this temperature.

The temperature is then lowered to the optimal annealing temperature, where the two primers can bind to the opposing DNA strands (Figure 8.1). This is the only temperature in a PCR cycle that can be varied widely. It is

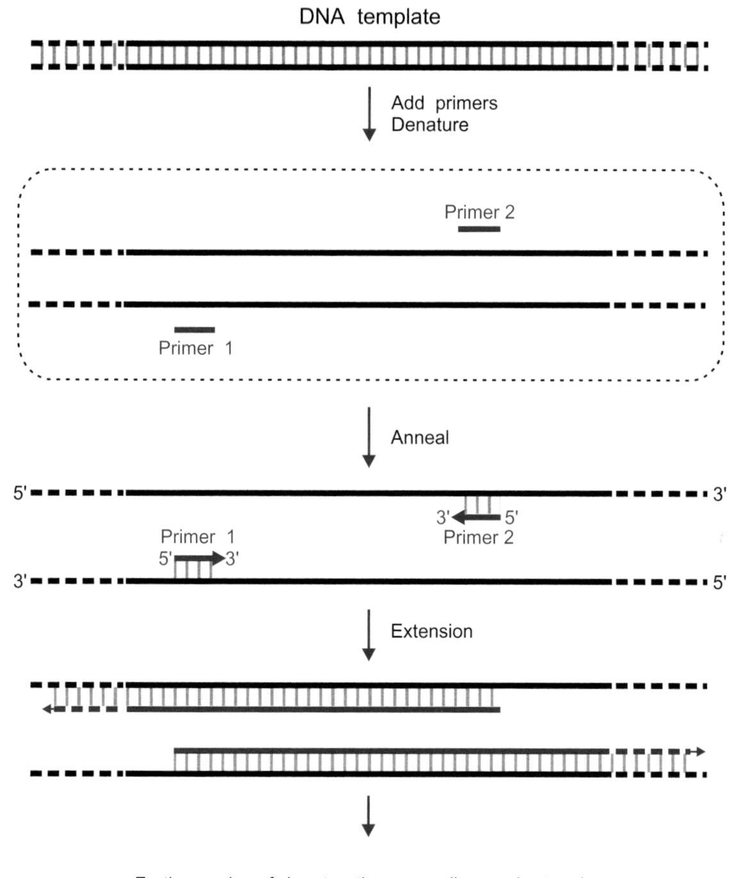

Figure 8.1 Polymerase chain reaction: first cycle

chosen for maximum binding of the primer to the correct template, and minimum binding to other sequences. If the annealing temperature is too low, the primers will bind at other positions on the template, resulting in false products or no detectable product at all. If the annealing temperature is too high, the primers may fail to bind at the correct site, resulting in no amplification. The temperature needed will depend on the sequence and length of the primers, as discussed later. Because the primers are short, and at relatively high molar concentrations, annealing is rapid, taking less than a minute.

The temperature is then raised to approximately 72°C, which is normally the optimum *extension* temperature for a PCR reaction. The *Taq* polymerase will now produce complementary DNA strands starting from the primers. The extension proceeds at approximately 1000 bases per minute.

Because the template in this case is many times larger than that, the polymerization will proceed until it is interrupted. This happens when the temperature is yet again raised to 94°C in order to start the next cycle in the PCR reaction, which is normally identical in temperature and duration to the previous ones. As we finish the first PCR cycle, we have two double-stranded DNA molecules for each one that we started with. Each one contains one strand of the original template, and one novel strand which is defined at one end, specifically, by the oligonucleotide primer and at the other end, non-specifically, by how far polymerization was able to proceed during the time we allowed for extension.

The advantage of *Taq* DNA polymerase over, say, *E coli* DNA polymerase will now become apparent. The *E coli* polymerase could have performed the extension, but it operates at 37°C, which means that you would not be able to increase stringency and would risk non-specific hybridization. Above all, however, you would lose all enzyme activity in the denaturation step, and would have to add fresh enzyme. *Taq* polymerase, by contrast, survives the denaturation step unscathed.

This second denaturation step creates four single-stranded template molecules, two of which are derived from the original template and two of which are our newly synthesized strands. At the following annealing step, one molecule of the complementary primer will bind to each of these single strands. As the temperature is again raised to 72°C, *Taq* polymerase will begin to extend the primers (Figure 8.2). Two of the four extensions, where the primer has again bound to the original chromosomal template, are identical to the first extension in that they terminate only when the temperature is raised. Note, however, that this does not apply to the two strands that are produced with the new strands as templates. These templates end abruptly where the opposite primers had bound.

If we envisage yet another cycle, then again the two original template strands will give rise to a long product, limited only by the duration of the extension reaction. Priming on all the other strands will yield a product that is defined and delimited by the primers at both ends. Each subsequent cycle will produce two new long strands, but the number of new short strands made will increase exponentially, so that eventually the reaction mixture will be completely dominated by the newly formed short DNA strands with one primer incorporated at each end. It follows from this that the ends of the new DNA are actually defined by the primers, unlike the intervening regions which are entirely defined by the original template (apart from any mistakes in the amplification).

This in turn means that modifications may be introduced through the use of modified primers. One particularly useful application of this principle is in the addition of restriction enzyme recognition sites to the ends of a PCR product for cloning purposes. In order to accomplish this (Figure 8.3), the 5′

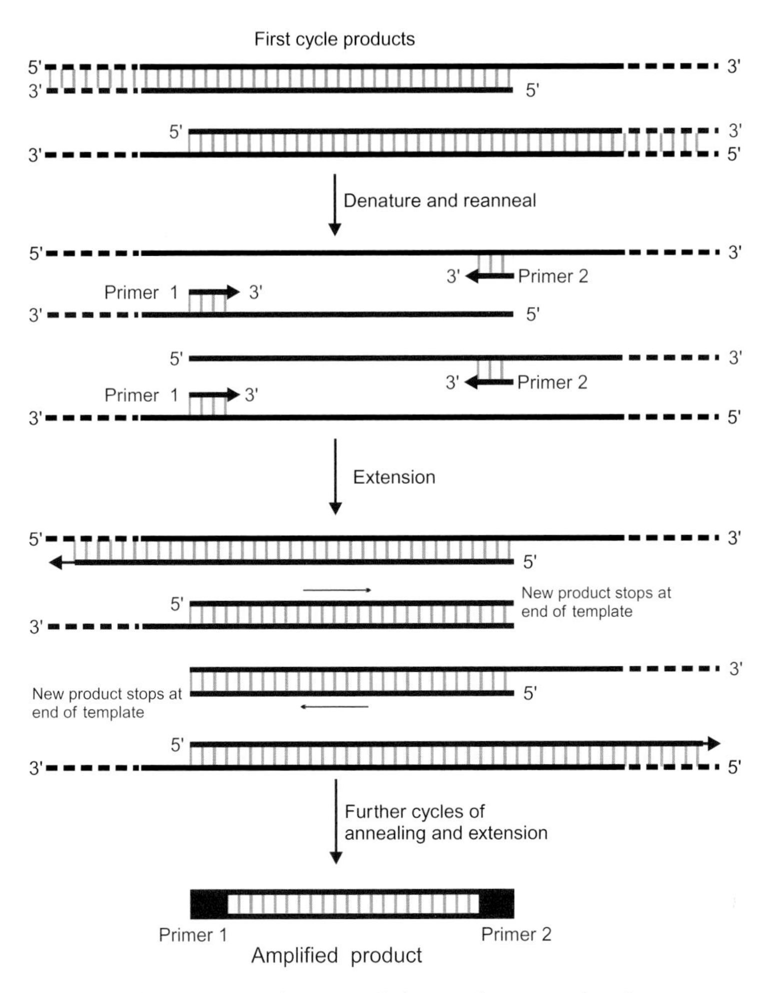

Figure 8.2 Polymerase chain reaction: second cycle

ends of the primer sequences are modified to incorporate the appropriate recognition site. In the first PCR cycle, the 5′ ends of the primer will not pair with the template, but that does not prevent annealing (provided there is sufficient similarity in the 3′ region of the primer) or extension (which is dependent on a perfect match of the final 3′ nucleotide). In subsequent rounds, the part of the primer carrying the restriction site (which is now incorporated into the product) will be accurately replicated, as the primers are now a perfect match with the template. The end product is a DNA fragment carrying the restriction site near the ends, so it can be cut with the restriction enzyme and ligated with an appropriate vector. The same principle can be employed to introduce other modifications into the product, including base changes. Attaching fluorescent dyes, or other labels, to one or both primers provides a good way of obtaining a labelled product.

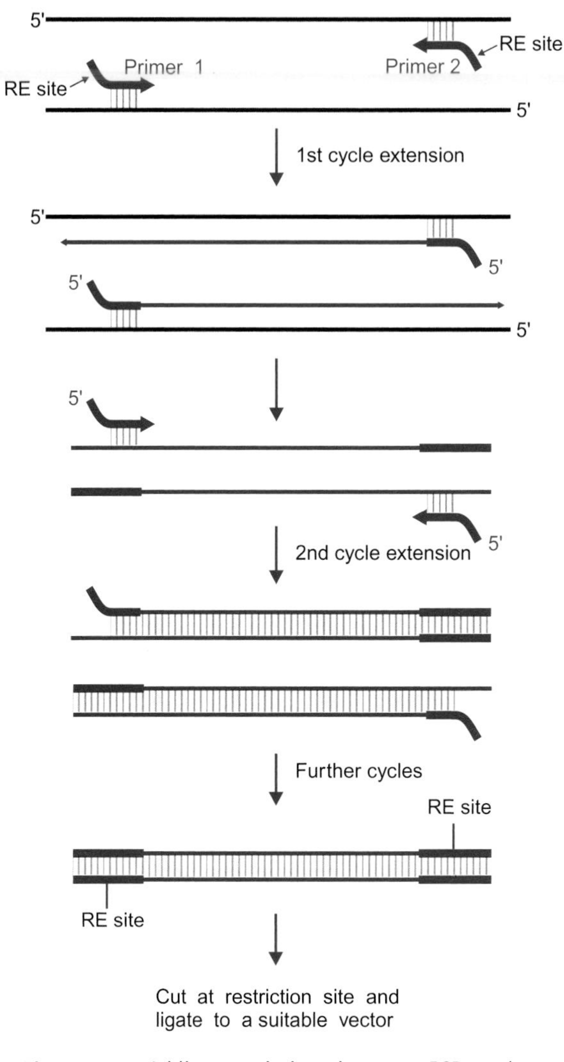

Figure 8.3 Adding restriction sites to a PCR product

8.2 PCR in practice

The launching of PCR as a revolutionary new technology was enabled by the development of the programmable thermocycler. These instruments are based on metal heating blocks with holes for the PCR tubes. The blocks are designed to switch between the programmed series of temperature steps with great speed and precision by a combination of heating and cooling systems. The use of small (0.2–0.5 ml), thin-walled, tubes helps to ensure a rapid change of temperature. Alternatively, for larger numbers of samples, microtitre plates are used.

Because the PCR reaction is performed in a small volume (typically 20–50 μl), and because of the high temperatures involved, it can be easily imagined that the water would quickly evaporate and end up on the inside of the lid rather than on the bottom of the tube. There are two ways of preventing this. The original one was to place a drop of mineral oil over the reaction. The less messy approach applied almost universally today is to heat the lids of the tubes to prevent condensation.

8.2.1 Optimization of the PCR reaction

The importance of the annealing temperature in PCR has already been discussed. If the temperature is too high, binding of the primers to the target will not be stable enough for amplification to take place. If it is too low, the system will become too tolerant of partial primer–target mismatches, and will therefore be non-specific. The annealing temperature will be affected by the sequence and length of the primers. Because G–C pairing is stronger than A–T pairing, the more Gs and Cs there are in the primer, the stronger it will bind (and therefore higher annealing temperatures can be used). See Chapter 7 for a discussion of other factors that influence the optimal annealing temperature. Although computer algorithms are available to predict the optimal annealing temperature for a primer, in practice some trial and error is often needed. Normally, the annealing temperature is chosen somewhere between 40 and 60°C, although for templates with a high GC content annealing temperatures as high as 72°C (the normal extension temperature, see below) may be used. Note also that, as with any other hybridization, we may want to use conditions that will allow priming from sites that are only partially matched to the primers, and for this purpose we would use a lower annealing temperature.

Another important factor is the concentration of magnesium ions, a necessary cofactor for the enzyme. Typically, a magnesium concentration of 1.5 mM is used. A higher magnesium concentration gives a higher yield, but also a lower specificity. Lowering the magnesium concentration increases specificity, but will decrease enzyme activity.

8.2.2 Primer design

The most important PCR parameter by far is the design of the primers. Assuming that we know the sequence of the target gene, we want to pick a region of a suitable length for amplification. For analytical purposes, 200–500 base pairs is adequate. If it is too small, it will be difficult to detect on an agarose gel; if it is too big, the amplification will be inefficient. (For other purposes, it is possible to amplify much longer sequences, but it requires special conditions.) We then select two sequences (usually 20–25 bp) either

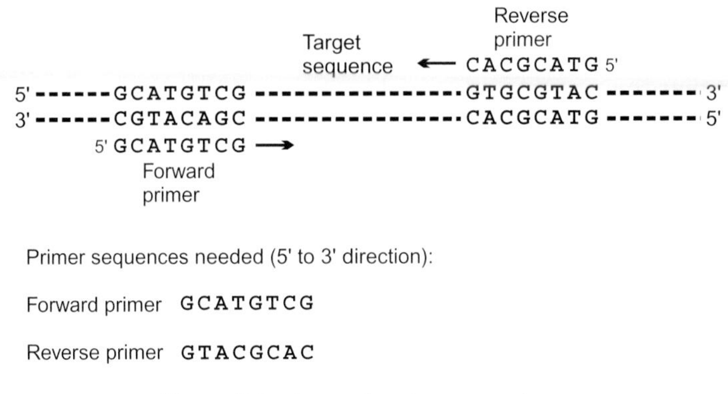

Figure 8.4 Forward and reverse primers

side of this region for binding our complementary primers. It is essential to get the orientation of the primers right. The 'left-hand' (forward) primer will be the same as the sequence of the 'top' strand – remember that double-stranded DNA is conventionally represented with the top strand in the 5′ to 3′ direction, reading from left to right; this is the strand that is shown if your DNA is shown in single-stranded format. The 'right-hand' (reverse) primer must be the sequence of the complementary strand, read from right to left (which is the 5′ to 3′ direction for the lower strand). This is illustrated in Figure 8.4 (using unrealistically short primers for clarity).

There are some further factors that affect the design of the primers. If an oligonucleotide contains sequences with complementarity to itself or the other primer, the result will be a partially double-stranded structure rather than annealing to the target sequence, because the primers are present in such great excess that they are much more likely to encounter a partially complementary primer molecule than a perfectly complementary template molecule. Thus, primer complementarity is to be avoided. It is also desirable for the two primers to have similar annealing temperatures, which means they should be similar in length and base composition. Computer programmes are available to check that the chosen sequences are suitable in both of these respects.

We also want to make sure, as far as possible, that each primer will anneal only to the chosen sequence, and not to any other region of the template mix. If we are working with DNA from an organism for which the complete genomic sequence is known, then we can search the database to check for other sequences that are complementary to our primers (see Chapter 9 for a description of how these searches are done). Ultimately, it is in part a matter of trial and error – even the best designed primers do not necessarily work well, in which case we need to change one or both primers.

What if we do not know the sequence of the target region? This problem can often be circumvented by some informed guesswork. For example, we may use sequence information from a more or less closely related organism to design our primers. The fact that the sequences will probably not be a perfect match can be accommodated in two ways. Firstly, we can use a lower annealing temperature, to allow the imperfectly matched primers to anneal, and secondly we can create degenerate pools of primers, i.e. we can incorporate ambiguities into the primers. This is analogous to the use of heterologous probes for library screening (see Chapter 7). When using heterologous primers in this way it is essential to remember that the ends of the amplified product represent the primer sequence, and are *not* derived from the target genome.

8.2.3 Analysis of PCR products

The normal way of analysing the products of a PCR reaction is to separate the samples in an agarose electrophoresis gel. This allows you to ascertain that only one fragment is obtained in each reaction, which is usually the objective. By comparing the size of the amplified fragment to a molecular weight standard, it is also possible to make sure that the molecular weight is the same as the predicted one (which is usually known). The assumption is that, if the fragment is the predicted size, then it probably corresponds to the predicted fragment – an assumption which occasionally leads you down the wrong track, which therefore often necessitates checking the identity of the fragment (see below).

On the other hand, you may get a fragment of a different size from the predicted one. Frequently, a specific amplification product (of the correct size) is found together with an abundant fragment of low molecular weight. These low-molecular-weight fragments are called *primer dimers*, and are caused by binding of the primer molecules to each other. They can usually be ignored, although the tendency of primers to form dimers can cause problems with a PCR, due (amongst other things) to competition between primer–primer and primer–template binding. Another effect may be an amplification that appears to be specific (as shown by the absence of a fragment of that size in control reactions), but which gives a product of a different size than the expected one. This may indicate that one, or both, of the primers is binding at a position other than the intended one, producing an artefactual product. This may also be manifested as a specifically amplified fragment together with a more or less complex mixture of non-specific ones. In these cases, greater specificity may be obtained by adjusting one or more of the PCR parameters (see the previous section).

Alternatively, *nested* PCR may be used to increase the specificity (and/or the sensitivity) of the amplification. In nested PCR (Figure 8.5), a small aliquot of the original reaction is transferred to a second, 'nested' PCR reaction. In the nested PCR reaction, one or both primers are replaced with a

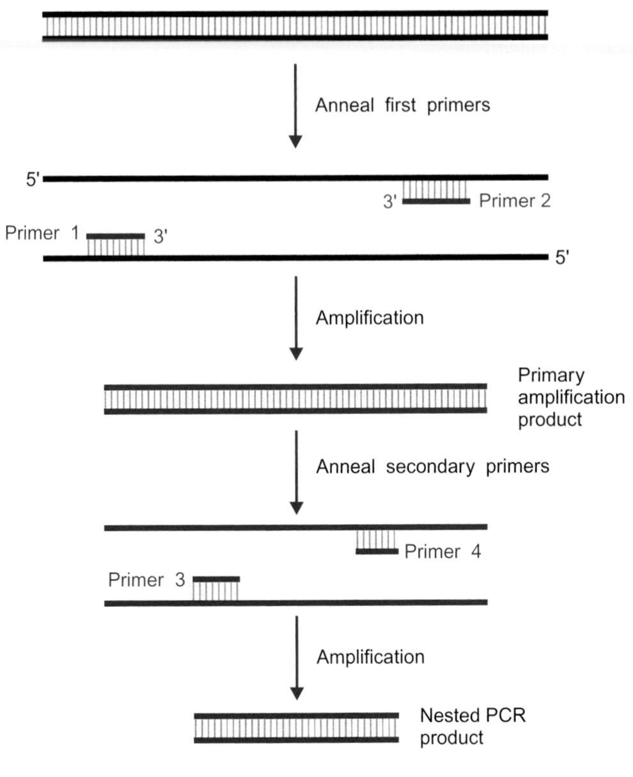

Figure 8.5 Nested PCR

second set of primers which will bind specifically within the desired amplification product. In this way, the undesirable products, which have typically been amplified because of a coincidental sequence similarity in the region of the original primers, will normally disappear.

In order to provide another level of certainly about the identity of a PCR product, in particular if the exact molecular weight is not known, the gel may be blotted and hybridized (see Chapter 7) with a probe complementary to the expected product. Alternatively, because of the remarkable recent progress in sequencing technology, many people find it faster, easier and cheaper to perform direct sequencing on the PCR product (see Chapter 9).

8.3 Cloning PCR products

Although PCR is commonly used merely to detect the presence of a specific sequence in different templates, it is also often employed for the amplification of such sequences as a convenient way of obtaining a specific product for cloning. This is especially important when the starting material is very

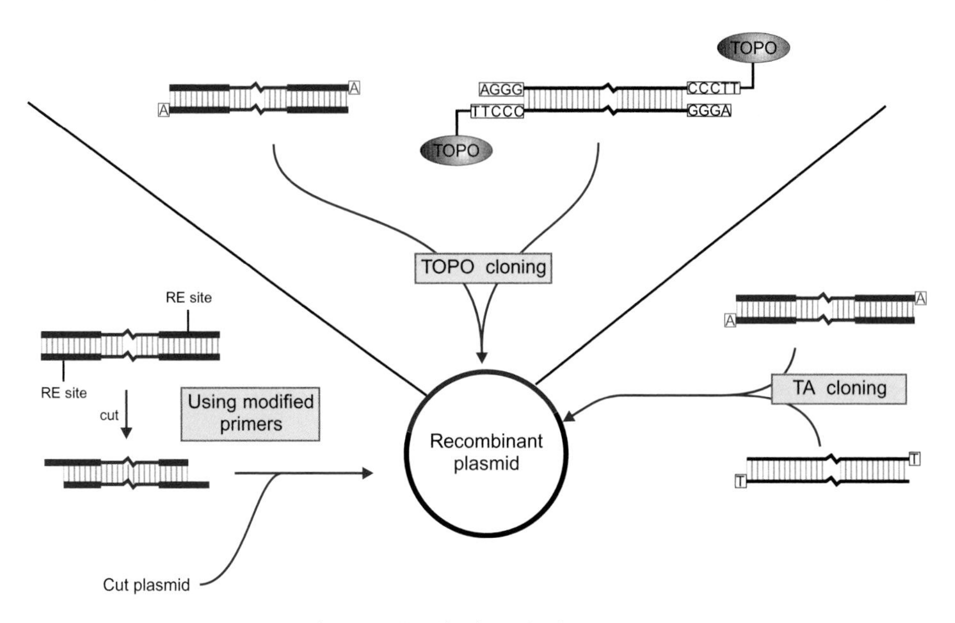

Figure 8.6 Cloning PCR fragments

scarce, as in such cases the conventional routes (for example constructing and screening a gene library) are impossible.

You might expect, from the description so far, that the PCR products would be blunt ended, and could therefore be cloned by normal blunt-ended cloning (see Chapter 4). In practice, this is often not very successful. Not only is blunt-ended cloning comparatively inefficient, compared with cloning sticky-ended fragments, but also *Taq* polymerase often tends to add, non-specifically, an adenosine residue to the 3′ ends of the product. The product is therefore not blunt-ended, but contains a tiny 3′ overhang, which is exploited in the so-called TA cloning method. In this method, a linearized vector plasmid is used which has been engineered to contain a thymidine overhang at each end. This produces short, but nonetheless sticky ends, which will anneal to the adenosine overhangs of the PCR products, allowing fairly efficient ligation (see Figure 8.6, and also Chapter 4). Note however that not all polymerases add adenosine residues to the product; the 'proof-reading' polymerases produce genuinely blunt-ended PCR products. If TA cloning is desired of such products, *Taq* polymerase may be added at the end of the amplification and allowed to proceed through one amplification cycle.

An alternative method has already been alluded to. This is to use modified primers that contain an additional restriction enzyme recognition site in its 5′ end. The resulting products can be cut and ligated to a plasmid that has been cut with the same enzyme, with great efficiency (Figure 8.6). This is particularly useful when cloning larger PCR fragments.

A further improvement to PCR cloning technology is the use of *Vaccinia* virus DNA topoisomerase I (TOPO); see also Chapter 4. *In vivo*, topoisomerases are involved in the supercoiling/relaxation of DNA. They will cleave DNA at specific sites, leaving a sticky end. The energy released by breaking the phosphodiester bond is stored in a covalent bond between the enzyme and one of the cleaved strands. The enzyme trapped in the sticky end will then rapidly and efficiently release its stored energy into the formation of a new phosphodiester fragment as soon as the sticky end encounters its complementary partner. Thus, TOPO has both endonuclease and ligase activity. Commercially available TOPO vectors offer sticky end overhangs, ranging from TA cloning to more complex sequences bound to the TOPO enzyme. This makes for rapid and ligase-free cloning.

8.4 Long-range PCR

We have already discussed the fact that *Taq* polymerase lacks proofreading activity and is thus unable to correct its own errors. These errors occur approximately once per 9000 nucleotides, on average. Such mistakes will then be perpetuated in all new molecules descended from the one containing the error. If direct sequencing is used, this is not really a problem, because the defects will be randomly distributed and the vast majority of molecules will be correct in any given position. However, if the PCR products are cloned it will potentially be a problem, because any error in the one molecule that happens to be cloned will be perpetuated in all the offspring of that one clone (Figure 8.7). For this reason, investigators normally sequence at least three independent PCR clones in order to ensure that they are identical.

Another consequence of the relative sloppiness of *Taq* polymerase is the fact that *Taq* polymerase can only efficiently amplify fragments of a few thousand base pairs. Both these problems can be solved by the introduction of thermostable DNA polymerases from other thermophilic organisms, such as *Pfu* and *Pwo* DNA polymerases. Unlike *Taq* polymerase, these enzymes do have proofreading activity. They are not necessarily as efficient, however. This has been ingeniously overcome by the introduction of proprietary mixtures of *Taq* and proofreading DNA polymerases. By combining proofreading and high yield, and by the ability for one polymerase to continue the extension where the other one stalls, these mixtures may allow the efficient amplification of DNA fragments as large as 50 kb.

8.5 Reverse-transcription PCR

In Chapter 6, we described the use of reverse transcriptase to obtain a cDNA copy of mRNA, and the construction of cDNA libraries. Combining reverse transcriptase (RT) with PCR, a procedure known as RT-PCR, extends the

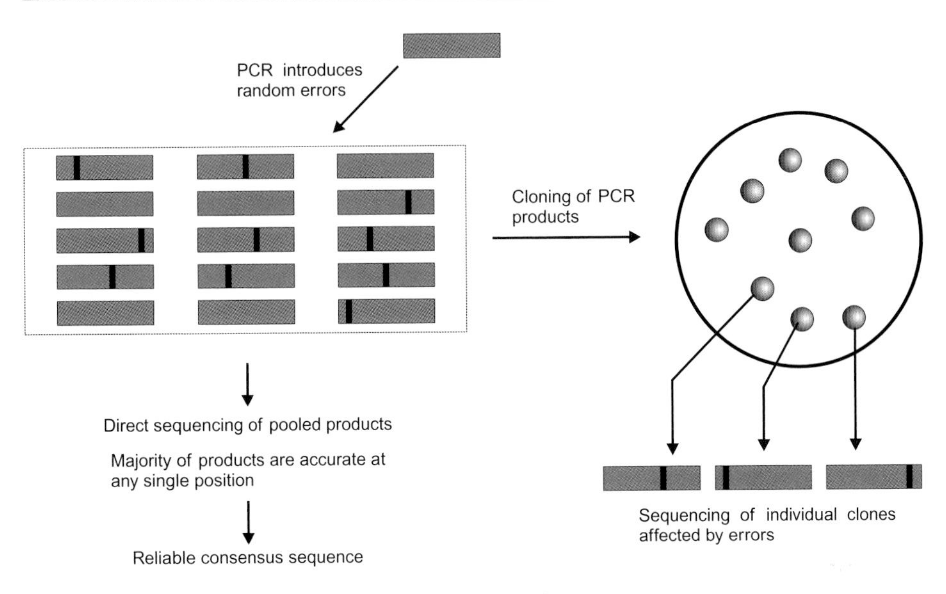

Figure 8.7 Detection of PCR errors

application of PCR into the analysis of gene expression, either qualitatively or quantitatively, as well as greatly facilitating the construction of cDNA libraries or the cloning of specific cDNAs. The use of RT-PCR for the analysis of gene expression is described in Chapter 10.

One problem with RT-PCR is that the initial mRNA preparation may be contaminated with genomic DNA. The PCR step can then result in amplification of the contaminating DNA. When working with eukaryotic material, this can often be overcome by designing the primers so that the amplicon spans at least one intron. In this way, amplification of any genomic DNA will either be prevented altogether (because the presence of the intervening intron makes the sequence too large to be amplified), or at least it will be readily distinguished from amplified cDNA (because of the different size of the product). Generally, it is preferable to remove all traces of DNA from the mRNA, usually by treatment with RNase-free DNase.

8.6 Rapid amplification of cDNA ends

The procedure described in Chapter 6 for generating cDNA suffers from the problem that the products are often incomplete, i.e. the ends of the mRNA molecules are under-represented in the clones generated. In Chapter 6 we considered some ways of at least partially overcoming this problem. PCR provides further enhancement of these techniques, by specifically amplifying the ends of cDNA molecules.

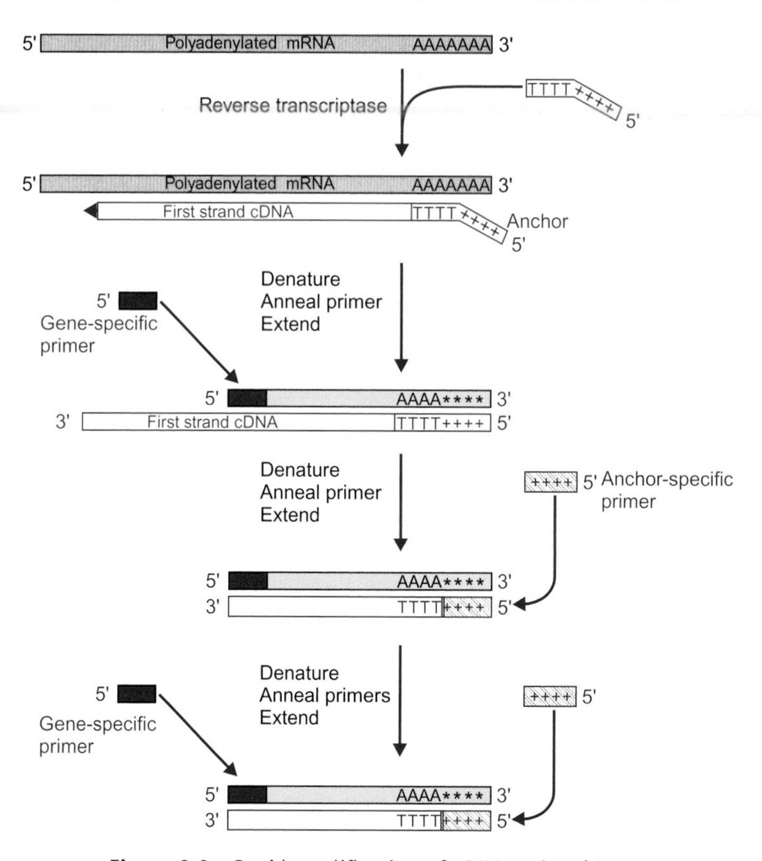

Figure 8.8 Rapid amplification of cDNA ends: 3′ RACE

The method known as *rapid amplification of cDNA ends* (RACE) is based on the amplification between a *gene-specific primer* (GSP) within a known area, and *a universal primer* at one end of the cDNA molecule. Note that the use of a gene-specific primer is only possible if you already know, or can in some way infer, the sequence of at least a small part of the gene.

The most straightforward type of RACE is 3′ RACE (Figure 8.8), which allows amplification of cDNA corresponding to the 3′ end of the mRNA. As in the description of cDNA synthesis in Chapter 6, this method takes advantage of the fact that eukaryotic RNA has a poly-A tail. This tail is used as the binding site for a complementary, universal oligo-dT primer, for synthesis of the first cDNA strand. For 3′ RACE, this primer has an additional sequence, known as an *anchor*, which provides a more effective site for subsequent amplification (and often includes a restriction site to assist in subsequent cloning). The second cDNA strand is initiated by a GSP. The resulting double-stranded DNA can then be amplified by the usual PCR procedures, using the original primers, or replacing the oligo-dT-anchor primer with one just containing the anchor sequence.

Normally, PCR would use two primers which are specific to the desired amplicon. In the basic 3′ RACE system, as shown in Figure 8.8, one primer (the GSP) is specific for the gene concerned, while the other primer (the oligo-dT-anchor primer) will act as primer for any polyadenylated mRNA. The specificity of the system can be enhanced by a nested PCR in which further amplification takes place using a second, nested primer (also specific for the gene in question) in place of the GSP.

In a similar way, 5′ RACE enables amplification of cDNA corresponding to the 5′ end of the mRNA. The key feature of this procedure is the incorporation of a universal tag at the 3′ end of all first-strand cDNA molecules. This is accomplished by the enzyme *terminal deoxynucleotidyl transferase*, usually known as terminal transferase, or TdT. This enzyme will non-specifically add deoxynucleotides to the 3′ end of a single DNA strand. By making only one dNTP available, for example deoxycytosine, a monomeric 5′ tail analogous to the 3′ poly-A tail can be added.

In the procedure shown in Figure 8.9, the mRNA is initially reverse transcribed using random primers, which are more likely to produce first-strand cDNA reaching the 5′ end of the mRNA than using oligo-dT primers (which start at the far 3′ end). This is a non-specific step, and will produce a mixture of single-stranded cDNA molecules, reflecting more or less all the mRNA present in the original material. The cDNA pool is then tailed. using terminal transferase, to provide a binding site for an oligo-dG primer coupled to an anchor sequence for second-strand cDNA synthesis. This is then followed by PCR amplification, using the anchor primer and a gene-specific primer. As with 3′ RACE, additional specificity is typically gained by using nested PCR, with a second specific primer replacing the first GSP.

Thus, even if a gene is novel, PCR can be used to obtain the entire cDNA sequence without ever using a vector-based library. This has made it possible for many laboratories to abandon cDNA library screening altogether, although it is currently experiencing a revival thanks to the development of array libraries from national genome facilities.

8.7 Quantitative PCR

It is difficult to derive reliable quantitative information from conventional PCR, for example by measuring the amount of product formed. There is no simple and reliable relationship between the amount of template you start with and the amount of product obtained, unless you can be certain that the PCR reaction is truly proceeding exponentially throughout. This will usually no longer be the case during the latter phase when the availability of reagents becomes limiting. This problem may be overcome, to a point, by adding a known amount of a second template that is almost identical (and amplifiable with the same primers) but still distinguishable on an electrophoresis gel.

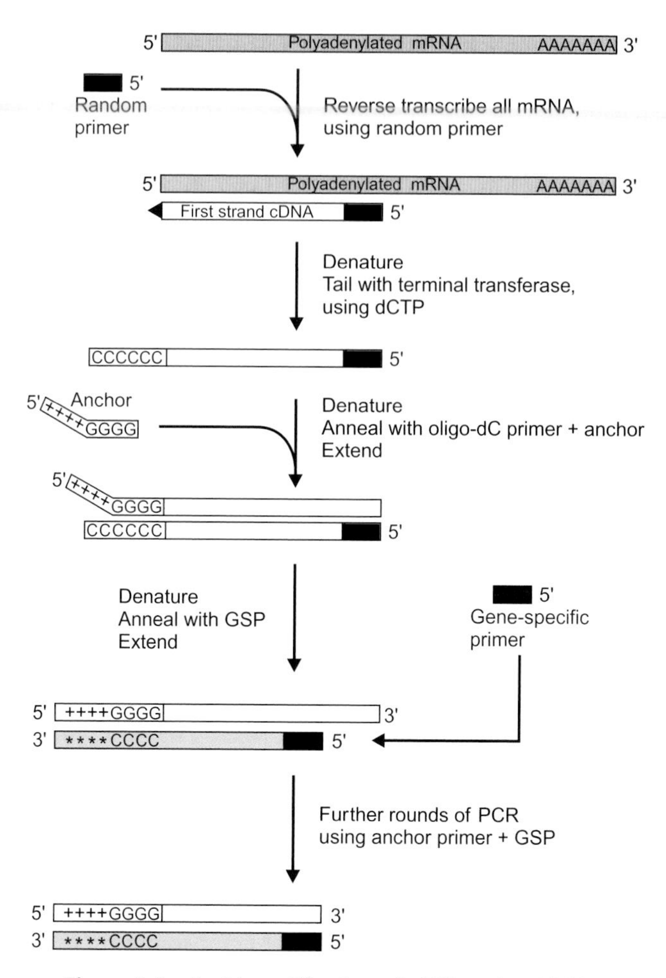

Figure 8.9 Rapid amplification of cDNA ends: 5′ RACE

Normally, this is produced by cloning the PCR product into a plasmid vector and deleting a small part of it. Adding a known quantity of this modified DNA to the PCR reaction will allow you to achieve much better quantitative accuracy by comparing the relative amounts of the two resulting PCR products. In practice, this has now been superseded by *real-time PCR,* a method that is much more accurate and also simpler. The previous limitation, affordability of equipment, has become less of an obstacle.

If you have a way of detecting the product without interrupting the series of amplifications, then you have an invaluable way of quantifying the procedure. This is the basis of real-time PCR. For real-time PCR, you need a way of detecting the product as it is formed, without having to stop the reaction and run the products out on a gel. The simplest method to understand is to add, to the PCR mixture, a dye such as SYBR™ Green (pronounced as

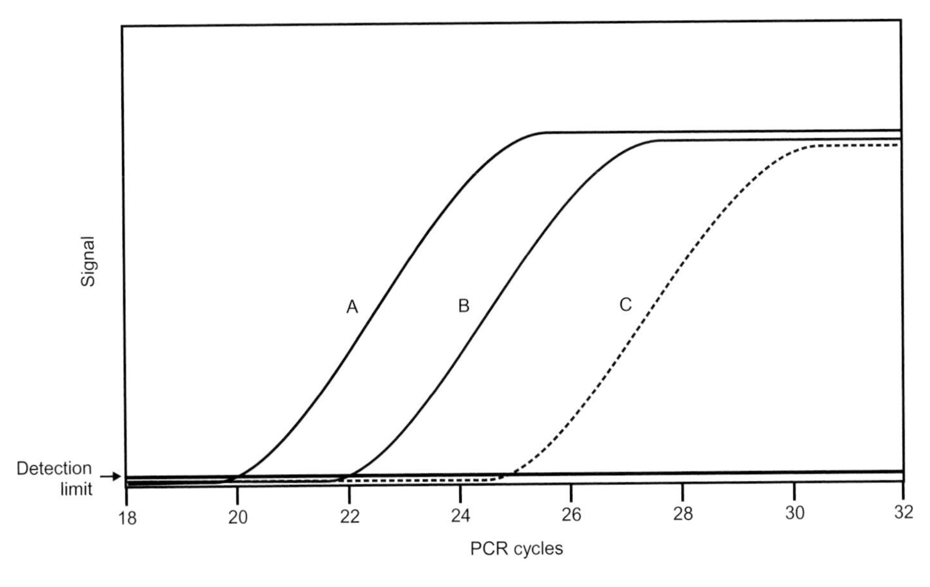

Figure 8.10 Real-time PCR

Cybergreen) that fluoresces when it binds to double-stranded DNA. This is the cheapest and simplest way of performing real-time PCR, because the dye is generic (and thus not reaction-specific) and you can use the same primers as in any normal PCR reaction. However, this method may lead to artefacts, as you will be unable to distinguish the signal produced if the wrong product is made. More specific methods are described below.

If you carry out the PCR reaction in a machine that not only performs the temperature cycling needed for a PCR reaction but also will detect the fluorescence of your samples, then you can monitor the progress of your PCR in real time. Initially, the template is single-stranded, and so there is no signal. As amplification proceeds, double-stranded product is formed, and eventually enough of this product is made to allow the resulting fluorescence to be detected by the machine (see Figure 8.10). The level of the fluorescence will increase over a number of cycles, and by extrapolation of the resulting curve back to zero you can determine the number of cycles needed for formation of a detectable amount of product. This value, known as the C_T value, is related to the initial amount of template: higher amounts of initial template will result in lower C_T values. In the figure, sample A produces a detectable signal after 20 cycles, so we say it has a C_T value of 20. This represents more template than in samples B ($C_T = 22$) or C ($C_T = 25$).

Since SYBR Green will bind to any double-stranded DNA, there is no guarantee (apart from the specificity of the primers) that the fluorescence detected is due to formation of a specific product. It is possible to circumvent this by programming the thermocycler to generate a melting temperature

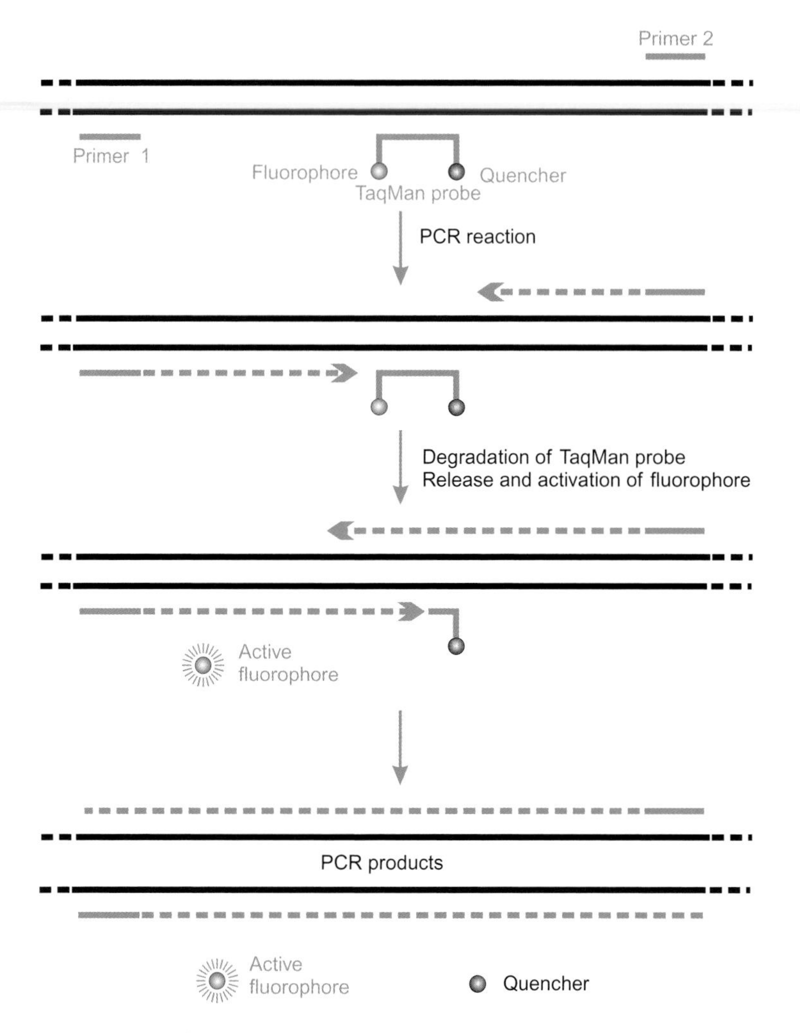

Figure 8.11 Real-time PCR: TaqMan reaction

curve of the product after the end of the amplification. The shape of this curve is a measure of the homogeneity of the sample.

The other commonly used methods achieve specificity through the use of a probe that hybridizes specifically to the desired product, the probe being labelled in such a way that fluorescence only occurs as a consequence of the PCR reaction. The most generally employed method is the so-called TaqMan method (Figure 8.11). Here, an oligonucleotide probe that is complementary to an internal sequence within one of the amplified strands is labelled with a fluorescent group at its 5′ end, and a *quencher* at its 3′ end, which quenches the fluorescence at the 5′ end so long as both fluorophore and quencher are within the same molecule. Thus, virtually no fluorescence is emitted at the

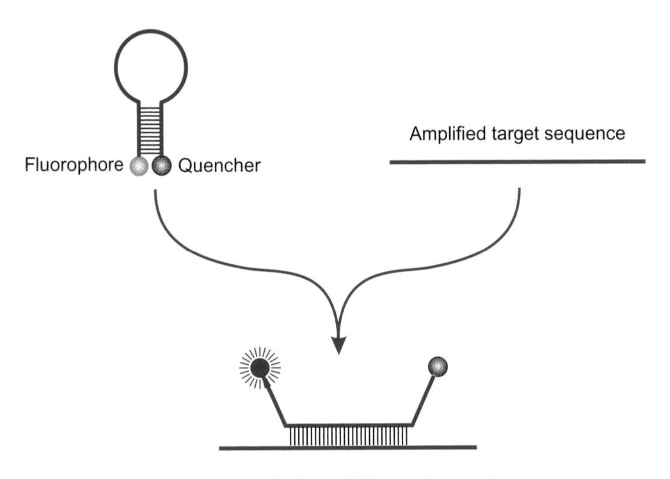

Figure 8.12 Real-time PCR: molecular beacons

beginning of the reaction. As the reaction proceeds, the oligonucleotide will bind to the increasing number of newly synthesized strands. When such a strand is copied, the intrinsic 5′–3′ exonuclease activity of the polymerase will cleave the fluorophore away from the probe, thus liberating it from the quencher and enabling it to fluoresce. The intensity of this fluorescence is a directly proportional measure of the amount of product that has been generated. Because the oligonucleotide sequence is complementary to an internal part of the amplicon independent of the primers, it provides high specificity. When using this method for amplifying cDNA, the oligonucleotide is often designed to span the junction between two exons. That way, any amplification of contaminating genomic DNA will not lead to the release of any fluorophore.

Another method, known as Molecular Beacons, also uses a probe containing a fluorophore and a quencher, but in this case instead of a linear probe (as in the TaqMan method) the probe is designed with complementary ends that form a hairpin structure, with the loop containing a sequence complementary to the target (Figure 8.12). Here, the quencher/fluorophore combination works in a different way from that used for TaqMan, in that fluorescence is only suppressed if the two are in close proximity. When the probe binds to an amplified DNA strand, the hairpin conformation is lost, and the fluorophore is activated, because it is no longer adjacent to the quencher. Since the detection is dependent on the relative stability of the hairpin and the target binding, it is more specific than TaqMan probes, and is especially useful for detecting single base changes (SNPs).

Real-time PCR is an important technique for estimating the level of expression of specific genes (see Chapter 10), when it is used in combination with reverse transcription. Its application for the detection of DNA sequences, including medical diagnostic applications (see Chapter 13), rests more in the convenience of detection of the product than in quantification. When analysing a number of samples, it is quicker than using agarose gels, and can be readily automated.

8.8 Applications of PCR

8.8.1 PCR cloning strategies

PCR can be used in several ways to provide alternatives to the screening of random libraries as described in Chapter 7. For example, analysis of the known sequences of a set of related genes may show that, although the genes are not similar enough to devise reliable hybridization probes, there may be some regions that are relatively highly conserved. It is possible to construct pairs of primers directed at these conserved regions, and use these to amplify the corresponding fragment from the genomic DNA of your target organism (Figure 8.13). Nested PCR is often used to add to the power of this approach. You can then sequence this product, to try to confirm that it is indeed derived from the correct gene, and you can use it as a probe for screening a gene library to isolate clones carrying the complete gene.

This approach needs to be treated carefully. One reason why part of a gene may be highly conserved is that it codes for an essential substrate-binding site, so any gene which uses the same substrate is likely to have a similar sequence. For example, many enzymes that use ATP as a substrate have a similar sequence that represents the ATP-binding site.

Alternatively, if you already know the sequence of the gene, then you can devise a reliable PCR system to amplify the whole gene or part of it, whether from genomic DNA or cDNA, and either clone or sequence the product. Cloning and/or sequencing a gene of which you already know the sequence is much more useful than it sounds. For example, you may want to know whether the sequence varies at all in different strains (see Chapter 13), or you may want to identify alternatively spliced transcripts of a eukaryotic gene. PCR amplification combined with direct sequencing provides a quick route to answering that question. It is also employed in medical genetics in the search for individual sequence differences which may cause disease.

Alternatively, you may have cloned a gene in one vector, and you then want to subclone it in a different plasmid – perhaps moving it into an expression vector, or removing extraneous DNA that was present in the original clone from a random gene library. You can easily devise a pair of primers that will produce exactly the product you want – down to adjusting the reading frame to fit in with your expression vector (see Chapter 11), or adding or removing restriction sites at the end of the fragment to aid subsequent manipulation.

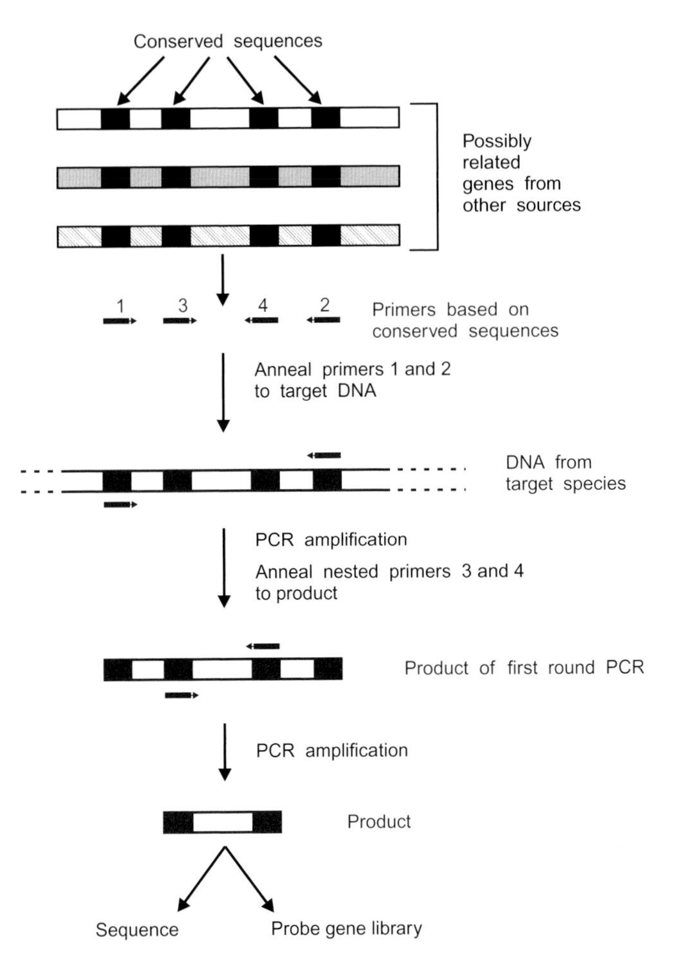

Figure 8.13 Use of conserved sequences to design PCR primers

In Chapter 11, we discuss the introduction of changes into the coding sequence of a gene (*site-directed mutagenesis*); PCR plays a major role in this as well. A further topic in Chapter 11 is the use of a procedure known as *assembly PCR* for assembling synthetic genes. There is virtually no end to the versatility of the procedures to which you can adapt PCR as an aid to cloning and re-cloning DNA fragments in a research laboratory.

8.8.2 Analysis of recombinant clones and rare events

The best way of characterizing a recombinant clone, to make sure you have indeed produced the structure that you intended, is to sequence it – or at least to sequence across the new junctions that have been made. For this, you can either directly sequence the appropriate part of the recombinant plasmid, or you can amplify across the inserted fragment and sequence the PCR product.

PCR can also be used to detect specific events that occur *in vivo*, without having to clone the affected region of the genome. For example, if we want to detect the transposition of an insertion sequence into a specific site, or if we want to test if our gene knockouts (see Chapter 15) have really inserted a foreign DNA fragment into that site, we can use PCR. We would use one primer directed at the insert and one at the genomic flanking sequence. We would only get a product if the event we are looking for has indeed occurred. Otherwise the two primers are binding to completely different bits of DNA.

However, this needs a word of caution, which applies to some extent in many applications of PCR. The technique is splendid for amplifying DNA where a genuine target exists. When there is no genuine target, there is the possibility of artefacts being created, due to the power of the technique for amplifying extremely rare events, such as the possibility of the polymerase 'jumping' from one DNA molecule to another. This may seem highly unlikely (although there are reports of it happening) – but it only needs to happen once, in the early stages, and the product will then be amplified effectively, leading to an incorrect conclusion.

A similar effect can occur if the template contains a stable hairpin (or stem-loop) structure. This will cause the polymerase to pause (or even to stop altogether). However while it is paused, it may encounter the other side of the hairpin and 'jump' across – yielding a product that lacks the hairpin. Although this rarely happens, the product will then amplify much more effectively than the original template, and the final product will be shorter than it should be – leading to the suggestion that your strain contains a deletion at that point.

8.8.3 Diagnostic applications

The polymerase chain reaction has many important practical applications in the real world outside molecular biology laboratories. Amongst these are forensic applications – the detection and identification of specific DNA fragments that can be traced to a particular individual – and uses in medical diagnostics, including the detection of mutations causing human genetic diseases. PCR can be employed for the detection of pathogenic microorganisms in a clinical specimen (such as sputum or blood). This is especially valuable for organisms which are difficult to culture, including many viruses and some bacteria. In addition, PCR tests (often in combination with specific gene probes) are available for the identification and speciation of pathogens that have been cultured. In these contexts, the problem of contamination has to be taken very seriously. The ability of PCR to amplify tiny amounts of DNA makes it an extremely sensitive test, but it also makes it extremely sensitive to low levels of contaminating DNA. In a diagnostic situation, extensive measures have to be taken to avoid contamination, notably the use of separate rooms and equipment for the preparation of the clean (pre-PCR amplification) material and the analysis of the post-PCR products (which contain large amounts of material that can act as template for further amplification). A battery of additional precautions must be taken, including the use of negative controls with each batch of samples. We will look at some of these applications further in later chapters.

9 Characterization of a Cloned Gene

Having considered the ways in which we can obtain specific genes or fragments of DNA – either by cloning or by PCR amplification – we can now look at how we characterize that gene. In this chapter, we will deal with the sequencing of individual genes, annotation of database entries, and some ways of confirming the function of the gene. In the following chapters, we will look at how we analyse the expression of individual genes, and ways in which genes and their expression can be manipulated. In Chapters 12-14, we will extend these topics to genome-wide studies.

9.1 DNA sequencing

9.1.1 Principles of DNA sequencing

Sequencing is the primary way of characterizing a macromolecule, whether it be determining the order of amino acids in a protein, or of bases in a nucleic acid. Protein sequencing was a very important tool before genes could be cloned and sequenced. With the advent of recombinant DNA technology, it was largely superseded by the much more efficient and economical method of DNA sequencing, with the sequence of the encoded protein being deduced from the sequence of the gene. However, the technology available for direct protein sequencing has improved dramatically over recent years. Although this is beyond the scope of this book, it is worth noting that protein sequencing is a viable alternative to cloning and sequencing an unknown gene, if you have the purified protein (and the necessary equipment), especially as it provides additional information, such as identifying post-translational modifications that are not available from the DNA sequence. At the same time, it remains overshadowed by DNA sequencing, especially as genome sequencing can now give you the predicted sequence of *all* the proteins that are

From Genes to Genomes, Second Edition, Jeremy W. Dale and Malcolm von Schantz

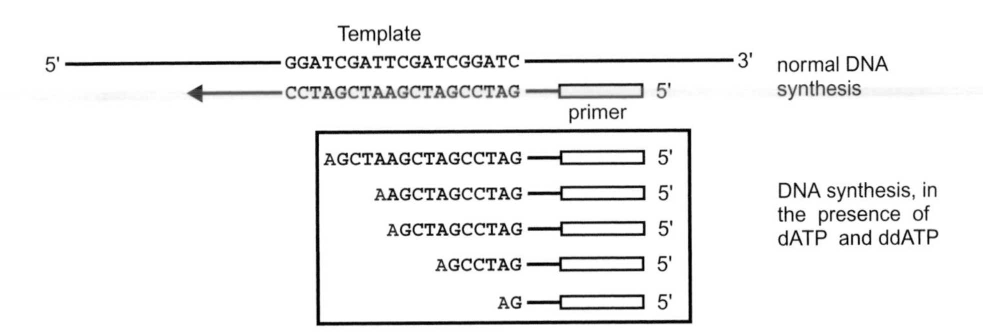

Similar reactions are carried out with ddGTP, ddCTP and ddTTP.
The fragments from the four reactions are separated on an acrylamide gel and detected by autoradiography

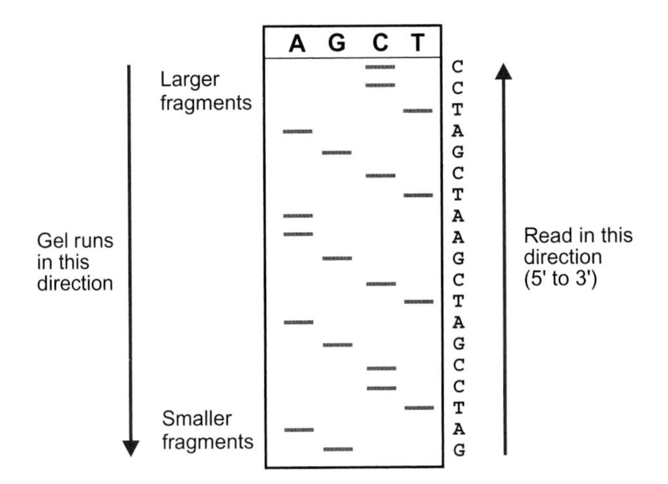

Figure 9.1 Determination of DNA sequence

potentially produced, not just one. Genome sequence assembly is covered in Chapter 12, but first we want to look at the principles of DNA sequencing as applied to individual genes.

The principle of the main procedure, known as dideoxy sequencing or the Sanger method, is illustrated in Figure 9.1. (There is an alternative procedure, known as the Maxam–Gilbert method, which works by selective chemical degradation. This method is now obsolete, and will not be described here. Another, newer, method known as *pyrosequencing* is described in Chapter 12.) To understand the dideoxy procedure it is necessary to remember two fundamental facts about DNA synthesis: firstly, synthesis of a DNA strand does not start from scratch. It requires a *primer* annealed to a *template* strand. Synthesis of the new strand works by adding bases, to the primer, that are

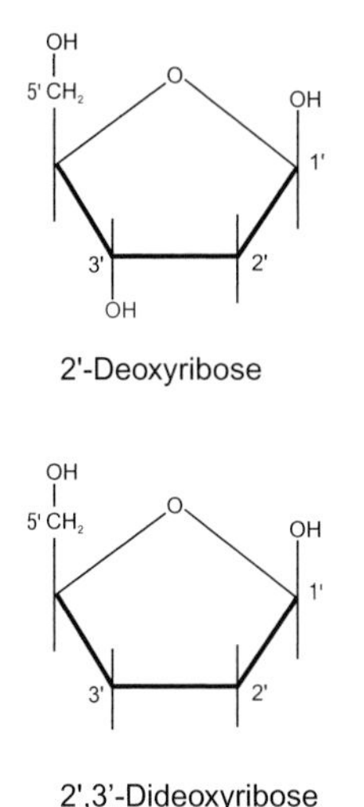

Figure 9.2 2′-Deoxyribose and 2′,3′-dideoxyribose

complementary to the template. Thus, by using a primer that binds to a specific position on the template, we can ensure that all the new DNA that is made starts from the same point. Secondly, the addition of bases to the growing strand occurs by formation of a covalent phosphodiester bond between the 5′-phosphate on the nucleotide to be added and the 3′-OH group on the existing molecule. The substrate for this reaction is a 5′-dNTP, i.e. a *deoxynucleotide* with three phosphates at the 5′-position of the deoxyribose sugar; two of these phosphates are eliminated in the reaction.

The sugar part of the natural substrate is more specifically a 2′-deoxyribose. This means that it does not have a hydroxyl group at the 2′-position (which distinguishes it from the ribose sugars that occur in RNA). However, it does have a 3′-OH group, which is necessary for the formation of the next phosphodiester bond when the next base is incorporated. What happens if we use, instead of the natural substrate, one in which there is no 3′-OH group, i.e. a 2′,3′-dideoxy derivative, a ddNTP (see Figure 9.2)? This can be incorporated into the DNA, by formation of a phosphodiester bond between its 5′-phosphate and the 3′-OH on the previous residue. However, that reaction produces a strand which does not have a 3′-OH at the end, and so no further bases can be added. DNA synthesis will therefore terminate at that point.

So if we replace one of the dNTPs with a dideoxy derivative – for example, if we use a mixture of dGTP, dCTP and dGTP (the normal substrates) but replace dATP with the 2′,3′-dideoxy derivative designated ddATP – then DNA synthesis will proceed only as far as the first A residue, after which it will stop (Figure 9.3). If we carry out a set of four such reactions, in each case replacing one of the dNTPs with its corresponding ddNTP, we would produce

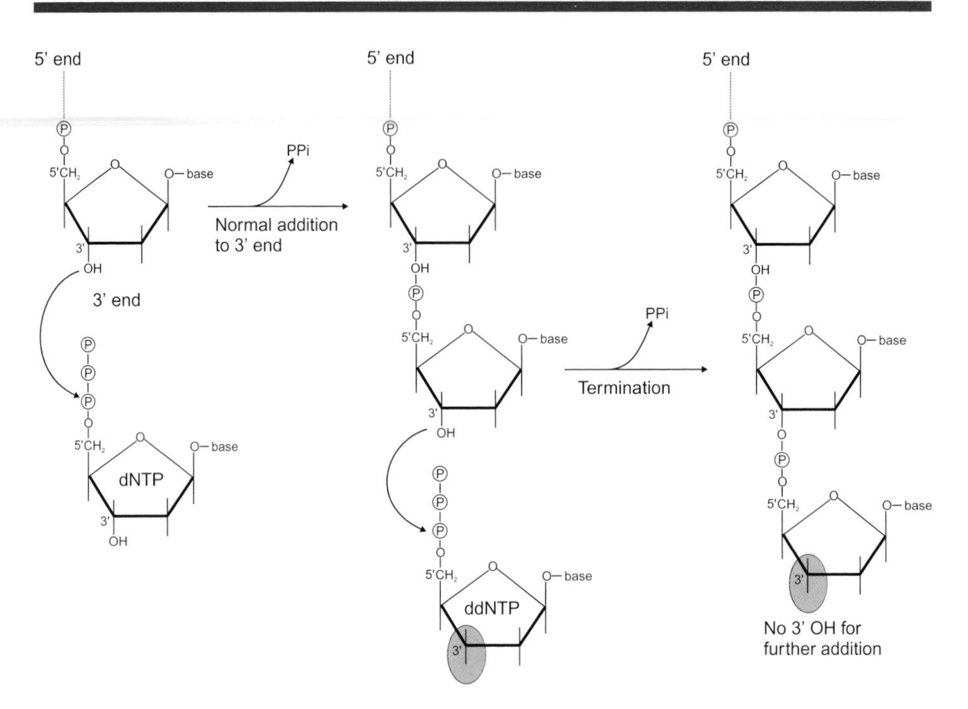

Figure 9.3 Chain termination by ddNTP

four molecules of different lengths, each proceeding just as far as the first occurrence of the relevant nucleotide in the sequence.

Just determining the first occurrence of each base would not be much use. However, instead of completely replacing, say, dATP with ddATP, we can use a mixture of the two substrates. So at the first T residue in the template, a few molecules of the new strand will have ddA added (and will therefore terminate), but most will have the normal A residue, and the reaction will be able to proceed. At the next T in the template, some more molecules will terminate, and so on. We will thus have a series of molecules of different chain lengths, each ending with ddA (see Figure 9.1). These can be separated by electrophoresis in a polyacrylamide gel, using denaturing conditions to prevent the DNA strands from folding up. The molecules are separated on the basis of their size, with the smaller molecules running faster through the gel. We therefore get a series of fragments, corresponding to the positions of A residues in the new strand (or T residues in the template). With a set of four reactions, each using a different ddNTP, we get a set of four lanes, from which the sequence can be read as shown in the figure.

For manual DNA sequencing, we would carry out the reaction with one of the dNTPs radioactively labelled, so that exposure of the gel to an X-ray film would result in a pattern of black bands on the film. Because the smallest fragments (which are closest to the primer) will migrate faster, they will be at the bottom of the gel; the autoradiogram is therefore read from bottom to top.

9.1.2 Automated sequencing

Today, virtually all DNA sequencing uses automated sequencers. Although the principle is the same, the method of detection is different. For automated sequencing, either the primer or the ddNTPs are labelled by incorporation of a fluorescent dye. Thus, rather than running the gel for a finite time, and reading the result, the machine uses a laser to read the fluorescence of the dye as the fragments pass a fixed point. Much longer sequences (up to 800–1000 bases) can be read from each track in this way, because the separation is not stopped at a specific point – instead, each fragment is allowed to proceed sequentially to the bottom of the gel where the resolution is the greatest. A further advantage is that the sequence read by the machine is fed automatically into a computer. This is not only much quicker than reading a gel manually and typing the resulting sequence into a computer, but also avoids the errors that are virtually inescapable with manual data entry. On the other hand, the computerized interpretation algorithm is more or less prone to some errors of its own, and may sometimes require manual checking.

If the primer is labelled, all the products carry the same dye, and so you still have to use four lanes. However, if the four ddNTPs are each labelled with different dyes which can be distinguished by the photometer, the sequencing reactions can be performed in a single tube and separated in a single lane, thus increasing the capacity of the machine. A further development on this is to replace the polyacrylamide gel, which needs to be cast anew for each run, with a reusable matrix-filled capillary.

Both manual and automated DNA sequencing methods suffer from problems that can give rise to errors in the sequence. These are often associated with the presence of certain combinations of bases in the DNA – for example runs of identical nucleotides, when it can be difficult to determine exactly how many bases there are in the run, stretches of C and G, which may cause such a high melting temperature that the sequencing is unable to proceed past them, and the presence of secondary structures in the DNA such as hairpin loops, which can interfere either with DNA synthesis (causing premature termination) or with the running of the fragment on the gel. Some of these potential errors can be identified and minimized by altering the reaction conditions, by sequencing a different overlapping fragment covering the problem region, or by determining the sequence of the complementary strand. A good complete sequence will therefore be derived by reading and assembling several overlapping sequences in each direction.

9.1.3 Extending the sequence

Although automated DNA sequencers allow much longer reads than manual methods, the length of sequence that can be obtained from a single run is still limited. As the fragments get larger, the degree of separation between them decreases, and the amount of product in each fragment is reduced, making the signal weaker. (In addition to the required termination due to incorporation of dideoxynucleotides, synthesis may pause or even stop for other reasons, such as a secondary structure in the DNA.) The sequence thus becomes progressively less reliable. Sometimes this is not a problem: if we are using sequencing to verify the structure of a recombinant plasmid that we have made, or if we are looking at the variation of a specific region between different strains, then the sequence of a few hundred bases may be quite sufficient, but if we want to sequence a whole gene, or especially a whole genome, then we will usually require a longer string of sequence data.

One strategy for extending the length of sequence determined is referred to as *walking* (Figure 9.4). Remember that the sequence depends on DNA synthesis starting from a specific primer. Usually, with an unknown cloned sequence, we would start with a primer directed at the cloning vector close to the point of insertion. Because most sequencing projects use the same cloning vector (either a pUC plasmid or an M13 vector, which have the same sequence flanking the cloning site), we can use the same primers for any clone. These are therefore known as *universal primers*.

The (forward) primer from one side of the insert will read the sequence in one direction, while the sequence of the complementary strand will be

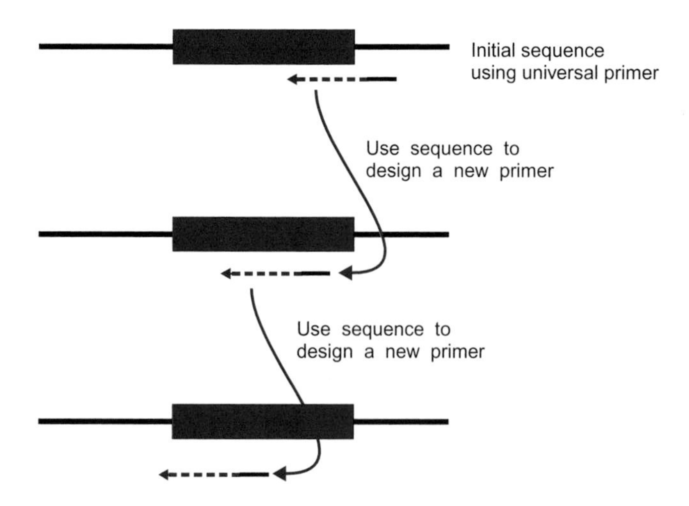

Figure 9.4 Extending a sequence by primer walking

obtained using the (reverse) primer from the other side of the insert. The sequence we obtain would thus start with a small amount of vector sequence (which has to be removed prior to analysis or to deposition in a databank, see below) and would then read through into the unknown insert. We can extend the length of sequence determined by using the first piece of data to design a new primer that would start synthesis further along the insert. That will produce a further length of sequence, which can then be used to design a third primer, and so on. This procedure is effective for relatively short sequences but becomes excessively tedious if long stretches of DNA are to be sequenced. It can, however, be a useful strategy for finishing larger sequences when other approaches have left short gaps (see below).

Transposons can also be used to facilitate the sequencing of a large insert. As discussed in Chapter 12, these are mobile genetic elements that insert randomly into the chromosome. By using a donor vector carrying a transposon in which a universal primer sequence is embedded, you can easily create a library of clones that can be sequenced for the rapid assembly of the full sequence of the insert.

9.1.4 Shotgun sequencing; contig assembly

For sequencing a longer fragment – say a cloned fragment of 10 kb – the best procedure is often to split it up into smaller fragments, each of which is a suitable size for sequencing in full. This is like producing a gene library. The insert from your recombinant vector is fragmented and cloned using a suitable vector. The bacteriophage vector M13 (see Chapter 5) was originally useful for this purpose, since it produces single-stranded versions of your fragments which gave cleaner results. However, improvements in the technology mean that good results are now obtained with double-stranded templates.

Note that it is *essential* to have overlapping fragments in this library, so mechanical fragmentation is often chosen (see Chapter 6). You then pick recombinant clones at random from this mini-library and sequence each one – an approach known as *shotgun* sequencing (see Figure 9.5). At the start, you will have no idea where each bit of sequence comes from in the original fragment – nor even which strand of the original it is derived from. However, once you have a number of such fragments sequenced, you can start to use computer algorithms to compare each bit of sequence to all the others. The computer will find any overlaps between the fragments, including comparing the complementary strand in case you have two fragments that overlap but are derived from different strands. Where there is an overlap, those fragments will be joined together to form a *contig*. As the project progresses, each contig will grow longer, and will then start to overlap other contigs. So the contigs themselves are joined together until eventually you have one

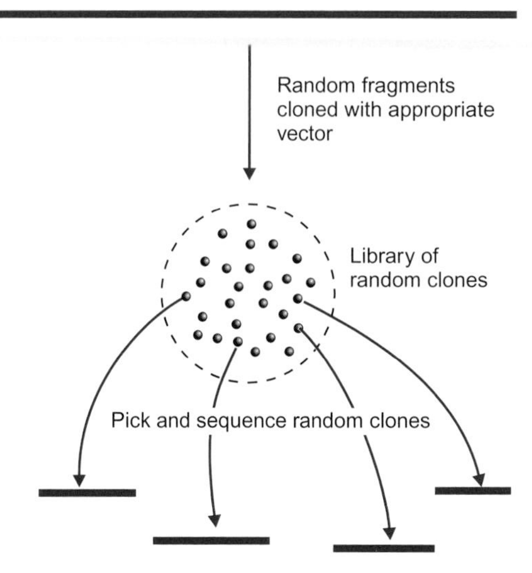

Figure 9.5 Shotgun cloning and sequencing

single contig covering the whole of the original piece of DNA, and all the sequenced fragments fit into that single contig (see Figure 9.6).

Of course, while initially each piece of sequence is new information, as you proceed, the sequencing becomes less and less productive. Towards the end, nearly all the clones sequenced will be completely contained within one of the existing contigs. Nevertheless, provided all parts of the fragment are equally represented in the library, it is quicker to continue with a shotgun approach rather than try to screen the library for the missing pieces. Sometimes, however, parts of the original fragment are under-represented in the library; they may be hard to clone for some reason. If this is the case, you will have a gap in your sequence that can be difficult to fill with a shotgun approach. If you have reason to believe that the gap is quite small, then you can use primer walking as described above to bridge the gap. Alternatively you can use the sequences you have determined to design PCR primers that will enable you to amplify a DNA fragment that spans the gap, and can then be sequenced. Other approaches for bridging gaps in longer sequences, such as genome sequences, are considered in Chapter 12.

For any sequencing project, whatever the size of the DNA to be sequenced, and whatever strategy is used, a diminishing returns effect is very marked. You may get 90 per cent of the sequence accurately determined quite quickly; the next 9 per cent may take as long again, and then the next 0.9 per cent a similar length of time. It is usually necessary to set some limits to this: how complete and how accurate do you need the sequence to be? Will 90 per cent do, or must you have 99 per cent or 99.9 per cent? Without some sort of

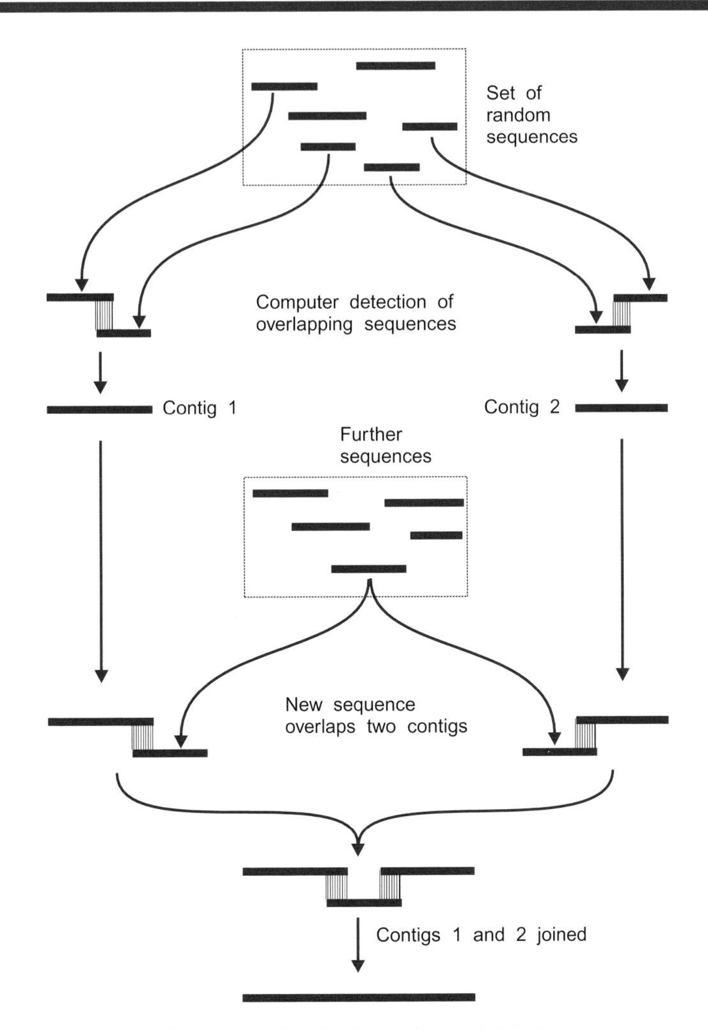

Figure 9.6 Contig formation and joining

compromise you can be drawn into a costly exercise trying to determine whether one difficult base in a sequence of a million bases is G or C.

9.2 Databank entries and annotation

Once we have determined a DNA sequence, it should be made publicly and freely accessible by submitting it to a databank (EMBL, GenBank, or DDBJ). This is done electronically via the internet, and the databanks have Web-based procedures that make this submission a simple automatic process. Box 9.1 lists the Web addresses for the databanks, and for the other databases and tools that are referred to in this and subsequent chapters. In practice, the databanks can be considered as one and the same, as they share their information on a daily basis, so you can just choose the one whose interface

Box 9.1 A selection of available on-line resources

Resource	Description	Web address
Nucleotide sequence databanks		
DDBJ (DNA Data Bank of Japan)		http://www.ddbj.nig.ac.jp/
GenBank		http://www.ncbi.nlm.nih.gov/Genbank/
European Molecular Biology Laboratory (EMBL)		http://www.ebi.ac.uk/embl/
Other databases		
UniProt Knowledgebase (UniProtKB	Merged protein sequence data from Swiss-Prot, TrEMBL and PIR-PSD	http://www.ebi.uniprot.org/
InterPro	Integrated resource for protein families, domains and functional sites	http://www.ebi.ac.uk/interpro/
PROSITE	Database of protein families and domains.	http://www.expasy.org/prosite/
Pfam	Multiple sequence alignments and HMMs for protein domains	http://www.sanger.ac.uk/Software/Pfam/
Prints	Compendium of protein fingerprints (motifs characterizing protein families)	http://www.bioinf.manchester.ac.uk/dbbrowser/PRINTS/
GDB Human Genome Database	Database for annotation of the human genome	http://www.genedb.org/
Online Mendelian Inheritance. in Man (OMIM)	Catalogue of human genes and genetic disorders	http://www.ncbi.nlm.nih.gov/OMIM
dbSNP	SNPs and short deletion and insertion polymorphisms	http://www.ncbi.nlm.nih.gov/SNP/
2D protein gels		http://www.mpiib-berlin.mpg.de/2D-PAGE/
imaGenes	Clone libraries, ESTs, plus a variety of tools	http://www.imagenes-bio.de/
Tools		
A wide variety of tools and databases are available at:		
NCBI		http://www.ncbi.nlm.nih.gov/Tools/
EBI (European Bioinformatics Institute)		http://www.ebi.ac.uk/
Sanger Institute		http://www.sanger.ac.uk/

Resource	Description	Web address
Some specific tools/sites:		
BLAST	Basic Local Alignment Search Tool	http://www.ebi.ac.uk/blast2/ http://www.ncbi.nlm.nih.gov/BLAST/
FASTA	Alternative to BLAST for database comparisons	http://www.ebi.ac.uk/fasta33/
Ensembl Genome Browser	Browse mammalian and other eukaryotic genome sequences	http://www.ebi.ac.uk/ensembl/
Artemis	Genome viewer and annotation tool that allows visualization of sequence features and the results of analyses	http://www.sanger.ac.uk/
ACT (Artemis Comparison Tool)	Downloadable DNA sequence comparison viewer	http://www.sanger.ac.uk/Software/ACT/
WebACT	On-line use of ACT for prokaryotic genome sequences	http://www.webact.org/WebACT/
ProtScale	Computes and represents a variety of profiles on a selected protein	http://www.expasy.org/tools/protscale.html
gMAP	Genome sequence comparison	http://www.ncbi.nlm.nih.gov/sutils/gmap.cgi

you prefer. It is only necessary to submit your sequence to (or to search in) one of them. An example of an EMBL databank entry for an individual sequence (human cDNA coding for thymidylate synthetase) is shown in Figure 9.7 (you may find when you retrieve a sequence that you get a more attractive display than this, but the information is the same). An increasing proportion of databank entries relate to whole genome sequences rather than individual genes like this one. We will consider the annotation and analysis of genome sequence data further in Chapter 12.

Entries in GenBank look slightly different, but contain the same information, and you should have little difficulty in switching from one to the other. Both are computer-readable, provided that the software has been set up to recognize the format.

The notes that we have added to the annotation should mainly be self-explanatory, but some aspects need additional comment. The *accession number* (AC) is important as it is the most convenient way of retrieving a specific sequence from the databank, and the one that is referred to in publications; the accession number is the same in the different databanks. It

Some elements of the annotation have been omitted, and sequence data are truncated

```
                        ID  X02308; SV 1; linear; mRNA; STD; HUM; 1536 BP.
Accession number        AC  X02308;
                        DT  07-NOV-1985 (Rel. 07, Created)
                        DT  12-SEP-1993 (Rel. 36, Last updated, Version 2)
Definition              DE  Human mRNA for thymidylate synthase (EC 2.1.1.45)

Keywords                KW  inverted repeat; synthetase; tandem repeat.

                        OS  Homo sapiens (human)
Species and             OC  Eukaryota; Metazoa; Chordata; Craniata; Vertebrata; Euteleostomi;
Classification          OC  Mammalia; Eutheria; Euarchontoglires; Primates; Haplorrhini;
                        OC  Catarrhini; Hominidae;Homo.

                        RP  1-1536
                        RX  PUBMED; 2987839.                       <- Link to the paper
Reference               RA  Takeishi K., Kaneda S., Ayusawa D., Shimizu K., Gotoh O., Seno T.;
                        RT  "Nucleotide sequence of a functional cDNA for human thymidylate synthase";
                        RL  Nucleic Acids Res. 13(6):2035-2043(1985).

                        DR  GDB; 678095.                           <- Links to human
                        DR  H-InvDB; HIT000320915.                    genome databases

                        FH  Key             Location/Qualifiers
Features table          FH
                        FT  source          1..1536
                        FT                  /organism="Homo sapiens"
                        FT                  /mol_type="mRNA"
                        FT                  /db_xref="taxon:9606"  <- link to taxonomy
                        FT  misc_feature    14..103
                        FT                  /note="triple tandemly repeated elements"
                        FT  misc_feature    35..103
                        FT                  /note="pot. stem-loop structure"
Coding sequence         FT  CDS             106..1047
                        FT                  /note="thymidylate synthase (aa 1-313)"
                        FT                  /db_xref="GDB:120465"                    ┐
                        FT                  /db_xref="GOA:P04818"                    │
                        FT                  /db_xref="HGNC:12441"                    │
                        FT                  /db_xref="InterPro:IPR000398"            ├ links to entries in
                        FT                  /db_xref="PDB:1HVY"                      │ other databases
                        FT                  /db_xref="PDB:1HW3"                      │
                        FT                  /db_xref="UniProtKB/Swiss-Prot:P04818"   ┘
                        FT                  /protein_id="CAA26178.1"
Translated sequence     FT                  /translation="MPVAGSELPRRPLPPAAQ ..."
                        FT  misc_feature    1231..1236
                        FT                  /note="pot. polyadenylation signal"
                        FT  misc_feature    1519..1524
                        FT                  /note="pot. polyadenylation signal"
                        FT  polyA_site      1536..1536
                        FT                  /note="polyadenylation site"

DNA sequence            SQ  Sequence 1536 BP; 390 A; 369 C; 399 G; 378 T; 0 other;
                            gggggggggg ggaccacttg gcctgcctcc gtcccgccgc ...
```

Figure 9.7 Sequence annotation (EMBL)

refers to a particular sequence submission, rather than to a specific gene or locus. There may therefore be several entries, with different accession numbers, relating to essentially the same sequence (for example from different strains or variants of the same species). If you want to look for different versions of a specific gene you may therefore need to search the databank by gene name, or alternatively use a known sequence and search for similarities (e.g. using BLAST, see below).

In addition to the sequence itself, there is a considerable amount of *annotation* that makes the sequence information much more useful. Some of the annotation is supplied with the submitted sequence, while other annotation is added by the databanks. Obviously, this must include information about the source of the sequenced DNA. In addition, the identification and extent of any open reading frame is a basic requirement (together with the computer prediction of the protein sequence), intron/exon boundaries (if applicable) and any further information on expression signals, motifs, structural elements, and so on, will enhance the value of the entry, as well as identification of the presumed or actual function of the gene. This information goes into the *features table* (lines starting with FT), and can be read by programs such as Artemis (see Chapter 12) to produce a visual display of the features of the sequence. How some of these features are identified is discussed below.

In addition to DNA sequences, there are databanks of protein sequences (see Box 9.1). These databanks, including the most popular one known as SWISS-PROT, have now been combined into a single databank, UniProt. You will see in Figure 9.7 that a reference is provided that identifies the corresponding entry in the UniProt database; clicking on the link will take you directly to that entry, which is shown in Figure 9.8. In addition to annotation that is similar to that described previously for DNA sequence entries, this entry contains information about the three-dimensional structure of the protein, and cross-references to a number of other databases, especially Pfam and Prosite, which will be considered later.

There are two types of protein sequence information. Some is derived from direct protein sequencing, but the majority (especially that arising from genome sequencing projects) is derived by computer translation of DNA sequences. Some care is needed when using computer-generated protein sequences, especially as there may be no direct evidence that this protein actually exists. Furthermore, it may be based on incorrectly predicted intron/exon boundaries, or alternative splicing may create different exon combinations, or the identification of the start site may be incorrect, or a variety of other factors (including post-translational modification or cleavage) may result in the protein within the cell being substantially different from the primary translation product. In particular, a base missing or incorrectly inserted will result in a shift of the reading frame, and the protein sequence beyond that point will bear no relationship to the real product.

```
Accession     ID  TYSY_HUMAN     STANDARD;     PRT;  312 AA.
number        AC  P04818;

              DT  13-AUG-1987, integrated into UniProtKB/Swiss-Prot.
              DT  17-OCT-2006, entry version 76.
              DE  Thymidylate synthase (EC 2.1.1.45) (TS) (TSase).
              GN  Name=TYMS; Synonyms=TS; ORFNames=OK/SW-cl.29;

References    RP  NUCLEOTIDE SEQUENCE [MRNA].
              RX  MEDLINE=85215597; PubMed=2987839;
              RA  Takeishi K., Kaneda S., Ayusawa D., Shimizu K., Gotoh O., Seno T.;
              RT  "Nucleotide sequence of a functional cDNA for human thymidylate synthase.";
              RL  Nucleic Acids Res. 13:2035-2043(1985).

              RP  PROTEIN SEQUENCE OF 1-24.
              RX  MEDLINE=85261174; PubMed=3839505;
              RA  Shimizu K., Ayusawa D., Takeishi K., Seno T.;
              RT  "Purification and NH2-terminal amino acid sequence of human thymidylate synthase ..."
              RL  J. Biochem. 97:845-850(1985).

              RP  X-RAY CRYSTALLOGRAPHY (3.0 ANGSTROMS).
              RX  MEDLINE=96110704; PubMed=8845352; DOI=10.1021/bi00050a007;
              RA  Schiffer C.A., Clifton I.J., Davisson V.J., Santi D.V., Stroud R.M.;
              RT  "Crystal structure of human thymidylate synthase: ..."
              RL  Biochemistry 34:16279-16287(1995).

Comments      CC  -!- CATALYTIC ACTIVITY: 5,10-methylenetetrahydrofolate + dUMP =  dihydrofolate + dTMP.
              CC  -!- PATHWAY: Nucleotide biosynthesis; deoxyribonucleotide biosynthesis.
              CC  -!- SUBUNIT: Homodimer.
              CC  -!- SIMILARITY: Belongs to the thymidylate synthase family.

Database      DR  EMBL; X02308; CAA26178.1; -; mRNA.
references    DR  PDB; 1HVY; X-ray; A/B/C/D=25-312.
              DR  PDB; 1HW3; X-ray; A=1-312.
              DR  Ensembl; ENSG00000176890; Homo sapiens.
              DR  HGNC; HGNC:12441; TYMS.
              DR  H-InvDB; HIX0017793; -.
              DR  MIM; 188350; gene.
              DR  ArrayExpress; P04818; -.
              DR  GO; GO:0009157; P:deoxyribonucleoside monophosphate biosynthesis; TAS.
              DR  InterPro; IPR000398; Thymidylat_synth.
              DR  Pfam; PF00303; Thymidylat_synt; 1.
              DR  PRINTS; PR00108; THYMDSNTHASE.
              DR  PROSITE; PS00091; THYMIDYLATE_SYNTHASE; 1.

Keywords      KW  3D-structure; Direct protein sequencing; Methyltransferase;
              KW  Nucleotide biosynthesis; Transferase.

Features      FT  INIT_MET     0       0
table         FT  CHAIN        1       312      Thymidylate synthase.
              FT  ACT_SITE     194     194
              FT  TURN         28      28
              FT  HELIX        29      42
              FT  STRAND       44      46
              FT  TURN         49      50
              FT  STRAND       54      65
              FT  TURN         67      68
              (etc)

Sequence      SQ  SEQUENCE  312 AA;  35585 MW;  A66F0E6D1973AB41 CRC64;
                  PVAGSELPRR PLPPAAQERD AEPRPPHGEL QYLGQIQHIL RCGVRKDDRT ...
```

Figure 9.8 Protein sequence annotation (UniProt)

9.3 Sequence analysis

9.3.1 Open reading frames

The analysis starts with the identification of the DNA region(s) that you can predict to code for expression of your target protein(s). To keep it simple at this stage, we will start with a discussion of sequences that do not have introns (such as bacterial sequences and cDNA) so we do not have to consider the identification of exons and introns. That will come later.

The key to identifying protein-coding sequences is to remember that mRNA has the potential to be translated into protein in any of three *reading frames*; which one is actually used will depend on where the ribosomes start. Because at this stage we may not know in which direction the DNA is transcribed, there are actually six possible reading frames altogether – three for an mRNA as read in one direction, and three others if the mRNA is produced in the other direction. How do we know which of these reading frames is used? Fortunately, we are provided with a clue. In those reading frames that are not used for translation, there are usually frequent stop codons. The same applies to any regions of the sequence that do not code for proteins. This is not just a fortunate coincidence – it prevents the ribosomes from accidentally producing useless, or even potentially damaging, products. On the other hand, any region of the sequence that does code for a protein must, obviously, be free of stop codons (until the stop codon that signifies the end of the protein). Such a sequence, without any stop codons, is known as an *open reading frame* (ORF). Therefore, we can use a computer to search the sequence for stop codons, and an ORF (without stop codons) can be predicted to code for a protein, in that reading frame. (In Chapter 12, we describe the use of a more sophisticated statistical approach, using hidden Markov models, or HMMs, which is especially relevant to the identification of ORFs in genome sequences.) Having identified an ORF, it is then straightforward to translate the DNA sequence into a protein sequence.

We would also look for a start codon. We can refine the concept of an ORF to mean the distance between a start codon and the first stop codon in the same reading frame. This is not actually as straightforward as it sounds. Although we regard ATG (using DNA rather than RNA nomenclature) as the 'normal' start codon, many organisms sometimes use other start codons as well, such as GTG, TTG or CTG – in some cases, especially in bacteria, these other start codons can be quite frequent. In such cases, we can also look for the presence of a ribosomal binding site adjacent to the start codon.

If we have some information about the nature of the target gene, or the protein it encodes – especially if we know the size of the protein – and if the piece of DNA we have cloned contains only that gene, then this is sufficient information to give us a putative identification of the gene. Ideally, there will be only one potential ORF of the required size. We can then confirm that identification by comparing the ORF, or the protein it encodes, with predicted proteins from other organisms. (We do not have to rely on the identified ORFs from other sequences; we can ask the computer to compare the product of our suspected ORF with all the possible products that would arise from translation of other complete sequences in all six reading frames.) If the predicted polypeptide from our query sequence is similar to polypeptides that might be made by other organisms, this suggests that it is a real coding sequence. Similarly we could look for known protein motifs in this sequence. The comparison of sequences is further considered later in this chapter.

9.3.2 Exon/intron boundaries

In eukaryotes, *introns* are spliced out from the primary transcript as the *exons* are joined into an mRNA molecule (Figure 9.9). This is not a consideration

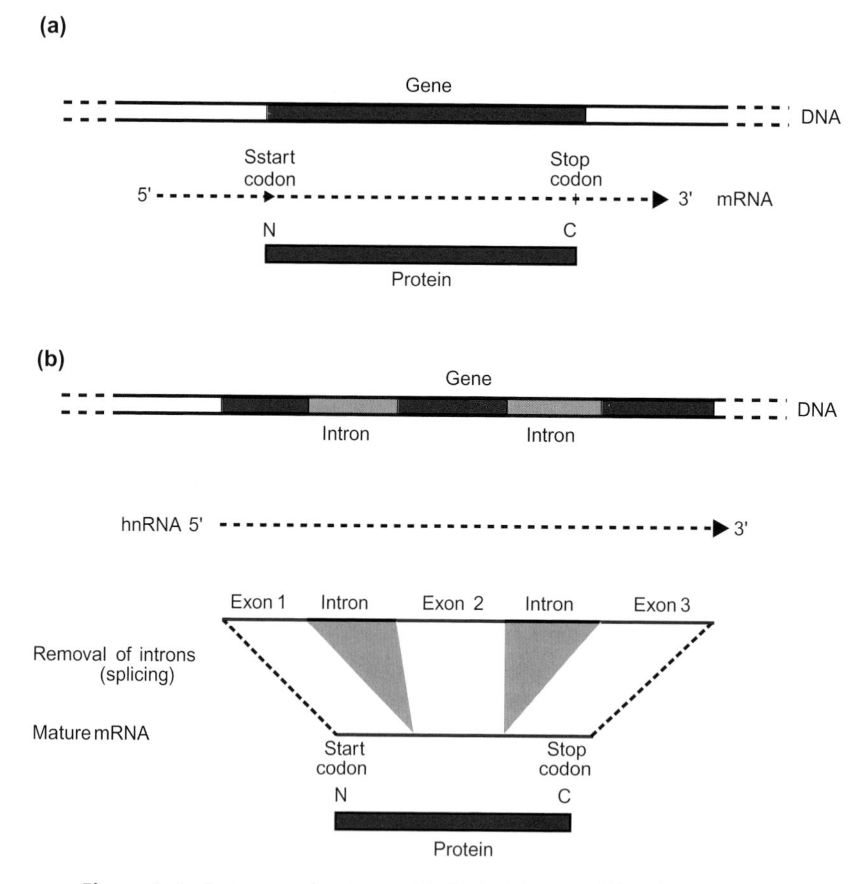

Figure 9.9 Introns and extrons. (a) Bacterial gene; (b) eukaryotic gene

when analysing cDNA sequences, such as that in Figure 9.7. The primary transcript, *heteronuclear* RNA (hnRNA) is so short-lived that it is unlikely to be represented in a cDNA library. Thus, the search for ORFs in eukaryotic cDNA sequences is similar to prokaryotes, with the very convenient exceptions that each transcription unit is neatly delineated – you know you will normally (if not always) find one ORF in the message, and you know that they are quite likely to be complete in each clone (if the clone itself is complete).

Genomic DNA sequences are, of course, a different matter. We can only start thinking of predicting the ORF if we can predict the mRNA sequence (if there is one in this particular region). Luckily, we can. In each eukaryotic species, intron/exon boundaries are surrounded by some more or less specific recurrent. The most obvious and conserved of these is the GU-AG motif. Most introns begin with GU (encoded by GT in the sense strand of the DNA) and finish with AG. Obviously, this is not the whole truth. Both of these combinations of two bases will theoretically occur with a frequency of 1:16, so a longer, more variable, sequence is required to define a splice site. It is therefore not reliable to simply inspect a sequence and identify consensus splice sites. This prediction becomes even more difficult when whole genome sequences are involved, but there are various computer programs that use statistical procedures (including HMMs – see Chapter 12) that can be trained to recognize intron/exon boundaries in a specific organism, based on comparison of known cDNA and genomic sequences.

The computer can then in effect excise the predicted introns and thus assemble a *predicted* cDNA sequence. This is not totally reliable without some corroborating evidence. Fortunately, such evidence is often at hand. The most obvious thing to do is to search a databank for matching full-length cDNA sequences, from the same species or from a different one. Additional evidence can be obtained from screening databanks of *expressed sequence tags* (ESTs) (see Chapter 14).

Figure 9.10 shows the annotation for the sequence of a genomic clone corresponding to the cDNA sequence shown in Figure 9.7. It will be seen that the sequence contains seven exons, separated by six introns, making the total length of the gene (as measured by the primary transcript) over 15 kb, or about 10 times the length of the cDNA.

9.3.3 Expression signals

An ORF is of no significance unless that region of the DNA is actually transcribed, and in the correct direction. Therefore the identification of potential transcription start (and stop) sites is important information relating to our sequenced gene. Although the identification of such signals by analysing the

Some elements of the annotation have been omitted, and sequence data are truncated

```
Accession      ID  D00596; SV 1; linear; genomic DNA; STD; HUM; 18596 BP.
number         AC  D00596;

               DT  17-JUL-1991 (Rel. 28, Created)
               DT  14-NOV-2006 (Rel. 89, Last updated, Version 3)
Definition     DE  Homo sapiens gene for thymidylate synthase, exons 1, 2, 3, 4, 5, 6, 7,
               DE  complete cds.

               RX  PUBMED; 2243092.                          (Link to the paper)
Reference      RA  Kaneda S., Nalbantoglu J., Takeishi K., Shimizu K., Gotoh O., Seno T., Ayusawa D.;
               RT  "Structural and functional analysis of the human thymidylate synthase gene.";
               RL  J. Biol. Chem. 265(33):20277-20284(1990).

               DR  GDB; 163670.                              (Links to human
               DR  GDB; 182340.                               genome databases)

Features       FH  Key         Location/Qualifiers
table          FT  source       1..18596
               FT              /organism="Homo sapiens"
               FT              /chromosome="18"
               FT              /map="18p11.32"
               FT              /mol_type="genomic DNA"
Primary        FT              /clone="lambdaHTS-1 and lambdaHTS-3"
transcript     FT  prim_transcript 822..16246
               FT              /note="thymidylate synthase mRNA and introns"
Coding         FT  CDS          join(1001..1205,2895..2968,5396..5570,11843..11944,
sequence       FT              13449..13624,14133..14204,15613..15750)
               FT              /translation="MPVAGSELPRRPLPPAAQ ..."
Location of    FT  exon         <1001..1205
exons and      FT              /number=1
introns        FT  intron       1206..2894
               FT              /number=1
               FT  exon         2895..2968
               FT              /number=2
               FT  intron       2969..5395
               FT              /number=2
               FT  exon         5396..5570
               FT              /number=3
               FT  intron       5571..11842
               FT              /number=3
               FT  exon         11843..11944
               FT              /number=4
               FT  intron       11945..13448
               FT              /number=4
               FT  exon         13449..13624
               FT              /number=5
               FT  intron       13625..14132
               FT              /number=5
               FT  exon         14133..14204
               FT              /number=6
               FT  intron       14205..15612
               FT              /number=6
               FT  exon         15613..>15750
               FT              /number=7

DNA            SQ  Sequence 18596 BP; 4521 A; 3991 C; 4479 G; 5605 T; 0 other;
sequence           cctgtagtcc cagctacgcg agaggctgag gcagcagaat tacttgaacc caggaggcgg ...   60
```

Figure 9.10 Sequence annotation (EMBL) of a genomic clone

sequence is not completely reliable, we can often achieve some indication of how the gene is expressed. Since Figure 9.7 refers to a cDNA sequence, transcriptional start and stop sites are not relevant. However a polyadenylation site is identified, together with potential polyadenylation signals – the 'potential' label indicating that it is a prediction that has not been confirmed by direct experimental evidence. For the corresponding genomic clone (Figure 9.10), the limits of the primary transcript are identified in the annotation.

In most bacteria, a high proportion of genes are transcribed by RNA polymerase-recognizing promoter sites which have two relatively conserved consensus regions: the –35 region (i.e. a region centred at 35 bases before the start of transcription) and a –10 region. The consensus sequences for these two regions are TTGACA and TATAAT respectively (see Chapter 2). However, these only represent a consensus. Very few promoters have exactly that sequence at either position, and the distance separating them can vary by a few bases as well. If we build in that degree of flexibility into our search for promoters, we will end up with a large number of sites, most of which are not genuine promoters. The binding of RNA polymerase is affected to a lesser extent by the sequence of a larger region, extending over perhaps 70 bases, as well as by other factors such as the supercoiling of that region of the chromosome.

In addition, the specificity of bacterial RNA polymerase can be changed by substitution of a different sigma factor, enabling it to recognize promoters with a markedly different structure. In organisms that are less well characterized than *E. coli*, the structure of these alternative promoters may be unknown, making promoter prediction impossible.

In eukaryotes, as usual the situation is even more complicated. The binding of RNA polymerase II is mediated by a number of canonical elements with similarity of structure, including the TATA-box, GC-box and CAAT-box. However, the spacing of these elements is not always consistent. Moreover, there is an element of species variation, so the parameters for searching for a transcription start site in a fruit fly are different from those needed to make the corresponding prediction in a human sequence. A number of websites are available for the prediction of transcription start sites in model organisms. These take advantages of neural networks that use information from known promoters in the same organism.

In addition to searching for possible promoters (i.e. sites at which RNA polymerase may bind directly), we can look for sites at which regulatory proteins can attach to the DNA to repress, or activate, transcription. Many of these proteins have quite well conserved, and characterized, recognition sequences, often known as 'boxes'. For example, in most bacteria iron uptake is regulated by proteins belonging to one of two families, related to either Fur (the ferric uptake regulator protein of *E. coli*) or DtxR (the diphtheria toxin repressor of *Corynebacterium diphtheriae*). These proteins, in the presence of Fe^{2+}, bind to specific DNA sites and repress transcription of the adjacent genes. Identification of a site such as the so-called 'Fur box', to which Fur binds, therefore provides an indication that the associated gene will be repressed in the presence of an adequate supply of iron, which in turn suggests that the function of that gene may be connected with iron uptake. There are a number of other known 'boxes', to which different regulatory proteins will bind, in both prokaryotes and eukaryotes – for example, the E-box (for enhancer) has been associated with an expression pattern following a circadian (24 h) rhythm. It is possible to screen DNA sequences for each of these boxes, providing evidence not only relating to the regulation of the relevant genes, but also a clue as to their possible function; the methods for searching for such binding sites, and other structural features of a protein, are considered later. As with putative promoter sites, it is essential to distinguish between computer predictions and direct evidence.

It is important to keep in mind that the promoter and its associated regulatory sequences extend in the 3′-direction beyond the transcription start site. Many of the latter are often found in the first intron, which is often very large in eukaryotic genes, and exert their function in the double-stranded DNA rather than in the single-stranded RNA transcript.

9.4 Sequence comparisons

9.4.1 DNA sequences

There is a vast, and rapidly growing, amount of sequence data available in the databanks. How can we find out if our sequence is the same as (or very similar to) some other sequence that has already been determined?

We can start by considering the simplest approach. Suppose we have 1 kb of sequence data (and we will for the moment ignore the possible complication of introns), and we want to compare it with another specific sequence. We can put our sequence (the *query sequence*) alongside the second sequence, and get the computer to count the number of bases that match. However, the two sequence fragments may be of different lengths, and may not start at the same place. So we (or rather the computer) must slide our query sequence along the sequence it is being compared with and see where the best match comes. However, there is another problem. Our sequence may have one or more gaps in it, compared with the second sequence (or vice versa). This can happen in several ways: there may be a genuine difference between the sequences (if for example we are comparing genes from different species), or one of the sequences might be wrong. Addition or removal of even a single base would render this simplistic approach non-viable. For example, our sequence may read at one point GGACT, while at the corresponding point in the databank sequence, which otherwise matches perfectly, the sequence is GG**G**ACT. This is a very small difference, but the consequence would be that part of our sequence would line up perfectly, while the rest would not match at all (or more strictly it would match just as well as any two random bits of DNA). In order to accommodate this problem, we can allow the computer to introduce gaps into one or both sequences.

We cannot allow the computer indefinite licence to introduce as many gaps as it likes, otherwise it would produce perfect matches between any two sequences, just by repeatedly sliding them apart until it found the next base that matched. The algorithm therefore incorporates a *gap penalty*; every time a gap is introduced, into either sequence, there is a reduction in the score. The gap penalty can be set at different levels, including different penalties for different lengths of gaps, although the software usually has a default value so we do not have to think about it.

There are several methods available for comparing DNA sequences, and some of them employ algorithms that essentially work as described above (although the description is highly simplified!). However, this requires a lot of computer power for longer sequences, if carried out literally as described. As the computer slides the two sequences along base by base, at each position it has to calculate a score with all possible combinations of gaps – even for a sequence of say 1 kb, there is a very large number of possible combinations. Usually we want to compare our query sequence with all sequences in the databank, not just one individual sequence.

Other programs therefore employ different algorithms that are designed to speed up the process. One approach is basically to split our query sequence into small fragments (*words*), using these words to decide how the two sequences will align best and then computing a score for the optimal alignment, including allowing for gaps. This is more or less the basis for one of the common search programs known as FASTA.

A somewhat different approach is used by one of the most widely used programs, known as BLAST (Basic Local Alignment Search Tool). At a highly simplified level, this works by finding very short matches (segment pairs) and then extending that match outwards until the score falls below a set value. Each matched pair of sequences above a certain length is then stored and reported (as high-scoring segment pairs, or HSPs), starting with those with the highest score. Figure 9.11 shows selected results of a database search using BLASTN (the version of BLAST used with DNA sequences), using the human thymidylate synthase cDNA (see Figure 9.7) as the *query sequence*. (Note that this is an edited version of the output, derived from several hundred reported matches.) The first column identifies the sequence that matches, including in this example the Division of the EMBL database. The most similar sequences, not surprisingly, are from various mammals, followed by birds, viruses, invertebrates and plants (EM_OV, EM_VI, EM_INV,

DB:ID	Source	Length	Score	Identity %	Positives %	*E*()
EM_OM:BC111139	*Bos taurus* thymidylate synthase, mRNA	1606	3101	83	83	8.5e − 175
EM_MUS:M13019	Mouse thymidylate synthase mRNA	961	3458	87	87	4.8e − 149
EM_RO:L12138	Rat thymidylate synthase mRNA	1598	3452	85	85	5.4e − 149
EM_OV:BX932834	*Gallus gallus* finished cDNA	1489	2917	79	79	8.5e − 125
EM_VI:AF148805	Human herpesvirus 8, complete genome	137969	1706	66	66	3.6e − 71
EM_INV:AY089595	*Drosophila melanogaster* cDNA	1343	1699	67	67	9.7e − 70
EM_INV:X51735	*Leishmania amazonensis* dihydrofolate reductase thymidylate synthase gene	1996	1509	63	63	2.5e − 61
EM_PL:AF073488	*Zea mays* dihydrofolate reductase-thymidylate synthase mRNA	1751	1462	64	64	3.8e − 59
EM_PL:AJ491794	*Pisum sativum* mRNA dihydrofolate reductase-thymidylate synthase	2143	1407	64	64	9.4e − 57
EM_PL:AK227728	*Arabidopsis thaliana* mRNA dihydrofolate reductase-thymidylate ...	1926	1397	63	63	3.0e − 56
EM_INV:AF494020	*Trypanosoma cruzi* dihydrofolate reductase-thymidylate synthase	1566	1284	62	62	4.6e − 51
EM_INV:X98123	*Plasmodium vivax* DHFR-TS gene	1872	1150	61	61	4.3e − 45
EM_PRO:J01710	*E. coli* thyA gene	1163	701	62	62	1.2e − 35
EM_PRO:X69661	*B. subtilis* thyB gene	896	766	63	63	4.3e − 34
EM_FUN:J04230	*Candida albicans* thymidylate synthase	1413	861	68	68	6.5e − 32
EM_INV:AC006708	*Caenorhabditis elegans* cosmid Y110A7A, complete sequence	70694	874	68	68	3.2e − 30
EM_PRO:U86637	*Neisseria gonorrhoeae* thymidylate synthase	1422	773	64	64	1.2e − 27
EM_FUN:AJ401221	*Agaricus bisporus* tms1 gene for thymidylate synthase, exons 1-2	943	763	64	64	7.9e − 27
EM_PRO:AE004091	*Pseudomonas aeruginosa*, genome	6264404	628	62	62	7.7e − 25
EM_FUN:J02706	*S. cerevisiae* thymidylate synthase	1813	760	63	63	2.8e − 24
EM_PRO:AE008838	*Salmonella typhimurium* LT2, section 142 of 220 of the complete genome	21583	609	61	61	7.5e − 23
EM_PRO:BX842580	*Mycobacterium tuberculosis* genome	346051	623	62	62	1.1e − 22
EM_INV:AC167560	*Culex pipiens* quinquefasciatus, complete sequence	118752	688	65	65	2.0e − 21
EM_FUN:AE017350	*Cryptococcus neoformans* chromosome 10, complete sequence	1085720	707	63	63	3.7e − 21
EM_PRO:AE016879	*Bacillus anthracis*, complete genome	5227293	726	60	60	1.3e − 19
EM_PRO:BX571856	*Staphylococcus aureus* subsp. aureus strain MRSA252, genome	2902619	689	60	60	2.5e − 17

Notes
Length = the length of each target sequence. In some cases this is a complete genome sequence (or part of it), rather than a specific clone
Score (or *bit score*) = a measure of the similarity of the target and query sequences
Identity = percentage of identical residues in the two sequences
Positives = the percentage of similar residues (as these are DNA sequences, this is the same is Identity)
E() = the probability of this match occurring by chance.

Figure 9.11 Comparison of DNA sequences: selected output from a BLASTN search of databanks with human thymidylate synthase cDNA as the query sequence

EM_PL), prokaryotes (EM_PRO) and fungi (EM_FUN). The entry in this column also includes the accession number, and in the on-line output it is an active link enabling you to obtain any of these sequences. The next column shows the definition of the sequence (truncated in this example).

The most significant column is that headed *E*( ), which provides an estimate of the probability of each match occurring by chance; thus a low *E* number, with

```
>EM_PRO:J01710; J01710 E.coli thyA gene coding for thymidylate synthase.
    Length = 1163

 Plus Strand HSPs:

 Score = 701 (111.2 bits), Expect = 1.2e – 35, Sum P(2) = 1.2e – 35
 Identities = 297/475 (62%), Positives = 297/475 (62%), Strand = Plus/Plus

Query:  578 TTGACCAACTG-CAAAGAGTGATTGACACCATC--AAAACCAACCCTGACGACAGAAGAA 634
            |||||||  |  | | | || | ||| |||| |  |||| | |||| ||     |  | |
Sbjct:  540 TTGACCAGATCACTACG-GT-ACTGA-ACCAGCTGAAAAACGACCCGGATTCGCGCCGCA 596

Query:  635 TCATCATGTGCGCTTGGAATCCAAGAGATCTTCCTCTGATGGCGCTGCCTCCATGCCATG 694
            | ||  | |  || |||||   | | || ||   |   ||||||||| | || |||||||
Sbjct:  597 TTATTGTTTCAGCGTGGAACGTAGGCGAACTGGATAAAATGGCGCTGGCACCGTGCCATG 656

Query:  695 CCCTCTGCCAGTTCTATGTGGTGAACAGTGAGCTGTCCTGCCAGCTGTACCAGAGATCGG 754
            |  ||| ||||||||||||||   || |  | || || |||||||| || ||| | ||
Sbjct:  657 CATTCTTCCAGTTCTATGTGGCAGACGGCAAACTCTCTTGCCAGCTTTATCAGCGCTCCT 716

Query:  755 GAGACATGGGCCTCGGTGTGCCTTTCAACATCGCCAGCTACGCCCTGCTCACGTACATGA 814
            | ||| |   ||||||  |||| |||||||| |||||||||||  |  |   | | ||||
Sbjct:  717 GTGACGTCTTCCTCGGCCTGCCGTTCAACATTGCCAGCTACGCGTTATTGGTGCATATGA 776

Query:  815 TTGCGCA-CATCACGGGCCTGAAGCCAGGTGACTTTATACACACTTTGGGAGATGCACAT 873
            | ||||| ||   ||  |  ||||   ||||| ||| |    ||    || ||  | |||
Sbjct:  777 TGGCGCAGCAGTGCGATCTGGAAGTG-GGTGATTTTGTCTGGACCGGTGGCGACACGCAT 835

Query:  874 ATTTACCTGAATCACATCGAGCCA-CTGAAAATTCAGCTTCAGC-GAGAACCCAGACCTT 931
             | |||   || || || || | | || |   | ||  || ||| | |||||  | ||
Sbjct:  836 CTGTACAGCAACCATATGGATCAAACTCATC-TGCAA-TTAAGCCGCGAACCGCGTCCGC 893

Query:  932 TCCCAAAGCTCAGGATTCTTCGAAAAGTTGAGAAAATTGATGACTTCAAAGCTGAAGACT 991
            | || ||| | |  ||    || |||   ||    ||    |||| |      |||||||
Sbjct:  894 TGCCGAAGTTGATTATCAAACGTAAACCCGAATCCATCTTCGACTACCGTTTCGAAGACT 953

Query:  992 TTCAGATTGAAGGGTACAATCCGCATCCAACTATTAAAATGGAAATGGCTGTTTA 1046
            || |||||||||| ||| ||||||||||    ||||||  |    ||||| | ||
Sbjct:  954 TTGAGATTGAAGGCTACGATCCGCATCCGGGCATTAAAGCGCCGGTGGCTATCTA 1008
```

Notes.

Plus Strand HSPs (High Scoring Pairs) indicates that the reported scores were obtained with the query sequence (the human cDNA) as entered, rather than the reverse complement (which would be indicated as Minus Strand)

Expect = 1.2e – 35, Sum P(2) = 1.2 e – 35. This shows the probability of the score(s) occurring by chance. This is a low value, suggesting that this match is highly significant, although the sequences are clearly far from identical

Strand = Plus/Plus. This indicates that the match shown contains the plus strands of both the query sequence (human cDNA) and subject (*E. coli*)

See Figure 9.11 for additional notes.

Figure 9.12 Results of a database search using BLASTN (DNA search); example of a reported match, from the search results shown in Figure 9.11

a high negative logarithm, indicates a strong match. The entries are arranged in descending order, so that the most significant matches are at the top.

The BLAST search will also return the pairwise alignment of the query sequence with each of the matches, although this is not necessarily the optimal alignment (see the section on CLUSTAL below). As an example, the alignment with the *E. coli thyA* gene is shown in Figure 9.12. Although the *E* value indicates that this is a highly significant match, you will see that the sequences are far from identical (62 per cent identity). In the next section, we will look at the match between these two sequences at the protein level.

9.4.2 Protein sequence comparisons

DNA sequence comparisons are inherently noisy, that is they are riddled with false positive matches. If you just take two random DNA sequences and put them side by side, there is a 25 per cent chance of a match at any specific position. Sometimes you have to use DNA sequence comparisons, for example if you are working with a non-coding sequence, but generally for searching databanks with an unknown sequence you will get much cleaner results with a protein sequence. A further advantage of searching at the protein level is that, owing to the degeneracy of the genetic code and the much lower evolutionary constraints on intron sequences, related proteins are generally more conserved than are the genes encoding for them. Note that all the predicted coding sequences in the DNA databanks are translated and incorporated in the protein sequence databank. Alternatively, using a variant of BLAST (BLASTX), you can input a DNA query sequence and the computer will translate it in all six reading frames (three in each direction) and compare all six to the sequences in a protein databank. Conversely, TBLASTN will compare a protein query to translated DNA sequences from the databank. Or you can even use TBLASTX, which compares a translated DNA sequence with translated sequences from a nucleotide databank. However, in this case, we will use BLASTP, for comparing a protein sequence query to a protein databank.

The initial basis for protein sequence comparisons is just the same as described above for nucleotide sequence comparisons, and the algorithms used are fundamentally the same. The major difference arises from the fact that you have 20 amino acids instead of four bases. This does not have to change anything – you could treat the 20 amino acids as all different from one another, and require a perfect match (*identity*) at any position. However, the computer offers you a more sophisticated scoring system, in which some pairs of amino acids are regarded as more different than others. This scoring system is based on a matrix of all the possible amino acid pairings, ranging from some pairs which score almost the same as a perfect match through to other pairs which are regarded as completely different. All of the scores in the matrix are used to calculate the overall score for the match, although the alignment presented will only mark matches above a selected cut-off score.

There are several such matrices to choose from, although here again the package you use will probably have a default in case you do not want to make a choice. Yet even if you accept the default, you should have some understanding of the basis of the matrix that is used. It is a common misconception that the alignment is calculated solely on the basis of the similarity between the amino acids. For example a biochemist might group valine, leucine and isoleucine together as they have large nonpolar (hydrophobic) side chains, phenylalanine, tyrosine and tryptophan (aromatic), glutamate with aspartate (acidic, negatively charged), lysine and arginine (basic, positively charged), and so on. A change from one amino acid to another within the same group (say lysine to arginine) would be expected to have less effect on the structure and function of the protein than a change between groups (say lysine to phenylalanine) – and so you would expect this similarity to be reflected in the scoring system.

That is indeed partially true, but in fact commonly used matrices such as the Dayhoff point accepted mutation (PAM) matrices are derived empirically from comparisons of related proteins from different sources, and the score is determined from the frequency with which specific changes in amino acid sequence occur. This partially reflects the similarity between the amino acids, but is also influenced by the nature of the genetic code. Where a pair of amino acids can be interchanged by only a single base change in the DNA, this will occur more frequently than a change that requires two or three base changes. There are a series of PAM matrices with different values; one example (the Dayhoff PAM250 matrix) is shown in Figure 9.13. If you are not familiar with the one-letter amino acid notation, see Box 9.2. Values of 0 indicate neutral changes, while increasing positive or negative values indicate increasingly acceptable or unacceptable mutations, respectively. Note that in some cases, a mismatch can give a positive score – and in particular a mismatch between tyrosine (Y) and phenylalanine (F) actually gives a higher score than a perfect match of most other amino acids. There are different versions of the PAM matrices, which vary in their sensitivity. Therefore the PAM250 matrix is suitable for distantly related sequences, while others such as PAM40 are better for more similar sequences.

Although the PAM matrices still perform well, they are no longer regarded as the best matrices for database searching. Most implementations of programs such as BLASTP use, as the default, an alternative family of matrices known as BLOSUM ('blossom'), or BLOcks SUbstitution Matrix. These are derived in a somewhat different way, which makes them more suitable for procedures based on local alignments. One example, BLOSUM45, is shown in Figure 9.13. The numbering of the matrix title works in the opposite direction from the PAM matrices, so the commonly used BLOSUM62 matrix is less sensitive for weak alignments than BLOSUM45. Yet other matrices are available; some implementations use the Gonnet matrix as the default. Generally speaking, at the start the best option is to use the default matrix, and only start to explore different matrices if you do not find the matches that you are looking for.

Box 9.2 Amino acid notations

	Three-letter notation	One-letter notation
Alanine	Ala	A
Arginine	Arg	R
Asparagine	Asn	N
Aspartate	Asp	D
Cysteine	Cys	C
Glutamate	Glu	E
Glutamine	Gln	Q
Glycine	Gly	G
Histidine	His	H
Isoleucine	Ile	I
Leucine	Leu	L
Lysine	Lys	K
Methionine	Met	M
Phenylalanine	Phe	F
Proline	Pro	P
Serine	Ser	S
Threonine	Thr	T
Tryptophan	Trp	W
Tyrosine	Tyr	Y
Valine	Val	V

Figure 9.14 shows selected results from the output from a BLASTP search of the UniProt database, using, as the query sequence, the amino acid sequence of the human thymidylate synthase, which is the gene that we employed for the nucleic acid search above (Figure 9.11). The general structure of the table is the same as before. Note that, although all of these proteins are labelled as thymidylate synthases (or as bifunctional DHFR-TS

Dayhoff PAM250 matrix

	C	S	T	P	A	G	N	D	E	Q	H	R	K	M	I	L	V	F	Y	W
C	12																			
S	0	2																		
T	−2	1	3																	
P	−3	1	0	6																
A	−2	1	1	1	2															
G	−3	1	0	−1	1	5														
N	−4	1	0	−1	0	0	2													
D	−5	0	0	−1	0	1	2	4												
E	−5	0	0	−1	0	0	1	3	4											
Q	−5	−1	−1	0	0	−1	1	2	2	4										
H	−3	−1	−1	0	−1	−2	2	1	1	3	6									
R	−4	0	−1	0	−2	−3	0	−1	−1	1	2	6								
K	−5	0	0	−1	−1	−2	1	0	0	1	0	3	5							
M	−5	−2	−1	−2	−1	−3	−2	−3	−2	−1	−2	0	0	6						
I	−2	−1	0	−2	−1	−3	−2	−2	−2	−2	−2	−2	−2	2	5					
L	−6	−3	−2	−3	−2	−4	−3	−4	−3	−2	−2	−3	−3	4	2	6				
V	−2	−1	0	−1	0	−1	−2	−2	−2	−2	−2	−2	−2	2	4	2	4			
F	−4	−3	−3	−5	−4	−5	−4	−6	−5	−5	−2	−4	−5	0	1	2	−1	9		
Y	0	−3	−3	−5	−3	−5	−2	−4	−4	−4	0	−4	−4	−2	−1	−1	−2	7	10	
W	−8	−2	−5	−6	−6	−7	−4	−7	−7	−5	−3	2	−3	−4	−5	−2	−6	0	0	17

BLOSUM45 matrix

	C	S	T	P	A	G	N	D	E	Q	H	R	K	M	I	L	V	F	Y	W
C	12																			
S	−1	4																		
T	−1	2	5																	
P	−4	−1	−1	9																
A	−1	1	0	−1	5															
G	−3	0	−2	−2	0	7														
N	−2	1	0	−2	−1	0	6													
D	−3	0	−1	−1	−2	−1	2	7												
E	−3	0	−1	0	−1	−2	0	2	6											
Q	−3	0	−1	−1	−1	−2	0	0	2	6										
H	−3	−1	−2	−2	−2	−2	1	0	0	1	10									
R	−3	−1	−1	−2	−2	−2	0	−1	0	1	0	7								
K	−3	−1	−1	−1	−1	−2	0	0	1	1	−1	3	5							
M	−2	−2	−1	−2	−1	−2	−2	−3	−2	0	0	−1	−1	6						
I	−3	−2	−1	−2	−1	−4	−2	−4	−3	−2	−3	−3	−3	2	5					
L	−2	−3	−1	−3	−1	−3	−3	−3	−2	−2	−2	−2	−3	2	2	5				
V	−1	−1	0	−3	0	−3	−3	−3	−3	−3	−3	−2	−2	1	3	1	5			
F	−2	−2	−1	−3	−2	−3	−2	−4	−3	−4	−2	−2	−3	0	0	1	0	8		
Y	−3	−2	−1	−3	−2	−3	−2	−2	−2	−1	2	−1	−1	0	0	0	−1	3	8	
W	−5	−4	−3	−3	−2	−2	−4	−4	−3	−2	−3	−2	−2	−2	−2	−2	−3	1	3	15

Figure 9.13 Examples of protein substitution matrices

DB:ID	Source	Length	Score	Identity %	Positives %	E()
TYSY_HUMAN	Thymidylate synthase (EC 2.1.1.45)	312	1675	100	100	2.30e −171
Q8VDV6_MOUSE	Tyms protein.	307	1482	89	93	6.60e − 151
TYSY_RAT	Thymidylate synthase (EC 2.1.1.45)	307	1475	88	93	3.70e − 150
TYSY_VZVD	Thymidylate synthase (EC 2.1.1.45)	301	1154	71	85	3.80e − 116
TYSY_DROME	Thymidylate synthase (EC 2.1.1.45)	321	1084	63	77	9.90e − 109
Q2F5K5_BOMMO	Thymidylate synthase isoform 1	298	1077	68	81	5.50e − 108
Q8L6F0_PEA	Dihydrofolate reductase-thymidylate synthase	530	1033	67	76	2.50e − 103
DRTS_ORYSA	Putative bifunctional dihydrofolate reductase-thymidylate synthase	494	1032	65	77	3.20e − 103
DRTS_SOYBN	Bifunctional dihydrofolate reductase-thymidylate synthase (DHFR-TS)	530	1031	66	76	4.10e − 103
TYSY_CANAL	Thymidylate synthase (EC 2.1.1.45)	314	614	65	74	1.60e − 95
TYSB_BACSU	Thymidylate synthase B (EC 2.1.1.45)	264	580	59	75	2.40e − 82
TYSY_ECOLI	Thymidylate synthase (EC 2.1.1.45)	264	523	54	69	2.20e − 75
TYSY_SALTY	Thymidylate synthase (EC 2.1.1.45) (TS) (TSase).	264	507	52	67	2.90e − 73
TYSY_MYCTU	Thymidylate synthase (EC 2.1.1.45) (TS) (TSase).	263	519	52	70	3.20e − 72
Q7Y4M9_BPT2	DTMP thymidylate synthase.	286	548	43	56	6.20e − 52

Notes
DB:ID Uniprot identity code. See Figure 9.16 for explanations
Identity = percentage of identical residues in the two sequences
Positives = the percentage of similar residues (according to the matrix used)
See Figure 9.11 for explanation of other headings

Figure 9.14 BLASTP search of Uniprot database for proteins resembling human thymidylate synthase (selected results only)

enzymes), many are derived from genome sequences and are labelled as such solely because of their similarity to other thymidylate synthases, without direct experimental evidence. With these examples, it is probably correct, but the potential for a circular argument has to be remembered.

As before, the program will also return a pairwise alignment of the sequences. If you compare the protein alignment, with the *E. coli* enzyme, shown in Figure 9.15, with the alignment of the corresponding DNA sequences in Figure 9.12, you will see that the protein comparison produces a much clearer similarity. You may wonder why the percentage identity (50–54 per cent) is lower than that for the DNA comparison (62 per cent), when we said that amino acid sequences tend to be more conserved. The reason lies in the noisiness of the DNA comparison, where many of the matches will be due to chance. Comparison of the *E* values shows that the protein sequence match is much more significant than the DNA alignment.

>UNIPROT:TYSY_ECOLI P0A884 Thymidylate synthase (EC 2.1.1.45) (TS) (TSase).
Length = 264

Score = 523 (189.2 bits), Expect = 2.2e − 75, Sum P(2) = 2.2e − 75
Identities = 102/187 (54%), Positives = 130/187 (69%)

```
Query:   126 EEGDLGPVYGFQWRHFGAEYRDMESDYSGQGVDQLQRVIDTIKTNPDDRRIIMCAWNPRD 185
             E GDLGPVYG QWR +            G+ +DQ+  V++ +K +PD RRII+ AWN  +
Sbjct:    86 ENGDLGPVYGKQWRAWPTP--------DGRHIDQITTVLNQLKNDPDSRRIIVSAWNVGE 137

Query:   186 LPLMALPPCHALCQFYVVNSELSCQLYQRSGDMGLGVPFNIASYALLTYMIAHITGLKPG 245
             L  MAL PCHA  QFYV + +LSCQLYQRS D+ LG+PFNIASYALL +M+A    L+ G
Sbjct:   138 LDKMALAPCHAFFQFYVADGKLSCQLYQRSCDVFLGLPFNIASYALLVHMMAQQCDLEVG 197

Query:   246 DFIHTLGDAHIYLNHIEPLKIQLQREPRPFPKLRILRKVEKIDDFKAEDFQIEGYNPHPT 305
             DF+ T GD H+Y NH++   +QL REPRP PKL I RK E I D++ EDF+IEGY+PHP
Sbjct:   198 DFVWTGGDTHLYSNHMDQTHLQLSREPRPLPKLIIKRKPESIFDYRFEDFEIEGYDPHPG 257

Query:   306 IKMEMAV 312
             IK  +A+
Sbjct:   258 IKAPVAI 264
```

Score = 265 (98.3 bits), Expect = 2.2e − 75, Sum P(2) = 2.2e − 75
Identities = 55/109 (50%), Positives = 71/109 (65%)

```
Query:    31 QYLGQIQHILRCGVRKDDRTGTGTLSVFGMQARYSLRDEFPLLTTKRVFWKGVLEELLWF 90
             QYL  +Q +L  G +K+DRTGTGTLS+FG Q R++L+D FPL+TTKR   + ++ ELLWF
Sbjct:     3 QYLELMQKVLDEGTQKNDRTGTGTLSIFGHQMRFNLQDGFPLVTTKRCHLRSIIHELLWF 62

Query:    91 IKGSTNAKELSSKGVKIWDANGSRDFLDSLGFSTREEGDLGPVYGFQWR 139
             ++G TN   L    V IWD     ++ D       E GDLGPVYG QWR
Sbjct:    63 LQGDTNIAYLHENNVTIWD-----EWAD-------ENGDLGPVYGKQWR 99
```

Notes
Query sequence is the human thymidylate synthase
The line between the query and subject shows the agreement between the two sequences
Identity = percentage of identical residues in the two sequences
Positives = the percentage of similar residues (shown as + in the consensus)
BLAST has reported two (partially overlappng) matched fragments, probably because of the gaps in the *E. coli* sequence needed for the alignment, which result in the score falling below the cutoff in that region.

Figure 9.15 Results of a database search using BLASTP (protein search): example of a reported match, from the results shown in Figure 9.14

The alignment reported by BLAST is not necessarily the best. For optimal alignment, and for alignment of multiple sequences, CLUSTAL (see below) should be used.

Note that all the cautionary comments discussed in relation to nucleic acid searches also apply to protein sequence comparisons: the list of matches is not definitive, the order shown is not necessarily a true reflection of the order of similarity, and you may get a different set of matches if you use a different program, such as FASTA, or set the variables (including, especially, the matrix) differently. FASTA and BLAST, although they work in different ways, have one thing in common: they are essentially shortcuts to enable database searches to be carried out rapidly without demanding too much computer power. The price to be paid is the possibility of missing something significant. A more sensitive method is provided by a method referred to as the Smith–Waterman algorithm, which is used by a program known as MPsrch.

9.4.3 Sequence alignments: CLUSTAL

As indicated in the above discussion, search methods such as FASTA and BLAST do not necessarily display the optimal alignment between your query sequence and the target. Furthermore, they only show pairwise alignments, whereas you are likely to want to see how your sequence lines up with a collection of other proteins. (Similar arguments apply to nucleic acid comparisons, but we will just look at protein alignments.) In order to produce optimized multiple alignments, the most commonly used program is CLUSTAL.

CLUSTAL starts by comparing each one of a number of defined sequences to produce a matrix of pairwise alignment scores. This can be done directly from the output from BLASTP, which allows you to feed selected matches into a version of CLUSTAL known as DBClustal. The two most similar sequences are then aligned, and a consensus generated. Each of the other sequences is then aligned in turn, in order of similarity, with a new consensus being generated at each step, until a multiple alignment of all the proteins is produced. The comments above about gap penalties and amino acid substitution matrices also apply to CLUSTAL alignments. An example of the output from CLUSTAL, using the selection of protein matches from Figure 9.14, is shown in Figure 9.16. It can be seen that a substantial number of amino acids (indicated by an asterisk) are conserved in all the sequences shown, whether from bacterial, plant or mammalian sources. The alignment produced should not be regarded as absolute – indeed, a visual inspection may show places where the alignment could obviously be improved, for example by introducing an extra gap into one of the sequences. Alignments of sequences can be used, cautiously, as aids towards the establishment or confirmation of evolutionary relationships. Some of the conserved amino acids that are identified by the alignment may have real significance, such as indicating a substrate-binding motif for example, and it is then likely that the aligned amino acids do actually represent an evolutionary relationship. However, in many cases, to be able to draw such a conclusion, you would need to look at the position of this amino acid in the three-dimensional structure of the proteins you are comparing.

CLUSTAL allows the production of a putative phylogenetic tree based on the sequence similarities of the proteins. An example is shown in Figure 9.17, where it can be seen that the results largely agree with what you would expect from the taxonomic relationships of the organisms. For example, the mouse and rat sequences are very close, while that from humans is less so, but still grouped with them. Similarly the plant sequences cluster together. However,

```
Human               PRPPHGELQYLGQIQHILRCGVRKDDRTGTGTLSVFG-MQARYSLR-DEFPLLTTKRVFW 80
Mouse               A-PQHGELQYLRQVEHILRCGFKKEDRTGTGTLSVFG-MQARYSLR-DEFPLLTTKRVFW 75
Rat                 A-PQHGELQYLRQVEHIMRCGFKKEDRTGTGTLSVFG-MQARYSLR-DEFPLLTTKRVFW 75
VZV                 GFTLTGELQYLKQVDDILRYGVRKRDRTGIGTLSLFG-MQARYNLR-NEFPLLTTKRVFW 69
Drosophila          APVNRDEMHYLDLLRHIIANGEQRMDRTEVGTLSVFG-SQMRFDMR-NSFPLLTTKRVFF 87
Silk moth           EIKSHDEYQYLKLIDQIIKSGDKRVDRTGVGTLSIFG-AMQRYSLHNNTLPLLTTKRVFT 66
Pea                 IFERHEEYMYLNLVQEIISQGTSKGDRTGTGTLSKFG-CQMRFNLR-RGFPLLTTKRVFW 298
Rice                IFERHEEYQYLNLVQDIIRNGAKKNDRTGTGTVSKFG-CQMRFNLR-RNFPLLTTKRVFW 262
Soybean             ISERHEEYLYLKLVQDIIAEGTTKGDRTGTGTLSKFG-CQMRFNLR-GNFPLLTTKKVFW 298
Candida albicans    S-PNTAEQAYLDLCKRIIDEGEHRPDRTGTGTKSLFAPPQLRFDLSNDTFPLLTTKKVFS 61
Yeast               DGKNKEEEQYLDLCKRIIDEGEFRPDRTGTGTLSLFAPPQLRFSLRDDTFPLLTTKKVFT 63
B. subtilis         ------MKQYKDFCRHVLEHGEKKGDRTGTGTISTFG-YQMRFNLR-EGFPMLTTKKLHF 52
E. coli             ------MKQYLELMQKVLDEGTQKNDRTGTGTLSIFG-HQMRFNLQ-DGFPLVTTKRCHL 52
S. typhimurium      ------MKQYLELMQKVLDEGTQKNDRTGTGTLSIFG-HQMRFNLQ-EGFPLVTTKRCHL 52
M. tuberculosis     ------MTPYEDLLRFVLETGTPKSDRTGTGTRSLFG-QQMRYDLS-AGFPLLTTKKVHF 52
Phage T2            ------MKQYQDLIKDIFENGYETDDRTGTGTIALFG-TKLRWDLS-KGFPAVTTKKLAW 52
                             *      ::  *    ***  ** : *.    *:.:    :* :***:

Human               KGVLEELLWFIKGSTNAKELS-------SKGVKIWDANGSRDFLDSLGFSTREEGDLGPV 133
Mouse               KGVLEELLWFIKGSTNAKELS-------SKGVRIWDANGSRDFLDSLGFSARQEGDLGPV 128
Rat                 KGVLEELLWFIKGSTNAKELS-------SKGVRIWDANGSRDFLDSLGFSARQEGDLGPV 128
VZV                 RAVVEELLWFIRGSTDSKELA-------AKDIHIWDIYGSSKFLNRNGFHKRHTGDLGPI 122
Drosophila          RAVAEELLWFVAGKTDAKLLQ-------AKNVHIWDGNSSREFLDKMGFTGRAVGDLGPV 140
Silk moth           RGVIEELLWMISGSTDSKALA-------AKGVHIWDANGSRNFLDNLGFTDREEGDLGPV 119
Pea                 RGVVEELLWFISGSTNAKVLQ-------EKGIHIWDDNASREFLDSIGLSEREEGDLGPV 351
Rice                RGVLEELLWFISGSTNAKVLQ-------EKGIHIWDGNASRQYLDSIGLTQREEGDLGPV 315
Soybean             RGVVEELLWFISGSTNAKVLQ-------EKGIHIWDGNASREYLDGVGLTEREEGDLGPV 351
Candida albicans    KGIIHELLWFVAGSTDAKILS-------EKGVKIWEGNGSREFLDKLGLTHRREGDLGPV 114
Yeast               RGIILELLWFLAGDTDANLLS-------EQGVKIWDGNGSREYLDKMGFKDRKVGDLGPV 116
B. subtilis         KSIAHELLWFLKGDTNVRYLQ-------ENGVRIWNEWAD------------ENGELGPV 93
E. coli             RSIIHELLWFLQGDTNIAYLH-------ENNVTIWDEWAD------------ENGDLGPV 93
S. typhimurium      RSIIHELLWFLQGDTNIAYLH-------ENNVTIWDEWAD------------ENGDLGPV 93
M. tuberculosis     KSVAYELLWFLRGDSNIGWLH-------EHGVTIWDEWAS------------DTGELGPI 93
Phage T2            KACIAELLWFLSGSTNVNDLRLIQHDSLIQGKTVWDENYENQAKDLG----YHSGELGPI 108
                    :.   ****:: *.::   *         :.  :*:   .              *:***:

Human               YGFQWRHFGAEYRDMESDYSGQGVDQLQRVIDTIKTNPDDRRIIMCAWNPRDLPLMALPP 193
Mouse               YGFQWRHFGAEYKDMDSDYSGQGVDQLQKVIDTIKTNPDDRRIIMCAWNPKDLPLMALPP 188
Rat                 YGFQWRHFGADYKDMDSDYSGQGVDQLQKVIDTIKTNPDDRRIIMCAWNPKDLPLMALPP 188
VZV                 YGFQWRHFGAEYKDCQSNYLQQGIDQLQTVIDTIKTNPESRRMIISSWNPKDIPLMVLPP 182
Drosophila          YGFQWRHFGAQYGTCDDDYSGKGIDQLRQVIDTIRNNPSDRRIIMSAWNPLDIPKMALPP 200
Silk moth           YGFQWRHSGAKYIDCQTDYTGQGIDQLQSVIDSIKNNPADRRMLICAWNSSDLKKMALPP 179
Pea                 YGFQWRHFGARYTNMHNDYAGQGVDQLLDVINKVKHNPDDRRIILSAWNPADLKLMALPP 411
Rice                YGFQWRHFGAEYTDMHADYVGKGFDQLMDVIDKIKNNPDDRRIILSAWNPTDLKKMALPP 375
Soybean             YGFQWRHFGARYTDMHHDYSGQGFDQLLDVINKIKRNPDDRRIILSAWNPVDLKLMALPP 411
Candida albicans    YGFQWRHFGAEYKDCDSDYTGQGFDQLQDVIKKLKTNPYDRRIIMSAWNPPDFAKMALPP 174
Yeast               YGFQWRHFGAKYKTCDDDYTGQGIDQLKQVIHKLKTNPYDRRIIMSAWNPADFDKMALPP 176
B. subtilis         YGSQWRSWRGA--------DGETIDQISRLIEDIKTNPNSRRLIVSAWNVGEIDKMALPP 145
E. coli             YGKQWRAWPTP--------DGRHIDQITTVLNQLKNDPDSRRIIVSAWNVGELDKMALAP 145
S. typhimurium      YGKQWRAWPTP--------DGRHIDQIATVLSQLKNDPDSRRIIVSAWNVGELDKMALAP 145
M. tuberculosis     YGVQWRSWPAP--------SGEHIDQISAALDLLRTDPDSRRIIVSAWNVGEIERMALPP 145
Phage T2            YGKQWRDFGGV-------------DQIVEVIDRIKKLPNDRRQIVSAWNPAELKYMALPP 155
                    ** ***                  **:   :  ::  * .** ::.:**  ::  *.*.*
```

Notes

VZV = Varicella Zoster Virus

* = identical residues in all sequences

: and . = similar residues in all sequences

Figure 9.16 Clustal alignments (partial) of selected thymidylate synthases

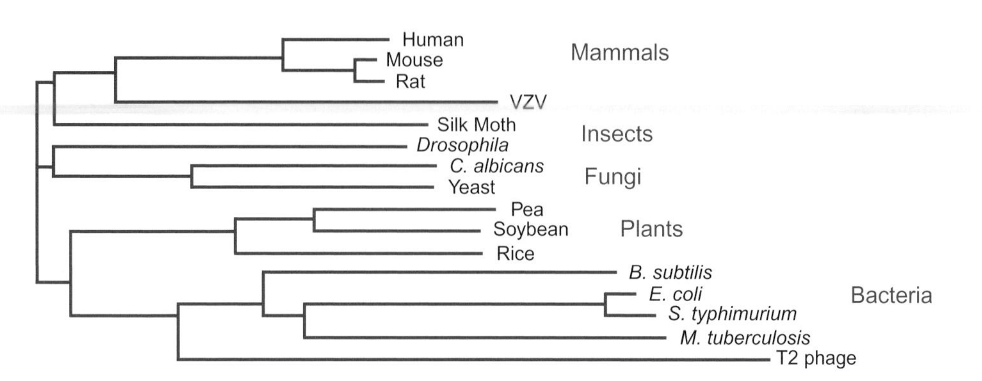

Figure 9.17 Phylogram based on CLUSTAL alignments of selected thymidylate synthases

the two insect examples are not grouped, and the two virus sequences (VZV and T2) seem to resemble their hosts rather than one another. However, it would be a mistake to read too much into this. A true phylogenetic tree would require analysis of a number of sequences, and you would need to use more sophisticated methods to discriminate between the many possible trees that could be drawn. We will look further at *molecular phylogeny* in Chapter 13.

9.5 Protein structure

9.5.1 Structure predictions

Having identified an ORF, and used the computer to translate it into a protein sequence, we can investigate a number of aspects of the structure of that protein. This is obviously particularly important where we do not already have clues to the function of our gene through its similarity with others that have already been characterized. A full consideration of all the possibilities is beyond the scope of this book, but it is worth a brief and selective overview, especially as an examination of protein structure can provide some leads as to its possible function.

For example, if we look at the amino acid composition of the protein, and especially at the occurrence and distribution of hydrophobic and hydrophilic amino acids, we should be able, just on this evidence, to detect if it is likely to be a membrane protein. Proteins that are embedded in a membrane will normally have substantial stretches of hydrophobic amino acids. Soluble proteins are more likely to be predominantly hydrophilic, although they can contain hydrophobic regions if the protein folds into a pocket that shields those hydrophobic regions from the aqueous environment. If we look more closely at the predicted structure (especially in combination with testing for specific motifs – see below) we may be able to guess whether it is involved in transport across the membrane, or energy generation, or acts as a receptor for signal transduction, or any of the other functions commonly associated with membrane proteins. As an example, Figure 9.18 shows a hydrophobicity plot for the human rhodopsin protein, a member of the so-called seven-transmembrane receptor family. This shows seven regions of marked hydrophobicity, with intervening regions that are more hydrophilic. The hydrophobic regions are embedded in, and span, the membrane, so that at one end of a transmembrane region the protein is exposed to the cytoplasmic environment, while at the other end the protein protrudes into the external environment. This is a commonly occurring theme in membrane proteins, especially in those proteins that function as transporters (that is they ferry material across the membrane) or as signal transducers (i.e. they respond to changes in the external environment and transmit a signal across the membrane to the interior of the cell).

Proteins do not, of course, exist merely as one-dimensional sequences of amino acids, but adopt higher orders of conformation. A variety of programs are available to predict, from the primary sequence of amino acids, which parts will adopt secondary structures such as alpha-helices and beta-sheets, examples also shown in Figure 9.18. However, these predictions are not entirely reliable, as the factors that determine the final structure of the protein are complex. Predicting how those elements of secondary structure will fold into the tertiary and higher orders of structure that are characteristic of the native protein is even more difficult. For this, in spite of the huge increase in computer capacity, we are still left with the need to produce crystals of the pure protein for X-ray crystallography.

However, as more and more proteins are characterized at this level, it becomes increasingly likely that the unknown protein we have predicted is a member of a family of proteins, some of which have been characterized structurally. By aligning the sequences, and looking for specific regions (motifs and domains) that are characteristic of families of proteins, we can therefore make much more reliable predictions as to the likely conformation, and function, of our unknown protein.

9.5.2 Protein motifs and domains

Although our protein may not show much overall similarity to any other characterized protein, we may find, if we look more closely, that it contains short sequences of amino acids that are very similar to parts of a number of other proteins. These conserved regions are known as *motifs*. The recognition of motifs provides yet another clue as to the function of our protein, or other aspects of its structure.

Some motifs occur because a wide range of enzymes, with otherwise disparate properties, may use the same substrate. For example, there is a wide range of enzymes that use ATP as a substrate. In many cases, the region of the enzyme that binds the ATP has a similar structure – although the remain-

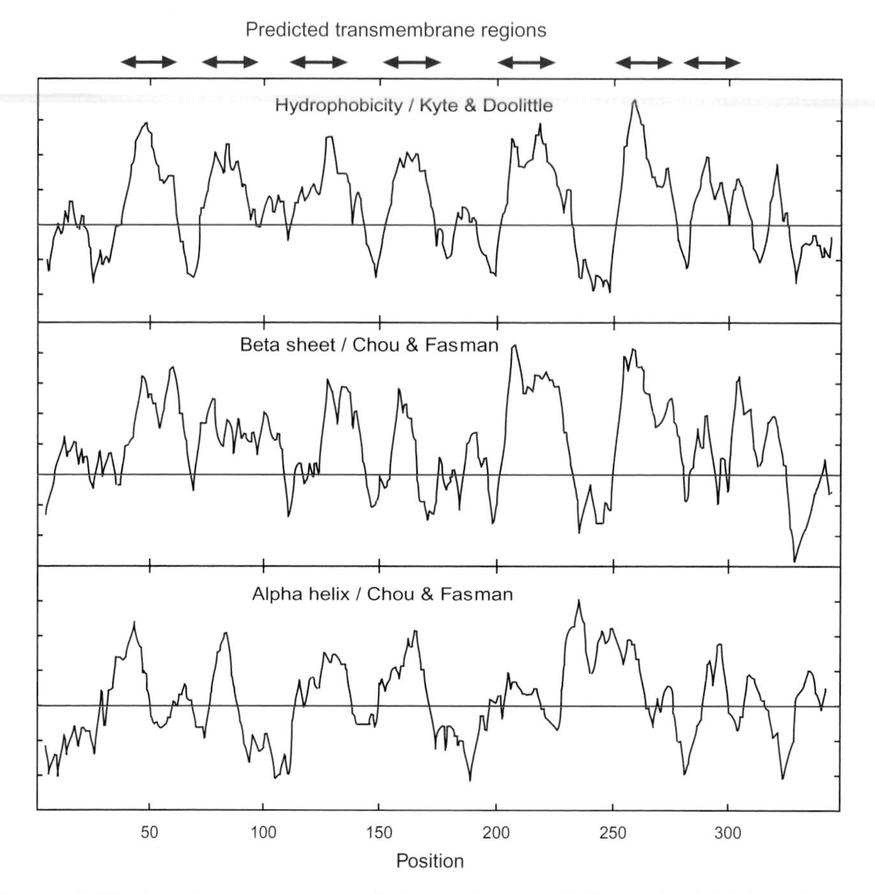

Figure 9.18 Protein structure predictions: Kyte–Doolittle hydrophobicity plot and Chou–Fasman secondary structure predictions for human rhodopsin (UniProt P08100). Analysis performed using ProtScale in the ExPASy server of the Swiss Institute of Bioinformatics (http://www.expasy.org/tools/protscale.htm)

der of the sequence may show no similarity. If we can recognize the presence of a putative ATP-binding site in our protein, then we can infer that it is probably an enzyme that uses ATP as a substrate.

As well as substrate-binding sites, there is a range of other motifs that indicate sites for structural modification such as lipid or sugar attachment, for secretion or for targeting to specific cellular compartments, or in the case of regulatory proteins, for DNA binding. There are libraries of known motifs available, so it is a simple matter to screen a protein sequence for any of these motifs. One of the longest-established libraries of patterns is PROSITE (see Box 9.1). Figure 9.19 shows the output from a web-based search for motifs present in protein Rv0194 (a protein from *M. tuberculosis* which is identified as a probable transporter protein; UniProt accession number O53645), and demonstrates the presence of two regions characteristic of the family of pro-

Edited output:

O53645 **O53645_MYCTU** (1194 aa)
PROBABLE DRUGS-TRANSPORT TRANSMEMBRANE ATP-BINDING PROTEIN ABC TRANSPORTER. *Mycobacterium tuberculosis*

PS50929 ABC_TM1F ABC transporter integral membrane type-1 fused domain profile :
21 - 301: **score** = 47.505
628 - 910: **score** = 45.116

PS50893 ABC_TRANSPORTER_2 ATP-binding cassette, ABC transporter-type domain profile:
334 - 568: **score** = 22.056
Predicted feature:
NP_BIND 367 374 ATP (Potential) [condition: [AG]-x(4)-G-K-[ST]]
942 - 1177: **score** = 21.382
Predicted feature:
NP_BIND 976 983 ATP (Potential) [condition: [AG]-x(4)-G-K-[ST]]

Figure 9.19 ProSite scan. The query protein was a probable ABC (ATP-binding cassette) transporter protein from *M. tuberculosis* (UniProt 053645); the analysis was done using Scan-Prosite, at http://expasy.org/tools/

teins known as ABC (ATP-binding cassette) transporters, and two potential ATP-binding motifs.

The overall reaction catalysed by some enzymes is actually a series of separate reactions, with different parts of the enzyme responsible for different steps. For example, acetyl-CoA carboxylase (catalysing the first step in fatty acid synthesis) has one component with a covalently attached biotin, a second component with (noncovalent) binding sites for ATP and CO_2 which carboxylates the biotin, and a third component, with an acetyl-CoA binding site, which transfers the carboxyl moiety from the biotin to the acetyl-CoA. In *E. coli*, these components are made as separate polypeptides, which associate with one another at the post-translational stage to form the complete holoenzyme. On the other hand, in mammalian cells the enzyme is made as a single polypeptide with different parts of the structure responsible for the different activities. We refer to these elements as *domains*. It is not uncommon to find enzymes that consist of several subunits in one organism but as a single polypeptide with several domains in another source – perhaps due to a fusion of genes that originally evolved separately.

Domains often form more or less structurally independent regions of a protein – in other words each domain folds up into its own secondary structure with little or no structural interaction with the other domains, other than a flexible loop connecting them. You can visualize such a structure, roughly, as a set of balls connected by bits of string. In a typical protein, some regions, especially those that are essential for the function of the enzyme, show relatively little variation from one species to another. They are referred to as

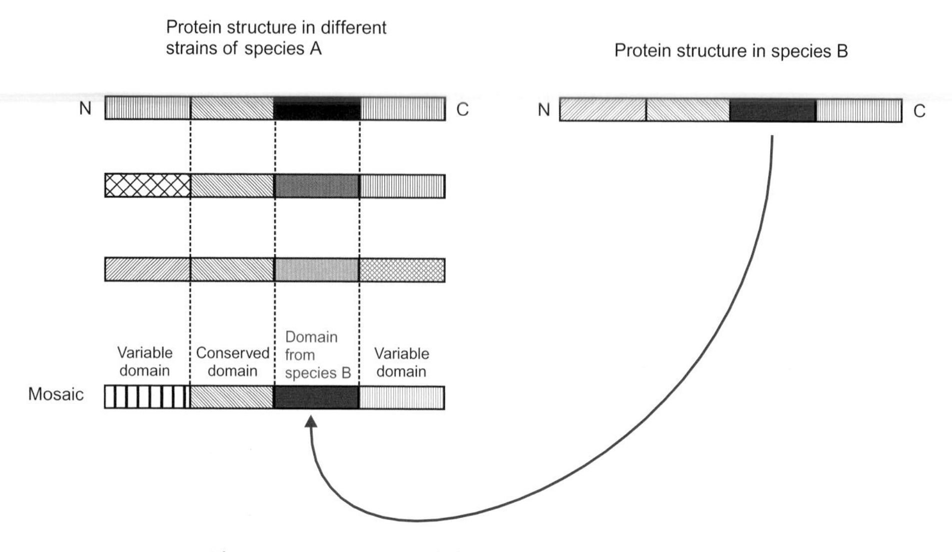

Figure 9.20 Conserved domains and domain shuffling

conserved domains. On the other hand, some regions are much more variable. These often include the terminal regions, and other sequences that form loops protruding from the main body of the folded protein. The latter are especially important in some viruses (notably HIV) where they are both *hypervariable* and *immunodominant*, i.e. they vary extensively and rapidly, and constitute the principal antigen 'seen' by the body's immune system. This is significant as it makes it extremely difficult to produce effective vaccines.

The domain structure of proteins can give rise to a high degree of variability through the formation of so-called *mosaic* proteins. This means that a protein in one strain contains a domain that is not related to the sequence found in other strains of the same species, but is closely related to the sequence found in a different organism (see Figure 9.20). The inference is that information has been transferred between organisms, so that a part of the gene is replaced with information from a different source. This contributes a further dimension to our overall understanding of the dynamic nature of the genome (sometimes referred to as *genome plasticity*).

A useful tool for the identification and comparison of domains is the Pfam database of protein domain families, available at several sites, including the Sanger Institute (see Box 9.1). This uses a set of multiple sequence alignments for each family. (Technically, these alignments are encoded using a statistical treatment known as hidden Markov models; HMMs have a variety of applications in bioinformatics. See Chapter 12 for further information.) Figure 9.21 shows part of the output from an analysis of the same protein as used in Figure 9.19. The graphic display is a composite of two versions, with different features displayed. The first shows the presence of two membrane

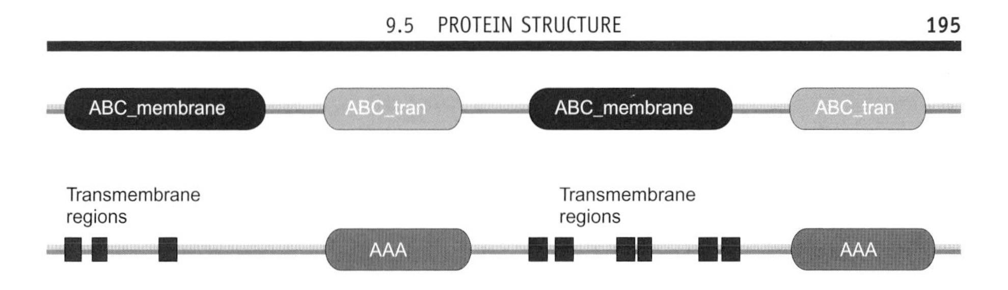

Description from UniProt for O53645_MYCTU probable drugs-transport transmembrane atp-binding protein abctransporter

Source	Domain	Start	End
tmhmm	Transmembrane	20	41
PfamA	ABC_membrane	20	289
tmhmm	Transmembrane	56	75
tmhmm	Transmembrane	144	166
PfamB	Pfam-B_52555 (hidden)	296	343
SMART	AAA	359	553
PfamA	ABC_tran	360	544
PfamB	Pfam-B_73432 (hidden)	545	619
tmhmm	Transmembrane	626	648
PfamA	ABC_membrane	627	898
tmhmm	Transmembrane	660	682
tmhmm	Transmembrane	741	763
tmhmm	Transmembrane	768	785
tmhmm	Transmembrane	847	869
tmhmm	Transmembrane	876	898
PfamB	Pfam-B_82283 (hidden)	902	958
SMART	AAA	968	1154
PfamA	ABC_tran	969	1153

Pfam entry ABC_membrane
Accession number: PF00664
ABC transporter transmembrane region
This family represents a unit of six transmembrane helices. Many members of the ABC transporter family (ABC_tran) have two such regions

Pfam entry ABC_tran
Accession number: PF00005
ABC transporter
ABC transporters for a large family of proteins responsible for translocation of a variety of compounds across biological membranes. ABC transporters are the largest family of proteins in many completely sequenced bacteria. ABC transporters are composed of two copies of this domain and two copies of a transmembrane domain ABC_membrane.

AAA
ATPases associated with a variety of cellular activities
SMART accession number: SM00382
Description: AAA - ATPases associated with a variety of cellular activities. This profile/alignment only detects a fraction of this vast family.

Transmembrane regions
Transmembrane helices indicate a protein has a membrane bound location. Prediction of individual transmembrane spans is quite accurate however care must be exercised and these regions should be verified by other means.

Figure 9.21 Protein families: Pfam database. The graphical display shows two runs, with different features hidden. The PfamB features, and other features not shown, are hidden in both cases

domains and two transporter domains, as identified by ProSite. The second shows predicted transmembrane regions, associated with the membrane domains, and two ATPase domains, corresponding to the two transporter domains. Clicking the links in the table produces more detailed information on each of these domains, a short extract of which is shown in the figure.

The description table also shows other sites (hidden in the graphic) which come from the PfamB database. The Pfam-A database is a high-quality, curated database (i.e. the quality is maintained manually). The Pfam-B database is computer-generated, and is of lower quality.

ProSite and Pfam originated in different ways, and are compiled differently, but it will be seen that functionally they now overlap considerably. If you look back to Figure 9.8, showing the annotation of a protein sequence in the UniProt database, you will see that there are links to both ProSite and Pfam. All UniProt entries have been analysed for the presence of motifs and domains in these databases, so you only need to click on the links to get this information.

9.6 Confirming gene function

From the sequence comparisons outlined above, we may be able to reach a conclusion as to the probable biochemical function of the protein encoded by our cloned gene. How do we confirm that this is really its activity? The obvious answer – to express the cloned gene in a suitable host and assay its activity – is often not possible, as many proteins do not have an enzymic function that is easily assayable *in vitro*. Even if it does, that still does not provide the full answer as to what its role is within the original organism. One of the most powerful strategies in such a situation is to make a specific deletion of that gene and determine the effect on the organism, using a procedure known variously as *allelic replacement, gene replacement* or *gene knock-out.*

9.6.1 Allelic replacement and gene knock-outs

Allelic replacement relies on the natural process of *homologous recombination*, which means that, when there are two identical pieces of DNA in the cell, enzymes within the cell may break the two DNA chains, cross them over and rejoin them. (This is a highly simplistic version of a more complex process, but it will do for our purposes.) We can exploit this process to replace a specific gene (or part of a gene) in the chromosome with a version that we have inactivated *in vitro*, thus destroying the function of that gene. This allows us to test the consequences of inactivating an individual gene, and thus make deductions about its function.

A typical procedure for allelic replacement in a bacterial host would be to manipulate the cloned gene so as to replace the central part of the gene with

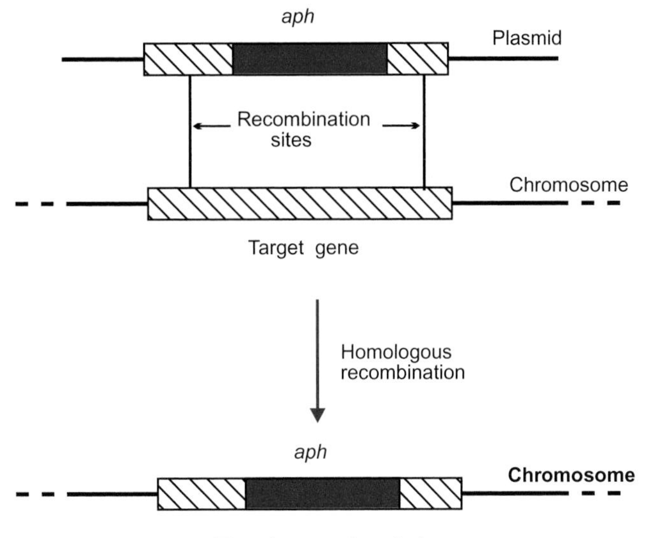

Figure 9.22 Gene disruption by allelic replacement. *aph* = aminoglycoside phosphotransferase, causing kanamycin resistance

an antibiotic resistance gene (see Figure 9.22). We would do this using a suicide plasmid (a plasmid that is unable to replicate in the chosen organism) so that, when we transform the bacteria with the construct, only those cells in which the resistance gene has become incorporated into the chromosome will become antibiotic resistant. We can select these on agar containing the antibiotic. If things go well, incorporation into the chromosome will have occurred by homologous recombination at the required position, thus inactivating the gene concerned.

It should be noted that replacement of the gene actually requires recombination at two positions, one either side of the gene. This is referred to as a double crossover event. A single crossover, in the homologous region at one side or the other of the construct, will produce resistant bacteria by incorporation of the entire plasmid into the chromosome rather than replacing the gene. Single crossovers may result in gene inactivation, depending on the details of the construct, but will usually be unstable, as further recombination may eliminate the plasmid again, restoring the original intact gene. We can select against single crossovers by incorporating a counter-selectable marker into the plasmid. In other words, we put a gene on the plasmid that will, if it is still present, confer a disadvantage on the cell. A gene known as *sacB* is commonly used for this purpose, in bacteria, as the presence of *sacB* renders the cells sensitive to sucrose in the medium. Plating the cells on a sucrose-containing medium will result in any cells containing *sacB* (which includes the single cross-overs but not the doubles) being unable to grow and form colonies.

An obvious limitation of gene replacement technology is that the inactivation of the gene concerned may be a lethal event. Nevertheless, it is a valuable approach for identifying the function of specific genes.

Although we have described gene replacement in terms of identifying gene function, its applications extend far beyond that. It can be used to inactivate genes that are necessary for the virulence of a pathogenic bacterium, thus producing attenuated strains that may be useful vaccine candidates. In addition, essentially identical procedures can be used to knock out genes in other organisms, including experimental animals (especially mice). Many strains of mice, lacking individual genes, have been produced in this way, and are invaluable for research purposes. This is considered further in Chapter 15, together with *RNA interference*, which provides an often simpler way of silencing a specific gene in eukaryotic hosts.

It should be noted also that the technique of gene replacement is not confined to the inactivation of the genes concerned. It can be readily adapted to the replacement of a gene by an altered version of the same gene (see the section on gene knock-in technology, in Chapter 15), thus conferring novel properties on the organism.

Other methods that are used for genome-wide identification of genes with specific functions are considered in Chapter 12.

9.6.2 Complementation

The above discussion makes the implicit assumption that the phenotypic consequences of the mutation are due solely to the effect of the loss of that gene. This is not always true. In particular, some mutations may show an effect known as *polarity*. This means that the mutation affects not only the altered gene, but also those adjacent to it. In bacteria, this can arise from the arrangement of genes into operons which are transcribed into a single mRNA. Mutation of one gene may interfere with transcription of the operon, and thus affect the expression of the genes downstream from it. Furthermore, genes and their products interact in many complex ways within the cell, so that disruption of one gene may have unexpected effects on the activity of other genes and their products.

It is therefore necessary to interpret the results arising from these experiments with care. The standard way of checking that the altered phenotype is a direct consequence of the inactivation of a specific gene is by *complementation*. This involves introducing into the mutant cell a fully active version of the affected gene (most simply, by a cloned version on a plasmid). If the alteration in the phenotype is indeed due solely to the loss of the affected gene and its product, then the mutation will be complemented by the plasmid, i.e. the wild-type phenotype will be restored. This is not entirely foolproof. For example if the original mutation disrupts the regulation of other genes, complementation may be successful in restoring the wild-type phenotype, even though the gene product is not directly responsible for the observed characteristics. Nevertheless, complementation does provide an element of confirmation of the consequences of the original mutation.

Complementation can be used in the same way to relate a cloned gene to the phenotype of mutant strains produced by other means, including mutants isolated by screening for specific altered phenotypes. For example if you have isolated a non-motile bacterial strain that is unable to produce flagella, you can introduce your cloned gene into that strain and test its motility. If motility, and the ability to produce flagella, is restored, then you have good evidence (subject to the caution above) that your cloned gene is the same one that is mutated in the host strain. The equivalent of complementation in eukaryotic model organisms is the phenotypic rescue of a mutation by transgenic delivery, see Chapter 15.

10 Analysis of Gene Expression

Gene expression can be divided into two main phases: *transcription* (copying DNA into RNA), and *translation* (production of a protein or polypeptide to the specifications of an mRNA template). For the purpose of this chapter, we will treat the polypeptide chain as the end product of gene expression; this ignores subsequent post-translational modification, including the folding of the polypeptide into the correct conformation for biological activity, as well as other modifications such as the specific addition of carbohydrate groups. In this chapter, we will discuss methods for the study of the products of single genes, and in Chapter 14 we will extend these to the study of complex samples on a genome-wide basis.

10.1 Analysing transcription

The first group of techniques for examining gene expression consists of various ways of assessing the amount of a specific transcript in a specific sample at a specific time. Other methods, described later in the chapter, are concerned with studying the transcriptional activity associated with the gene in question. Two points concerning the measurements of mRNA levels should be noted. Firstly, the amount of a specific mRNA in a cell at a point in time is influenced not only by the level of transcriptional activity, but also by the stability of that message. A gene that is transcribed at a low level but gives rise to a stable product may result in a higher amount of mRNA than a transcriptionally more active gene with an unstable message. Secondly, the amount of mRNA present does not necessarily correlate with the amount of protein made. This is most obvious with bacterial cells, where a single polycistronic message may be translated into many different polypeptides; although these are all translated from the same mRNA template, the levels

From Genes to Genomes, Second Edition, Jeremy W. Dale and Malcolm von Schantz

of the different proteins made can be widely different – translation efficiencies can vary considerably.

10.1.1 Northern blots

The oldest methods for detecting a specific mRNA rely on hybridization to specific, labelled, probes. A common manifestation of this is known as a *northern* blot. A northern blot involves electrophoretic separation of a purified extract of RNA from living cells, followed by immobilization onto a membrane and hybridization with a specific probe, as shown in Figure 10.1. This enables us to detect the presence of a specific mRNA, assuming that the probe we use is sufficiently specific, and to estimate its size. By careful calibration, for example by comparison of the strength of the hybridization signal with that of other genes that are expressed at known levels, we can get an idea of the *relative* amount of mRNA present. In Figure 10.1, we can see that sample B gave a weaker signal than sample A, and therefore contained correspondingly less of the specific mRNA, while in sample C expression of the gene was not detected. In Figure 10.1, we can see that there is a large amount of ribosomal RNA present, because the extract contained total RNA. If we are working with a eukaryotic organism, the mRNA can be enriched for poly-A containing RNA, using an oligo(dT) column (see Chapter 6). This will remove the ribosomal RNA.

However, there are some limitations to this technique. Firstly, although we can get a *relative* quantitative estimate of the strengths of the signals, it is generally not possible to obtain an *absolute* measurement of the amount of specific mRNA. Secondly, the technique is not very sensitive, and requires fairly large amounts of RNA (normally at least a microgram of poly-A-

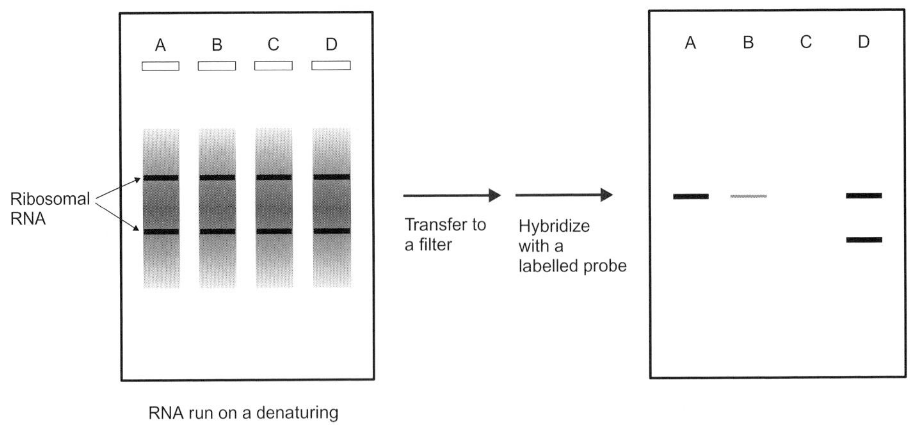

Figure 10.1 Northern blotting

enriched mRNA or ten micrograms of total RNA). It may not be possible to obtain sufficient quantities of starting material. For example, if we want to look at the expression of specific genes in a pathogenic bacterium when it is growing within an infected host (rather than the artificial situation of laboratory culture), or levels of mRNA in material obtained by biopsy, it is likely to be extremely difficult to get large enough quantities of RNA to produce a northern blot.

For these reasons, PCR-based techniques (specifically quantitative RT-PCR, see later in this chapter) are now more popular. On the other hand, northern blotting has some lasting advantages. Above all, it remains a major method for determining the size of a specific transcript, and for detecting the presence of different transcripts for a specific gene (e.g. in Figure 10.1, we can see the presence of a second, smaller, transcript in sample D). This is becoming more important as the human genome project teaches us that the number of genes is considerably smaller than the number of gene products. An important part of the explanation for this is that many genes are differentially spliced, and northern blotting is an important tool for detecting this.

10.1.2 RNase protection assay

This method is based on *solution hybridization* rather than filter hybridization. The target is a specific RNA species in a complex RNA mixture purified from a biological sample. The probe is a radiolabelled antisense RNA. Such RNA probes can be produced quite easily from plasmid cloning vectors, such as the pGEM series (Chapter 5) that are equipped with binding sites for RNA polymerases (T3, T7 and/or Sp6). In the example shown (Figure 10.2), the gene (or part of it) has been inserted into the vector in the correct orientation for transcription from the T7 promoter. Addition of T7 RNA polymerase, and the NTP substrates, will therefore produce the normal (sense strand) mRNA. However the gene is in the opposite orientation with respect to the SP6

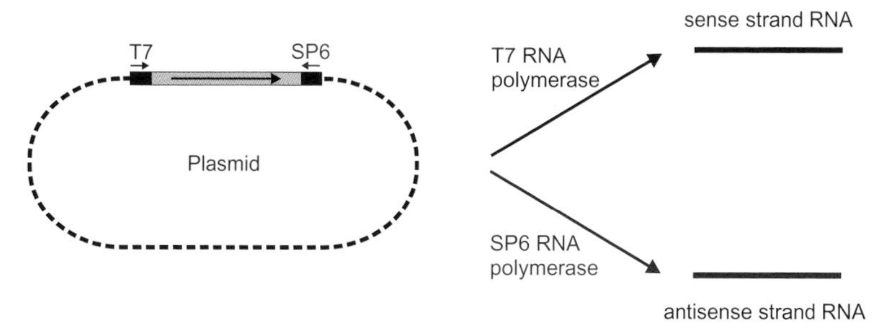

Figure 10.2 Production of antisense RNA

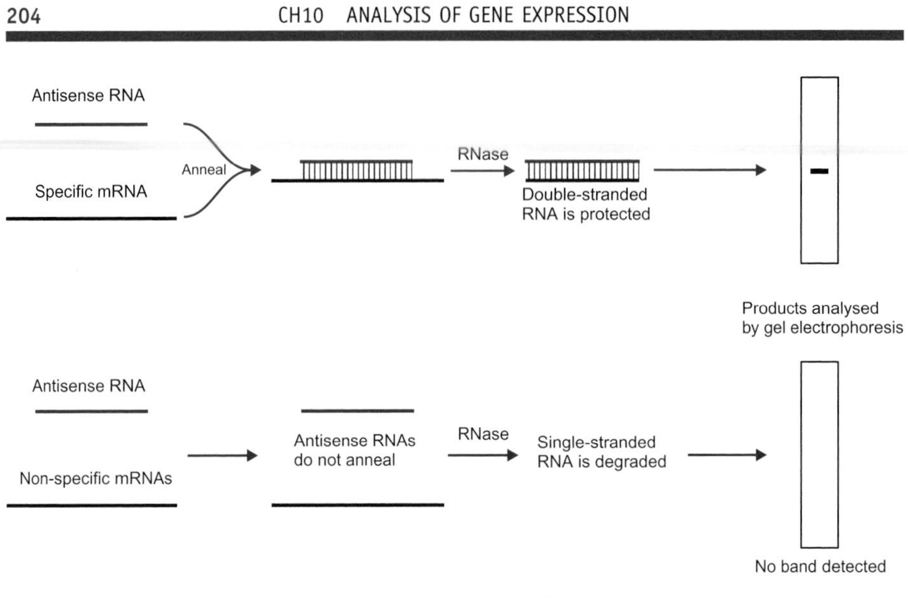

Figure 10.3 RNase protection assay

promoter. Therefore, if you use SP6 RNA polymerase it will copy the opposite, antisense strand. The resulting RNA (the antisense RNA) will be complementary to the true (sense strand) mRNA, and will therefore hybridize to it.

When you add the (labelled) antisense RNA to your mRNA preparation, it will anneal to the specific RNA you want to measure, forming double-stranded RNA, while non-specific mRNA molecules remain single-stranded. When ribonuclease (RNase) is added, it will degrade all the single-stranded RNA, while the double-stranded hybrids formed by annealing the antisense RNA probe to the specific mRNA are protected (Figure 10.3). After RNase digestion, the hybridization mixture is separated in a polyacrylamide gel. The gel is then dried and left to expose an X-ray film.

This assay has some distinct advantages over northern blotting. It is more sensitive, it is more quantitative, and it is more tolerant of partial degradation of the RNA samples. The reason for the latter is that only part of the RNA molecule needs to be intact to take part in the hybridization. This is simultaneously a disadvantage compared with northern blotting. It does not allow you to determine the size of the transcript(s), because the fragment you see in the gel is only defined by the length of the probe. The other disadvantage is that it demands more work than northern blotting in the production of the probe.

10.1.3 Reverse transcription-PCR

A much more sensitive way of detecting specific mRNA is to adapt the polymerase chain reaction (PCR, see Chapter 8) in order to be able to amplify the specific message. Because this requires copying the mRNA into cDNA

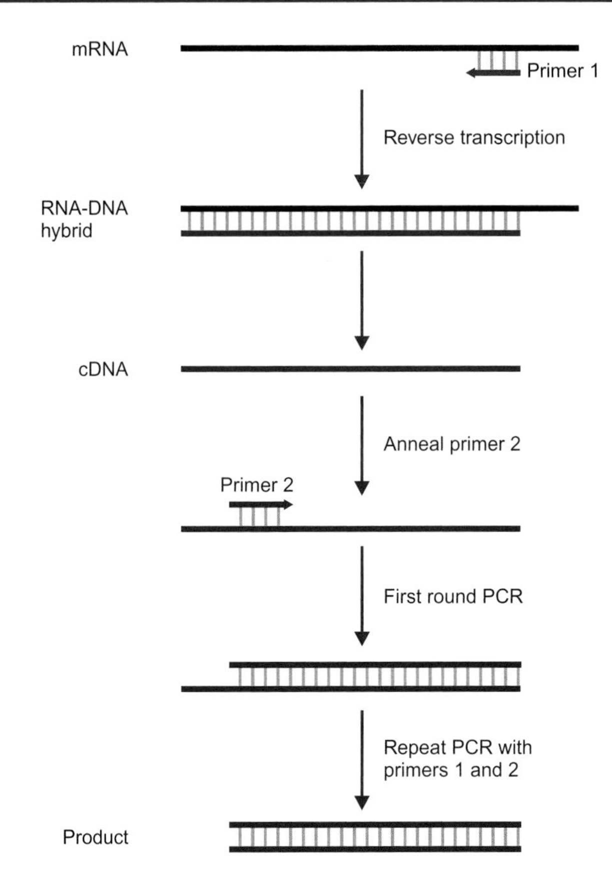

Figure 10.4 Reverse transcription PCR

using the enzyme reverse transcriptase (see Chapter 6), it is described as reverse transcription PCR (RT-PCR). The principle of the method is illustrated in Figure 10.4.

The sensitivity of this procedure is such that it is possible to detect a specific mRNA species in a single cell. This has obvious advantages, in being able to detect low abundance mRNA (i.e. mRNA that is present at extremely low levels), or to analyse gene expression in cells that are difficult to obtain in large numbers. Examples are the analysis of gene expression in bacteria from lesions in an infected animal, or in cells in tumours, in order to find out which genes are expressed under those circumstances. Culturing the bacteria or tumour cells is no use, as the gene expression would change. RT-PCR enables us to analyse gene expression in the limited number of cells recoverable from the lesion.

The sensitivity of RT-PCR confers a further, rather less obvious, advantage. If we want to look at tissue-specific expression, or at expression in cells under different physiological conditions (whether bacterial or eukaryotic), we have to try to ensure that all the cells from which we extract the mRNA are indeed expressing the same repertoire of genes. The relatively large amounts of mRNA required for a northern blot means that we have to use mRNA extracted from a considerable number of cells, and this population is more likely than not to be heterogeneous. RT-PCR requires much less mRNA, and hence far fewer cells. It is therefore much more likely that all the cells are in the same physiological state, and we will thus get a much more accurate picture of the true profile of gene transcription in those cells. Reporter genes (see below) provide an alternative way of addressing this question.

For analysis of expression of a specific gene, we need to be able to do more than merely detect the presence or absence of the relevant mRNA. We need to be able to assess how much of it is present. In Chapter 8, we saw how quantitative PCR results can be obtained using various techniques collectively known as *real-time PCR*, in which the PCR reaction is adapted so that product formation is accompanied by an increase in fluorescence which can be monitored continuously. The number of cycles needed for formation of a detectable amount of product, known as the C_T value, is related to the initial amount of template; higher amounts of initial template will result in lower C_T values. For quantitative assessment of the amount of a specific mRNA in a sample, you first need to reverse transcribe it into DNA before carrying out the real-time PCR; this is then known as quantitative RT-PCR. (Do *not* be tempted to abbreviate 'real-time' to RT, or you will get really confused! If you must have an abbreviation, use qRT-PCR, where q stands for 'quantitative'.)

If you are comparing the levels of mRNA in different samples, you have to make sure that you start with the same number of cells, or more specifically with the same overall amount of mRNA. Standardizing the amount of *total RNA* (rather than specifically mRNA) is not completely reliable, as the ribosomal content of the cells may vary considerably. The best way of standardizing the template is to carry out a parallel RT-PCR using another gene which is known to be constitutively expressed (or at least expressed at the same level under the different conditions you are testing). This enables testing for differences in the specific expression of an individual gene under different conditions. Can you use the same method to compare the expression of different genes? Subject to certain safeguards, yes. The problem to be dealt with is that PCR is not equally efficient with all combinations of primers and templates. Even if the amounts of template are the same, a more efficient PCR will result in earlier detection (a lower C_T value) than for another gene with a less efficient PCR. Some preparatory work is therefore necessary to ensure that the templates are amplified equally efficiently, possibly using different primers to change the efficiency so that they match. (It should be noted that this consideration also applies to the comparison of your gene with the standard template as referred to above.)

10.1.4 *In situ* hybridization

In this method, the target is found immobilized within a cell on a microscope slide. The target may be either DNA or RNA. The former is often used for gross chromosomal mapping of genes. A labelled probe is hybridized to cells with metaphase chromosomes. By using a differentiating counterstain, the investigator can identify the localization of the gene to a specific region of a specific chromosome (see Chapter 13 for a more detailed account). What we are concerned with in the context of gene expression is RNA *in situ* hybridization which, in contrast, can be used to identify those cells (for example in a tissue section) which produce mRNA for a specific gene. *In situ* hybridization is also used for the detection of viruses. In this case, the target is either DNA or RNA depending on the pathogen.

Although it is possible to obtain semi-quantitative information by counting the number of grains (when using radioactive labels), quantification is not the strength of this method. Rather, its strength is to tell us which particular cell types express a specific transcript. By using a panel of RNA samples from different tissues, other methods such as northern blot hybridization, RNase protection assay and RT-PCR can tell us in which tissue(s) expression takes place. *In situ* hybridization allows us to refine this analysis even further than either of these methods. In addition, the size of the sample required is minimal.

10.1.5 Primer extension assay

As part of the study of gene expression, we are likely to want to know not only how much of a specific mRNA is made, but where does it start and stop. Studying the 3′ end of a (processed) eukaryotic mRNA species is rarely a problem. Most cDNA libraries are constructed by oligo-dT priming of poly-A-enriched mRNA, so by definition this is the part of the mRNA molecule that is always employed for cDNA synthesis (although, occasionally, the primer ends up binding to an adenosine-rich stretch within the mRNA molecule instead). However, as mentioned earlier (Chapter 6), the 5′ end is not always incorporated into the cDNA. The reverse-transcription of the full mRNA may not occur because of degradation of the template, or because of secondary structures, or the presence of RNA-binding proteins that stop the reverse transcriptase. Moreover, studying the sequence of a cDNA will not necessarily tell us if its 5′ end goes all the way to the predicted transcription start site. We can establish whether the whole open reading frame for the protein product is present, but that is pretty much it. Comparing the length of your cDNA sequence with the size of the full transcript as seen on a northern blot will give you an idea of whether you are dealing with a gene with multiple transcripts (which may vary in which exons are included or in

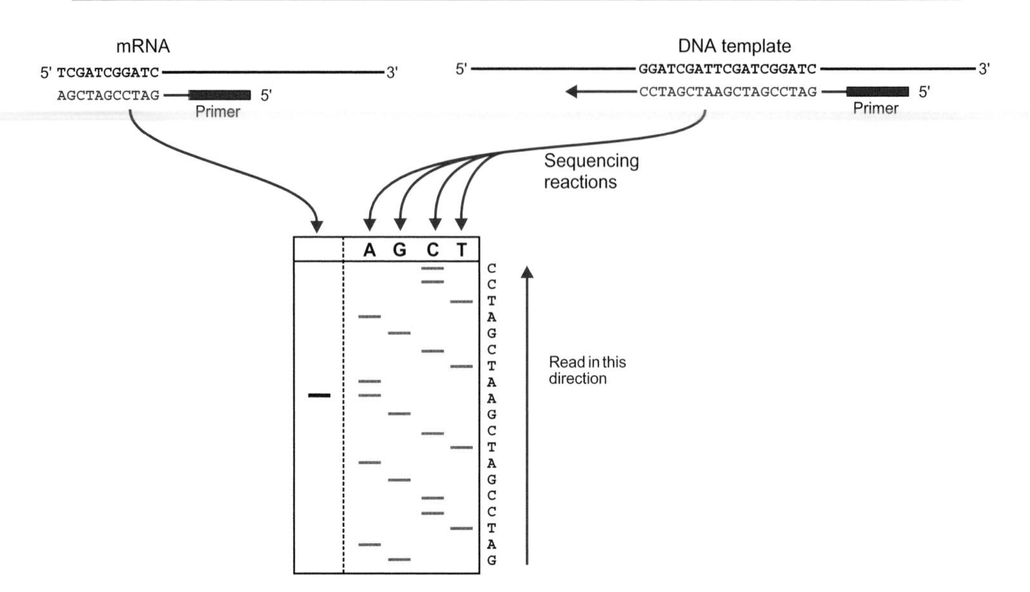

Figure 10.5 Determination of 5′ end of mRNA: primer extension

the length of their untranslated regions), and the approximate size of these transcript(s).

The primer extension assay is used for the exact mapping of the 5′ terminus of an mRNA. Poly-A RNA is purified from a tissue or cell type known to be expressing the gene in question at as high a level as possible. This complex mixture of all present mRNA species is then mixed with a radiolabelled oligonucleotide primer complementary to a region close to the 5′ end of the sequence, which will need to be as specific as possible for the transcript you are studying. Reverse transcriptase is then added, and used to extend the primer to the very 5′ end of the mRNA, at which point the polymerization automatically terminates. You will then need to run out the new radioactive cDNA strand in a sequencing gel for optimum resolution, and by running alongside a sequencing reaction of the same gene you are studying, starting with the same primer, you will be able to identify the transcription start site precisely (see Figure 10.5). Fluorescent labels can now be used to perform the same procedure in an automated sequencer.

Mapping the ends of bacterial transcripts is more difficult. Bacterial mRNA is often very short-lived *in vivo*, and is usually not poly-adenylated – and hence more difficult to isolate. In addition, the existence of operons means that the 5′ end may be some distance away from the gene you are investigating. However, the primer extension assay can also be used for identifying the 5′ end of a bacterial transcript, but often only after a potential promoter has been identified by the methods described below.

10.2 Methods for studying the promoter

10.2.1 Reporter genes

There are a variety of situations where even the most sophisticated of the above approaches is inadequate or inconvenient for following the nature of gene expression. Reporter gene technology allows us to examine gene expression within the cell, without disrupting it. This is invaluable, for example, in the study of cellular differentiation in a multicellular organism. Reporter gene studies are also very important in the analysis of a promoter and its transcription factors (see below), and are a crucial preamble to the construction of a transgenic organism (Chapter 15).

Reporter gene analysis involves the use of gene cloning technology to make a construct in which a gene that codes for a readily detectable product (the reporter) is attached to the promoter, or other regulatory elements, from the gene being investigated. Examples of commonly used reporters include β-galactosidase (detected using a chromogenic or fluorogenic substrate), green fluorescent protein (also known as GFP; a naturally fluorescent protein originating from a jellyfish) or firefly luciferase (where enzymatic activity is detected by the emission of visible light). The simplest procedure is then to introduce this construct, carrying a reporter gene driven by a promoter, into the organism in question. Cells in which the promoter is activated will show detectable and quantifiable expression of the reporter. Even if expression occurs only transiently, so that detection of the mRNA may be difficult, the greater stability of the reporter protein may allow the identification of the time at which expression is activated. (It follows from this, however, that it is less easy to determine when expression is switched *off.*)

A further application of reporter proteins, for ascertaining the localization of a protein within the cell, is described later in this chapter.

10.2.2 Locating the promoter

Reporter genes provide a useful way of locating the sequences that are necessary for promoter activity, and for assessing the regulatory functions associated with upstream sequences (see later). In order to locate the promoter of a specific gene, you can start by cloning fragments of DNA around and upstream from the transcription start site, using a vector that carries a promoterless reporter gene (this is known as a *promoter probe vector* – see Figure 10.6). Expression of the reporter gene will only occur if an active promoter is inserted at the cloning site. Or you can do genome-wide screening for promoters with specific function, for example by making a library of random DNA fragments in a promoter probe vector, and selecting those clones which exhibit expression of the reporter under the chosen conditions.

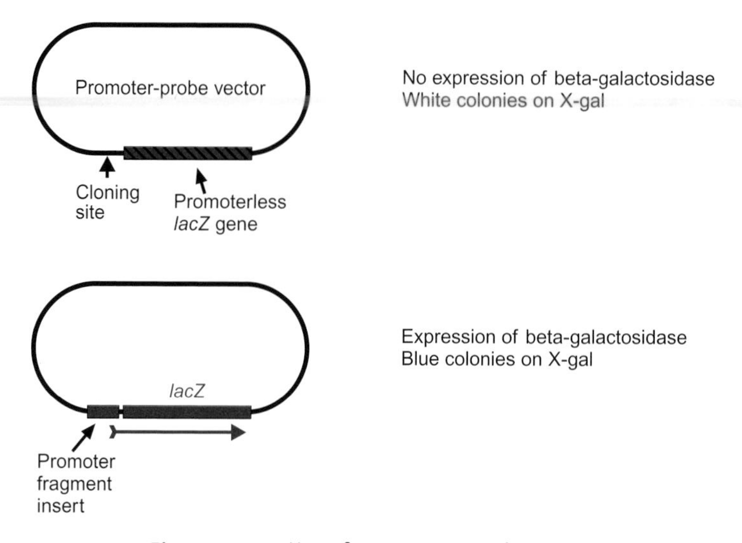

Figure 10.6 Use of a promoter–probe vector

In Chapter 12, we describe the related technique of *enhancer trapping*, which involves the use of a transposable element carrying a reporter gene with a weak promoter. Other approaches described later in this chapter include the yeast one-hybrid system, which is analogous to the promoter–probe approach in enabling the identification of DNA sequences that interact with known transcription factors. It is also used to identify DNA-binding proteins that interact with a defined DNA sequence, and it is in that context that we consider it in this chapter.

One disadvantage of using reporter genes on a plasmid is that the system is not free from artefacts. Fragments of DNA, taken out of context, may show quite different effects in their ability to promote transcription. Some promoters do not work properly (or their regulation is quite different) when located on a plasmid; conversely, some DNA fragments that produce positive signals on a promoter–probe vector are subsequently found not to be 'real' promoters. In prokaryotes, reporter genes, used in either of the above ways, often give more reliable results when inserted into a specific chromosomal position rather than in the artificial environment of a plasmid.

One important example of the use of the promoter–probe approach is the IVET (*in vivo* expression technology) method for identifying potential virulence genes in pathogenic bacteria (see Figure 10.7). In the original version of this procedure, used for identifying *in vivo* expressed genes in *Salmonella typhimurium*, the vector contained two promoterless genes (*purA* and *lacZ*) and the host strain was a *purA* mutant of *S. typhimurium* (i.e. it had a defect in purine biosynthesis, which makes it unable to grow in experimental animals). Random fragments of *S. typhimurium* DNA were inserted in the

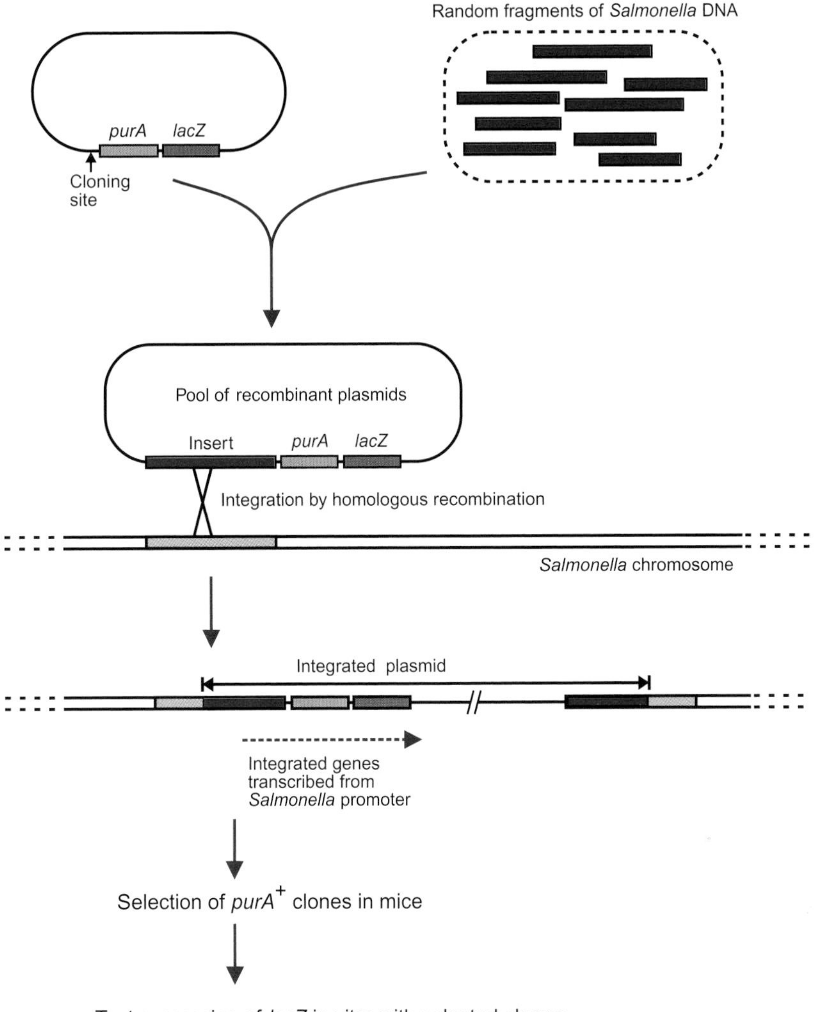

Figure 10.7 *In vivo* expression technology

vector to produce a gene library, and the constructs were integrated into the chromosome by homologous recombination (see Chapter 12). When animals were challenged with the mixture of clones, only those clones with a promoter that was active in this situation were able to express the *purA* gene – thus providing direct selection for promoters that were active during infection. The promoterless *lacZ* reporter gene allowed confirmation of the results; most of the recovered bacteria contained promoters that were also active in the laboratory, but about 5% were specifically activated during infection. Another elegant method for identifying virulence genes, known as *signature-tagged mutagenesis*, is described in Chapter 12.

The techniques above will identify a DNA fragment that has promoter activity, but will not tell you which parts of that fragment are responsible. You can extend the approach to provide a more precise identification of the sequences that are necessary for promoter activity. At the simplest level, this could involve making a series of deletions from one or both ends of the fragment, and seeing how much you can remove without affecting promoter activity. Other techniques that we will describe later (Chapter 11) can be used to alter specific bases within the promoter in order to assess the sequence requirements for initiation of transcription. This approach can also be used to identify regulatory sequences that affect the activity of the promoter, usually through binding of regulatory proteins. Subsequent sections in this chapter examine other methods for investigating the DNA binding of RNA polymerase, transcription factors and regulatory proteins.

10.2.3 Using reporter genes to study regulatory RNA elements

Reporter gene assays based on joining a promoter to a reporter gene have been very useful both for dissecting the components of the promoter and for quantifying its activation in different situations. However, of course it has limitations. If you join your promoter to a gene that encodes a completely different protein, your readout will be a function of the intrinsic properties of that mRNA and of that protein. These include the stability of the mRNA and the protein, as well as the efficiency of translation. However, as long as you compare like with like – the activation of the promoter, *as measured by its effect on reporter gene levels*, under different circumstances – you can make a convincing case that your observations are comparable as a measure of promoter activation, because you are keeping all other parameters constant.

You may, however (especially in eukaryotic systems) wish to study other parameters that affect the levels of the transcript encoded by your gene, such as the 5′-untranslated region (UTR) and its effect on translation efficiency, or the 3′-UTR and its effect on mRNA stability. In these cases, you will wish to keep all other factors constant, including the promoter. Thus, you choose a vector that contains your reporter gene under the control of a constitutively active promoter without cellular or organismal specificity, such as that of the large T antigen of the SV40 virus. You can then slot the whole or part of your 5′-/3′-UTR on either side of your reporter gene. By comparing the amount of reporter gene product in different constructs, and in the native vector, you are able to gauge the positive or negative impact of different elements on mRNA *translatability*.

10.3 Regulatory elements and DNA-binding proteins

The most direct, precise and reliable way of locating a promoter, or any other DNA sequence that influences transcription through the binding of specific proteins (which includes many but not all regulatory elements) is to detect the binding of those proteins to specific DNA fragments. Three methods are worth describing briefly: yeast one-hybrid assays, DNase footprinting, and gel retardation assays.

10.3.1 Yeast one-hybrid assays

The purpose of the yeast one-hybrid system is to identify proteins that interact with a defined DNA sequence, or conversely to locate DNA regions that will bind to known transcription factors. In the example illustrated in Figure 10.8, tandem copies of the DNA sequence under investigation (known as the *bait*) are inserted upstream from a reporter gene. This construct is then integrated into the yeast genome. In the absence of transcription factors that bind to the bait, the reporter gene will not be expressed. However, if you transform the recombinant yeast strain with a cDNA expression library, any

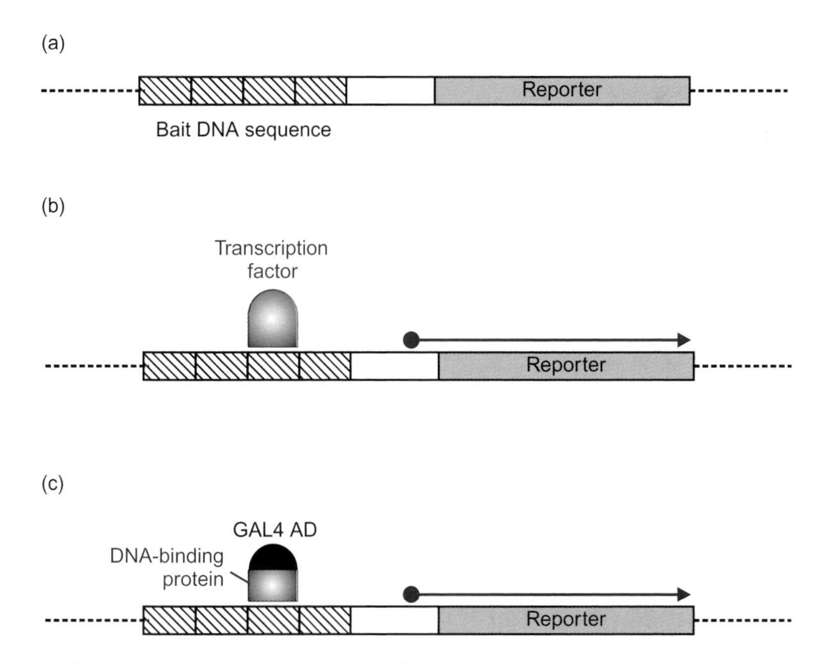

Figure 10.8 Yeast one-hybrid system. (a) Absence of transcription factor: no expression of reporter; (b) presence of transcription factor: expression of reporter; (c) presence of GAL4 AD/DNA-binding protein fusion: expression of reporter

clones that express a transcription factor that can interact with the bait will activate transcription of the reporter gene (Figure 10.8b).

The versatility of the system can be extended by exploiting the bipartite nature of eukaryotic transcription factors. This means that in general they consist of two domains: one part of the protein (the DNA-binding domain) makes specific contacts with certain DNA sequences, while the other part (the activation domain) is responsible for activating transcription. These domains can be separated, and function independently, so the activation domain (AD) can be joined to other DNA-binding proteins, when it will activate transcription from other sites according to the specificity of the new DNA-binding domain. The best-known example is the yeast activator GAL4, which normally controls the transcription of genes involved in galactose metabolism. If the DNA-binding domain of GAL4 is removed and replaced with another DNA-binding domain, the hybrid protein will activate transcription of a different set of genes. To take advantage of this, the cDNA library is made using a special vector containing the GAL4 AD so that the products will be expressed as fusion proteins with the GAL4 AD. If the product binds to the bait sequence, the GAL4 AD will activate transcription of the reporter (Figure 10.8c). Since the activation is provided by the GAL4 AD, this system can detect a wider range of DNA-binding proteins, and is not limited to those that are themselves transcription factors.

An extension of this concept, the yeast two-hybrid system, is used to study protein–protein interactions, and is considered in Chapter 14.

10.3.2 DNase I footprinting

In *DNase I footprinting* (or just *footprinting*), you identify which part of the suspected promoter sequence binds to transcription factor proteins. Nuclei are isolated from whole cells (whether a cultured cell line or a whole tissue), and proteins isolated from the nuclei. These proteins, or a specific protein purified from the mixture, are then mixed with the DNA fragment you wish to analyse. A parallel control reaction contains DNA without protein. In both reactions, the DNA is radiolabelled at the 3′ end. You then add the nuclease DNase I to both reactions. DNase I will cleave all the phosphodiester bonds in the DNA chain that it can reach. The reaction is not allowed to go to completion; thus, as a result, the DNA in your control reaction will be fragmented randomly into a continuous ladder of DNA fragments of all possible sizes between full size and one base. However, in the other reaction, DNase I will be unable to access its substrate wherever a *cis*-acting DNA element has been protected by a bound *trans*-acting protein. Thus, if you run these two samples out on a sequencing gel side by side, you will find that the continuous ladder as seen in the control is interrupted in these protected regions (Figure 10.9). Since DNase digestion is random, there will of course be a lot

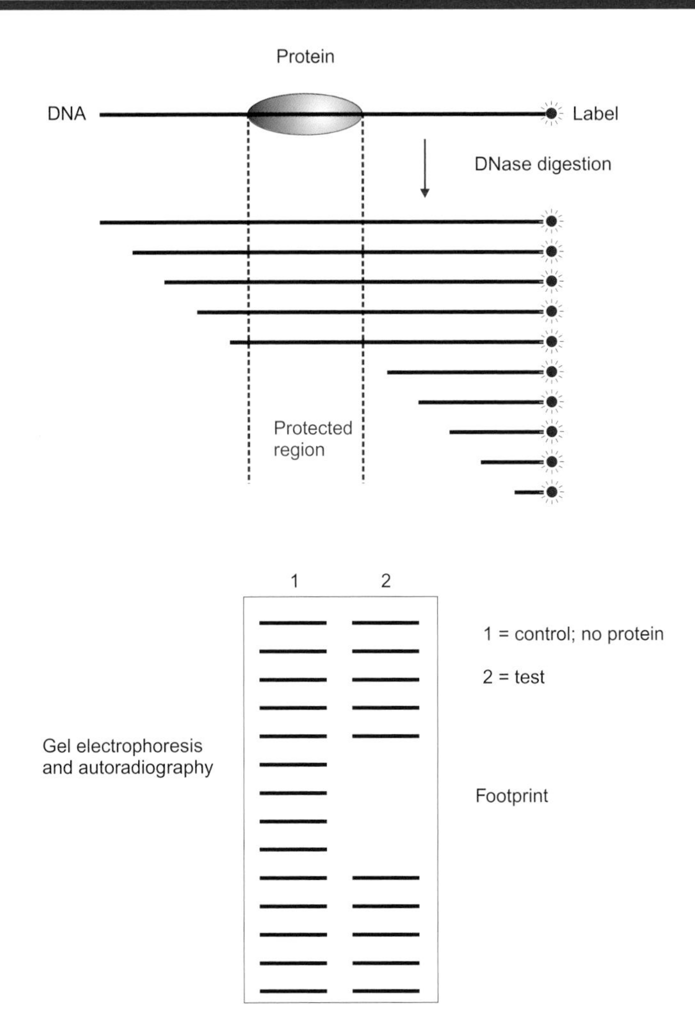

Figure 10.9 DNase I footprinting

of other products as well as those shown – but as you are using autoradiography to detect the label attached to the DNA, you will only detect those fragments that contain the 3′ end of the original DNA molecule. By running out a sequencing reaction as well, you will be able to tell exactly which regions of DNA were protected. Note, however, that the extent of the DNA that is protected by a specific protein is usually larger than the region that is necessary to make specific contacts with the protein.

10.3.3 Gel retardation assays

Another way of determining if a particular sequence of eukaryotic DNA is involved in transcription factor binding is a *gel retardation* or *gel shift* assay.

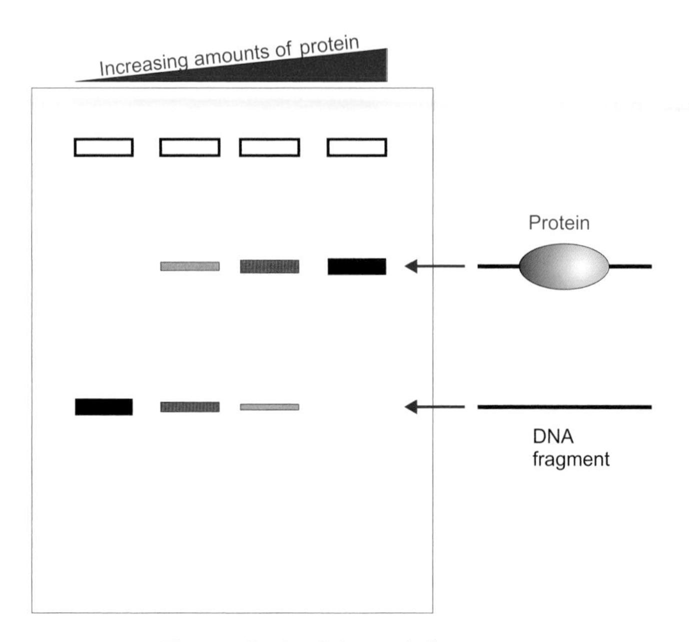

Figure 10.10 Gel retardation assay

Here, the DNA fragment is mixed with nuclear proteins. If one or more of these bind to the DNA fragment, its mobility in a polyacrylamide or agarose gel will be decreased. This can be easily detected by comparing it to a control sample without added protein (Figure 10.10). To investigate the specificity of the reaction, we can add different amounts of protein to ensure that there is a linear dose relationship.

Both footprinting and gel retardation assays are applicable to any protein that binds to specific DNA sequences, and are not limited to those proteins that regulate transcription of the DNA.

10.3.4 Run-on assays

Above, we have discussed several methods for determining the amount of an mRNA species in a sample 'frozen' in time (what is called, not always accurately, 'steady-state levels'). These, as we have seen, have the disadvantage of only showing us the net effect of promoter activation and mRNA degradation at one given moment. Using reporter gene assays, we are able to start disentangling the actual effects of the promoter and of the other regulatory elements, but they have the disadvantage of being 'contaminated' by the fact that the readout is a different protein. All of these methods are very useful within these limitations, but sometimes you want to obtain a more accurate readout. This is found in the *run-on assay*, also known as *nuclear* or *transcriptional* run-on assay (also, confusingly, called run-off assay!). This assay quantifies the actual transcription of the specific gene in a sample. Nuclei are harvested either from cultured cells or from whole tissue and supplied with radiolabelled UTP. They are then left to proceed with transcription, for a short period of time, of all active genes according to their degree of activation in the situation when the nuclei were harvested. The radiolabelled mRNA created by transcription of your gene of interest can then be specifically quantified by immobilizing it through hybridization to 'cold' complementary DNA.

10.4 Translational analysis

The methods for studying protein expression were developed earlier than DNA cloning methods. For example, it was possible to sequence proteins (if they could be purified in large enough amounts) before gene sequencing was developed. When gene cloning techniques became available, the characterization of DNA and mRNA became much easier than sequencing their protein products. However, seasoned protein biochemists have gleefully witnessed the renaissance of their craft, now renamed proteomics (see Chapter 14). Not least the realization that the human genome contains many fewer genes than there are proteins has reminded molecular biologists that gene products must be studied on both levels (to say nothing of the actual effect on the cell or organism as a whole) in order to understand how a gene works.

10.4.1 Western blots

The conventional way of analysing the proteins produced by a cell involves electrophoresis through a polyacrylamide gel in the presence of sodium dodecyl sulfate (SDS polyacrylamide gel electrophoresis, or SDS-PAGE). In general, there are too many proteins present to give a clear picture of the complete protein profile if you simply stain the gel with a general protein stain, although much higher resolution can be obtained by two-dimensional gel electrophoresis (see Chapter 14). In a one-dimensional SDS-PAGE, you will usually only be able to see the major proteins, and even then it is not always easy to identify any specific ones.

The generally applicable technique for the detection of a specific protein, and analysis of its expression, relies on the use of specific antibodies, in combination with SDS-PAGE. Following separation of the cell extract by SDS-PAGE, the proteins are transferred to a membrane by running a perpendicular current through the gel into the membrane, and a specific protein or proteins detected using *antibodies* (see Chapter 7). These antibodies can be either labelled themselves, or, much more commonly, detected by a second labelled

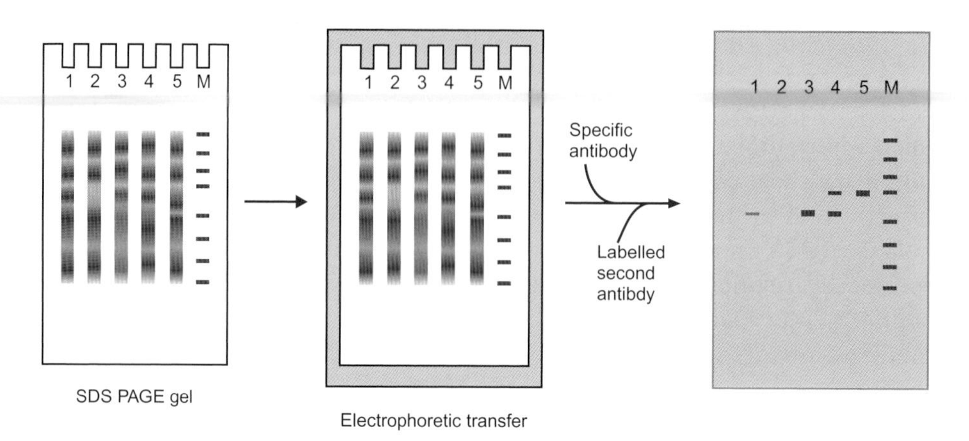

Figure 10.11 Western blotting. M = standard marker proteins

antibody. For example if your primary antibody is a mouse antibody that recognizes the required protein, the binding of that antibody can be detected by a labelled rabbit anti-mouse antibody. The analogy with a Southern blot led to this technique being termed *western blotting*. The technique will identify, not just the presence or absence of a protein that reacts with the antigen, but also its size and an estimate of relative levels of expression. In Figure 10.11, we see that expression was not detected in sample 2, and only weakly in sample 1, while the protein detected in track 5 was a different size, perhaps due to post-translational modification; both proteins were present in track 4.

The major considerations in the use of antibodies for molecular biology were covered in Chapter 7 (see Box 7.2). The main point to remember is that proteins in an SDS-PAGE gel are denatured, so the antibody must be capable of recognizing a linear epitope. For recombinant proteins, you also require an antibody that binds to a non-glycosylated epitope. The complex mixture of antibodies in a conventional antiserum is virtually guaranteed to contain some suitable antibodies, but an individual monoclonal antibody may fail one or both criteria.

10.4.2 Immunocytochemistry and immunohistochemistry

This method (differently named depending on whether we are working with cells or tissues) relates to western blotting the same way *in situ* hybridization relates to northern blotting. The blotting method allows us to determine whether the probe (either an antibody or a nucleic acid probe) is specific, and if it is, how many polypeptides or transcripts it recognizes, and what size they are. The cytological or histological method, by contrast, allows us to determine which cell type the gene product is localized to. Apart from being much easier to perform and less prone to artefacts than *in situ* hybridization, the study of the translated protein product has an added bonus. The *intracellular* localization of an mRNA, as determined by *in situ* hybridization, is rarely particularly revealing – we are only told the story of what happens until the transcript is translated into a protein, which is usually (but not always) in the vicinity of the nucleus. However, the destination of the translated protein shows much greater variation. Some are cytoplasmic, some are membrane-bound; some gather at one pole of the cell, and others re-enter the nucleus. The localization of a protein can give us important clues to its function, and indeed the incorrect transport of a protein can give us important information about a disease phenotype.

Reporters such as GFP can also be used to study the fate of specific proteins within the cell. By fusing the *gfp* gene to that coding for the protein in question, we can easily see where that protein ends up (for example, in the cytoplasmic membrane) simply by detecting the fluorescence.

11 Products from Native and Manipulated Cloned Genes

In the preceding chapters, we looked at ways of characterizing individual genes and analysing their expression. Genetic techniques can be used for modification as well as analysis, and in this chapter we focus on ways in which genes can be manipulated for product formation – whether of the natural product of that gene, modified versions of the product or, ultimately, entirely novel proteins. In a later chapter, we will consider ways in which cells or organisms can be genetically modified.

We use proteins, or smaller polypeptides, in many ways – ranging from enzymes that can be added to washing powders to hormones that are used for treating medical conditions such as diabetes. Some of these proteins can be extracted easily, although at some cost, from starting material that is readily available, e.g. plant material or microbial cultures. However, often the potential source of such proteins is scarce or difficult to obtain, which severely limits the application of this approach – or may even make it impossible. This applies especially to proteins and polypeptides from human sources, which can be invaluable for treating specific diseases. The classic example here is of human growth hormone (somatotropin), which is used for treating a condition known as pituitary dwarfism, where a child's growth is affected by a deficiency in the production of this hormone by the pituitary gland. For a long time, the only source of this hormone was from pituitary glands removed from the bodies of people who had died from a variety of other causes. The limitations on supply can be imagined; the safety implications, in terms of the potential transmission of disease by this route, were only partly appreciated at the time. Other consequences were only realized more recently when it was discovered that spongiform encephalopathies (CJD) could be transmitted by this material.

From Genes to Genomes, Second Edition, Jeremy W. Dale and Malcolm von Schantz

Many bacteria and other microorganisms (especially yeasts) can be grown easily and cheaply in essentially unlimited quantity. The expression of the gene coding for the required protein in a microbial host can therefore enable us to obtain that product in large amounts – especially as we can manipulate the gene to maximize its expression. Making a therapeutic product in this way also removes the possibility of transmitting infectious agents from another human or other animal, so it is much safer than the 'natural' product. It should also be emphasized that (exempting any post-translational modifications which will be discussed later in this chapter) the purified product obtained from recombinant bacteria is exactly the same as the pure product from its original source. The only way in which the product differs is that it is likely to be purer and will be free from infectious agents.

The realization that the new genetic technologies would provide a way of making products that had hitherto been expensive or impossible to produce was a major driving force behind the commercial side of the biotechnology revolution. The research implications are even more dramatic. Some proteins, especially those responsible for regulating cellular activities and for communication between cells, are produced naturally in tiny amounts. It is likely to be effectively impossible to extract and purify such a protein in sufficient quantity to be able to characterize its structure or to analyse the way it works. Yet in principle, as long as you can identify and clone the gene responsible, it will be possible to express that gene in a microbial host and obtain substantial quantities of the product.

11.1 Factors affecting expression of cloned genes

Expressing a foreign gene in a bacterial cell is not entirely straightforward. The expression signals that control transcription and translation can be quite different from one organism to another, especially if you are moving a gene from a eukaryotic source into a bacterial host. Expression can also be affected by the base composition of the gene, and by the codon usage. These factors were covered in Chapter 2, but are summarized here for ease of reference (see Figure 11.1). We are generally assuming to start with that the host organism is a bacterium; the expression of foreign genes in eukaryotic cells is considered later on.

Transcription The main factor is the promoter site, i.e. the region where the RNA polymerase binds to initiate RNA synthesis (Figure 11.2). This is the principal step where bacteria control which DNA regions are to be expressed, and how strong that expression should be. The structure of the RNA polymerase (and especially of the sigma factors that determine the specificity)

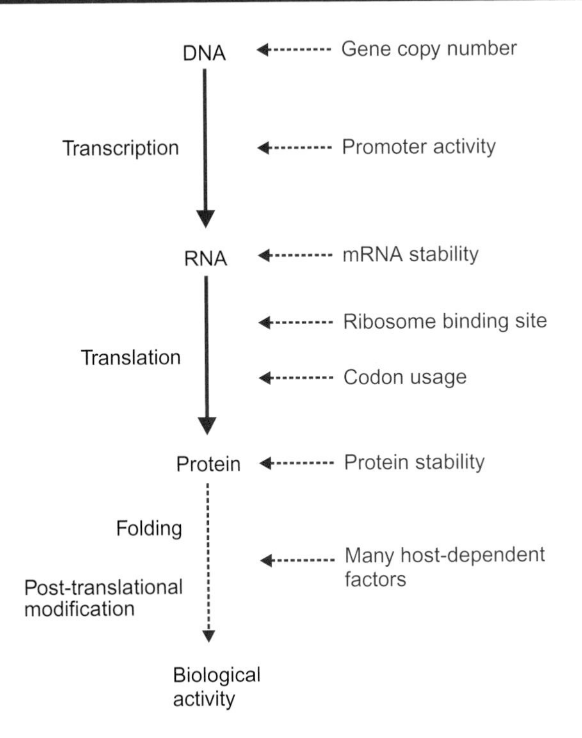

Figure 11.1 Gene expression

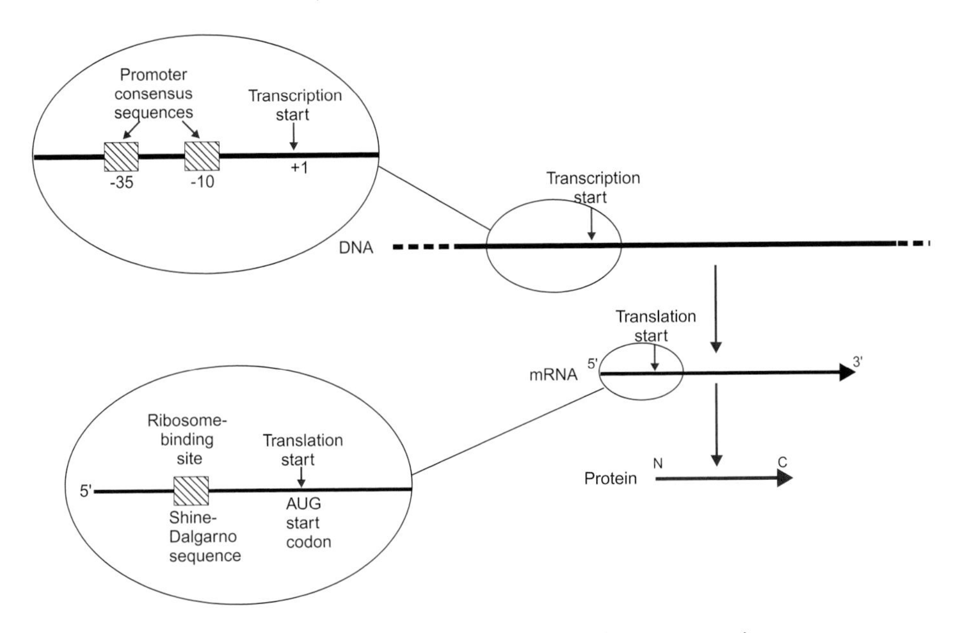

Figure 11.2 Principal factors in bacterial gene expression

and the sequences of the promoter regions have evolved together to produce this careful regulation of expression. It is therefore not surprising that a gene from one species may not be expressed when inserted into a different host.

Transcription signals in eukaryotes are substantially different from bacterial promoters, but that can usually be overcome by replacing them. For gene expression, you will usually require cDNA rather than genomic DNA (to remove introns). You have to provide a bacterial promoter, and any other necessary bacterial expression signals, using an *expression vector*. This concept was introduced in Chapter 5, and is considered again later in this chapter.

Transcription can also be affected by the base composition of the DNA. If the cloned DNA comes from an organism with a higher G + C content than that in the new host, the stronger association of the two strands may hinder transcription, which requires local strand separation and unwinding of the helix.

It should also be borne in mind that in principle it is the *amount* of mRNA available for translation that is important regardless of the rate with which it is produced. Thus, the level of a specific mRNA is determined by its stability as well as by its rate of synthesis. In bacteria, a transcript is typically very short-lived (with a half-life measured in minutes), but translation normally starts as soon as a ribosome-binding site (see below) is available. Differences in stability between mRNAs are therefore less important in bacteria than in eukaryotes. Furthermore, it is easier to manipulate the level of transcription than to alter the stability of a message, so we would generally focus on the former.

Translation initiation In bacteria, ribosomes normally bind to a region of the mRNA adjacent to the start codon, facilitated by a sequence (the Shine–Dalgarno sequence) that is (partially) complementary to the 3′ end of the 16S rRNA (Figure 11.2). This structure is reasonably well conserved between bacteria, so you could expect to get some degree of translation initiation if you move a gene from one bacterium to another (assuming that transcription occurs). However the structure may not be optimal; even within the same organism, genes may not have an optimum ribosome binding site; for example, the distance between the Shine–Dalgarno sequence and the start codon can vary, which can affect the efficacy of translation initiation, so you may not obtain maximum gene expression unless you manipulate this region.

Ribosome binding in eukaryotes works in a different way. Thus, if you choose a bacterial host for expression of eukaryotic genes, you will usually have to provide translational signals as well as a promoter.

The nature of the start codon may also have to be considered. In the standard genetic code, AUG is used to signal the start of translation. However, in some organisms, a proportion of genes (sometimes the majority) use alternative start codons such as GUG, UUG or CUG. Trying to express such a gene in an organism with a stricter preference for an AUG start may limit or prevent translation, or give a product with a different N-terminus.

Codon usage In the standard genetic code, which is basically conserved between all living organisms, there are many sets of *synonymous codons*, i.e. several codons that code for the same amino acid. For example, any one of six different codons in the mRNA will result in incorporation of leucine into the polypeptide (see Chapter 7, Box 7.1). These different codons are not completely equivalent. The cell will use some codons more readily than others, depending on the availability and specificity of tRNA molecules that recognize these codons. Any occurrence of the less readily used codons will slow down translation. The sequence of the genes in that organism will therefore have evolved to match the availability of tRNAs (and/or vice versa), and some synonymous codons will occur more frequently than their synonyms – which we refer to as *codon bias*. In a different organism, the tRNA population is different, and the codon usage will also be different.

In nature, codon bias is much more pronounced for highly expressed genes, such as those coding for enzymes of the central metabolic pathways. There is more evolutionary pressure favouring the optimization of codon usage of such genes than there is for genes coding for proteins that are produced at a lower level. This is reflected in the behaviour of cloned genes. If we take a gene from one organism and try to express it in another, with different codon usage, we will probably get some expression (assuming everything else is working), but the extent of that expression is likely to be limited. Expression may be enhanced by altering the sequence of the gene to improve the codon usage for that host.

In extreme cases, you may actually fail to get expression at all. The host organism may be virtually unable to recognize certain codons in your cloned gene (and these would be absent from natural genes in that host); in effect, these are acting as additional stop codons, and will cause premature termination of protein synthesis (remember, the standard stop codons terminate translation because there is no tRNA that can bind to them). This does not happen in *E. coli*, which is capable of recognizing (to varying extents) all the codons in the standard code. However, some other hosts have a more pronounced bias in their codon usage, with some codons occurring extremely rarely.

The converse can occur, too. In some organisms, a codon that is a stop signal in the standard code is actually used by that organism to code for a specific amino acid. For example, in some bacteria UGA codes for tryptophan rather than being a stop codon. If you try to express such a gene in *E. coli*, you are likely to get premature termination at the UGA codon. It is possible to overcome this by using *E. coli* host strains that have been engineered to produce additional types of tRNA.

G + C content Organisms vary very substantially in the base composition of their DNA (ranging, approximately, from 30 per cent G + C to 70 per cent G + C). Therefore, the composition of the cloned gene may be very different from that of the new host. We have alluded to the possible effect this might have on transcription; it is also likely to be reflected in differences in codon usage. There are a wide range of other effects which could occur, which might affect not only the expression of the gene but also other properties of the recombinant. These include the structure and stability of the mRNA, supercoiling of DNA, and replication and stability of the recombinant vector itself.

Nature of the protein product We also have to remember that the nature of the protein product may influence the amount of protein that is recovered. Most obviously (as with mRNA) the stability of the protein can have a major effect: an inherently unstable protein will not be recovered in high yield, even if we have maximized the rate of production. Amongst other factors, the location of the product can have an important influence of the levels obtained. If the protein remains in the cytoplasm, it may become insoluble at high rates of synthesis, leading to the production of *inclusion bodies*, or aggregates of insoluble product. It can be difficult to recover active soluble protein from such inclusion bodies. If the protein becomes inserted into the cytoplasmic membrane, it is likely to have a deleterious effect on the functions of the membrane of an organism to which it is alien; expression of such proteins may be lethal, even at quite low levels.

For high levels of protein production from bacterial cultures, it is often advantageous to express the product in a form that can be secreted into the culture medium, e.g. by attaching *signal sequences*. Because the volume of the supernatant is very much larger than the combined volume of the cytoplasm of all the cells in a culture, even in a very dense culture, problems of product insolubility are very much less likely to arise. Unfortunately, attaching secretion signals may not work. Proteins that are not naturally secreted may not pass properly through the cytoplasmic membrane, even when directed to do so by a secretion signal.

We also need to bear in mind that our aim is usually the production of a biologically active protein, rather than a simple polypeptide chain. Biological activity often depends on a variety of post-translational effects, ranging from folding of the polypeptide into the correct three-dimensional structure, to post-translational cleavage or modifications such as glycosylation. This can be a difficult barrier to surmount when attempting to express a protein in a foreign host, and is often a major reason for using a non-bacterial host for expression of eukaryotic proteins.

11.2 Expression of cloned genes in bacteria

11.2.1 Transcriptional fusions

When attempting to express a foreign gene in a bacterial host, the first parameter to be considered is the requirement for a fully functional promoter attached to the cloned gene. In principle, this can be addressed by simply adding a known, characterized promoter in a separate cloning step. A more convenient procedure is to use a ready-prepared vector (usually a plasmid) that already carries a suitable promoter adjacent to the cloning site. This forms a simple type of *expression vector* (Figure 11.3), as already introduced in Chapter 5. Insertion of the cloned fragment at the cloning site (in the correct orientation) will then put that gene under the control of the promoter carried by the vector. In other words, transcription initiated at the promoter site will continue through the cloned gene; since this means a fusion of the gene and the promoter into a single transcriptional unit, it is referred to as a *transcriptional fusion* (cf. translational fusions, below). This is very similar to the concept of a reporter gene (Chapter 10), but for a different purpose. A reporter gene is used to study the activity of the promoter. Here we know what the promoter does, and we are using it to stimulate transcription of the cloned gene.

In *E. coli* there are a wide range of suitable promoters that can be used in this way. Amongst naturally occurring promoters, the *lac* and *trp* promoters that normally drive expression of the *lac* (lactose utilization) and *trp* (tryptophan synthesis) operons, respectively, and the P_L promoter, from bacteriophage lambda, are examples of commonly used promoters, the P_L promoter being one of the strongest natural promoters in *E. coli* (or its associated genetic elements).

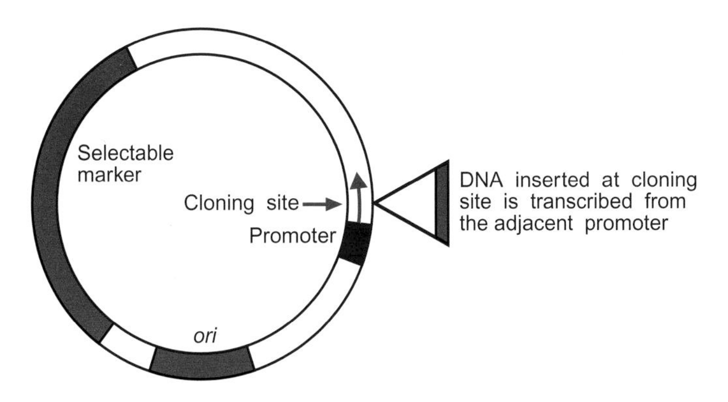

Figure 11.3 Basic features of an expression vector

Box 11.1 Comparison of selected *E. coli* promoters

Promoter	-35 region	-10 region
Consensus	TTGACA	TATAAT
lac	TTTACA	TATATT
trp	TTGACA	TTAACT
P_L	TTGACA	GATACT

However none of these three natural promoters actually represents a perfect match with the consensus sequence obtained by comparing the sequence of a large number of *E. coli* promoters (see Box 11.1).

The –10 and –35 regions are not the sole determinants of the strength of a promoter; a region of about 70 base pairs makes contact with the RNA polymerase and influences the strength of that contact. Nevertheless, these two regions (and the distance separating them) are the most highly conserved and play the major role. Since none of these sequences is a perfect match with the consensus, we can infer that further manipulation of the sequence could produce even stronger promoters, and this is in fact the case. One artificial promoter that is commonly used is known as the *tac* promoter, since it represents a hybrid between the *trp* and *lac* promoters, and it is capable of higher levels of transcription than either of the original natural promoters.

Not all promoters conform to the consensus shown. In particular, some bacteriophage RNA polymerases (notably those from T7, T3 and SP6) have a virtually absolute requirement for different sequences of bases, each specific to that enzyme. These polymerases will not initiate transcription from the usual promoters, nor will *E. coli* RNA polymerase recognize one of these special promoters. We met these enzymes in the discussion of cloning vectors (Chapter 5); they can be very useful in ensuring that we only get expression of our cloned gene if and when we need it. The ways in which we can control that expression are considered later.

In addition to the choice of promoter in our expression vector, we need also to consider the plasmid copy number. Most routine cloning vectors for use in *E. coli* are referred to as 'multi-copy' plasmids – but this can conceal a wide variation in the actual copy number. The early cloning vectors (such as pBR322) are normally present in 15–20 copies per cell (although under special conditions they can be amplified to up to 1000 copies per cell). Subsequent development of these vectors removed some control elements, so that many of the currently used vectors (such as the pUC series) have hundreds of copies per cell. More copies of the plasmid means more copies of the cloned gene, and hence (usually) more product formation. However, the relationship may not be linear; you will expect to get more product from a

200-copy plasmid than from one with only 20 copies, but not necessarily 10 times as much (and indeed excessive amounts of product may kill or at least seriously inhibit the growth of the cells that produce it).

So if you want to maximize gene expression, not only should you optimize the promoter, but you should also use a high copy number plasmid. Under these conditions (assuming translation works perfectly), you may get a bacterial clone in which your product represents up to 50 per cent of the total protein of the cell. For commercial production, such high yields not only mean more product per litre of culture (which is of course important), but in addition the proportion of contaminating protein (and other material) that has to be removed is lower – thus reducing the costs of downstream processing. However, expressing large quantities of protein, and maintaining very high copy number plasmids, brings problems which we need to consider.

11.2.2 Stability: conditional expression

There is a downside to such high levels of product formation. Producing vast quantities of a protein that is of no use to the cell is inevitably going to result in a reduction in growth rate, because of the amount of resources that are being diverted in a manner that is non-productive, from the host cell's perspective. This applies even if the protein itself has no damaging effects. Slower growth rates will reduce the efficiency of the process. If the protein is directly damaging, the problem becomes much more acute.

A bigger problem is that the slower growth of the producing cells means that there is a very strong selective pressure in favour of any of a wide range of potential mutants that are non-productive. These include cells that have lost the plasmid altogether, as well as any mutation to the plasmid that reduces or prevents product formation. This can be a problem whether you are growing a few millilitres in the laboratory or many thousands of litres in an industrial fermenter. However, on a laboratory scale it is possible to include antibiotic selection in your culture, to ensure that any mutants that have lost the plasmid will be unable to grow. On a large scale, this is not only an expensive solution, but also the disposal of large volumes of antibiotic-containing waste is a problem.

One way of ameliorating this situation is to use *controllable* promoters, i.e. promoters whose activity can be altered by changes in the culture conditions. The promoters listed above provide examples of such control. For example the *lac* promoter is naturally expressed only if *E. coli* is growing on lactose as a carbon and energy source. [Normally bacterial geneticists use IPTG (see Chapter 5) instead of lactose, as it is both more convenient and is not broken down by the cell's β-galactosidase; it is known as a *gratuitous inducer*.] In the absence of an inducer, a repressor protein binds to a DNA sequence known as the *operator* (which in this case overlaps with the *lac*

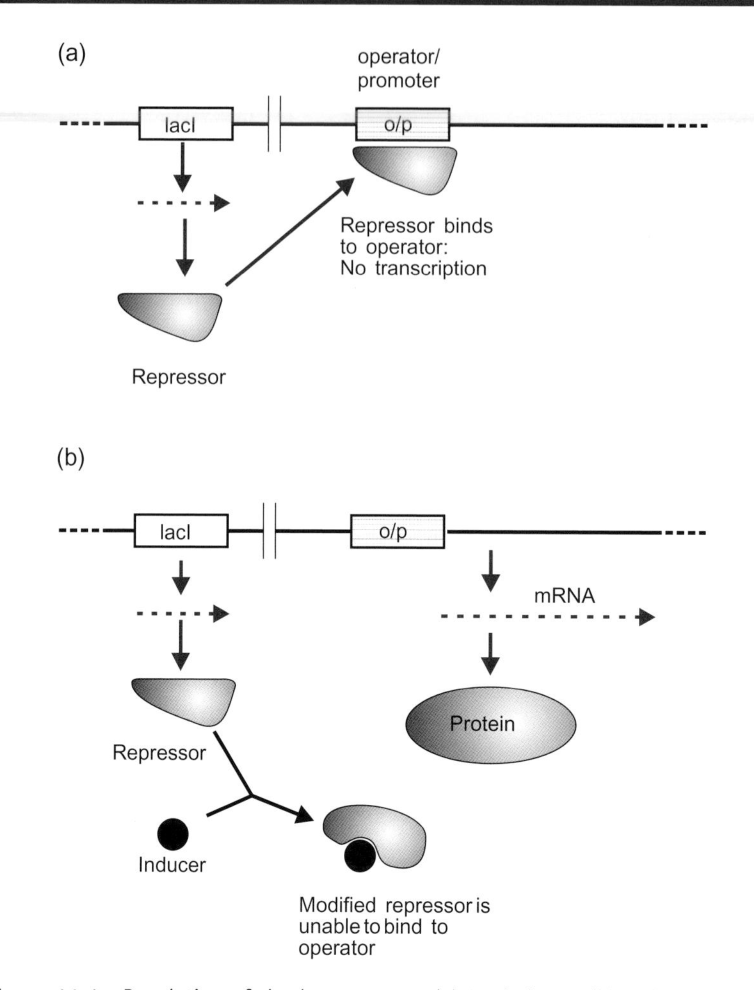

Figure 11.4 Regulation of the *lac* promoter. (a) No inducer; (b) inducer present

promoter) and prevents transcription from the *lac* promoter. The inducing agent binds to the repressor protein, altering its conformation so that it no longer binds to the operator (Figure 11.4). We can grow the culture to an appropriate density in the absence of an inducer, so that we will not get high levels of gene expression and hence the selective pressure will not exist. Then, when we have enough cells, we add the inducing agent and switch on gene expression.

However, we have to remember that we are using a multi-copy vector. *E. coli* produces enough of the repressor protein to switch off the single copy of the promoter that it has in the chromosome, but this is not enough to switch off several hundred copies of this promoter. We refer to the repressor being *titrated out* by the presence of so many copies of the operator. So we have also to increase production of the repressor protein. The gene that codes for the repressor (the *lacI* gene) is not actually part of the *lac* operon; it has its own promoter. So one way of increasing production of the LacI repressor protein is by using a mutated version of the *lacI* gene which has a more active promoter (an *up-promoter* mutant). This altered *lacI* gene is known as *lacI*^{q}. Alternatively, we can put the *lacI* gene onto the plasmid itself, so subjecting it to the same gene dosage effect and therefore increasing the production of LacI. Commonly, we would do both – i.e. put a *lacI*^{q} gene onto the plasmid.

Adding IPTG to a laboratory culture is fine, but on a commercial scale it is still not an ideal solution. Adding IPTG to an industrial-scale fermenter would be very expensive. An example of an alternative strategy is to use a promoter such as the *trp* promoter, which controls transcription of the tryptophan operon. This is subject to repression by tryptophan; *E. coli* switches off expression of the tryptophan operon when the enzymes encoded by it are not needed, i.e. if there is a plentiful supply of tryptophan. It is possible to monitor and control the availability of tryptophan so that there is an adequate supply during the growth phase, and then limit the supply of tryptophan when expression is required. You may be puzzled by this. If we stop supplying tryptophan, how is the cell going to make the protein that we want, which will probably contain some tryptophan residues? However, it is possible to supply a low level of tryptophan which is not enough to switch off the *trp* promoter, but will still enable production of the required protein, and we can then feed the culture continuously with a low level of tryptophan so that protein production will continue.

This may solve the problem in part, by removing the selective pressure imposed by excessive product formation. However, there is still some element of selection imposed by the presence of so many copies of the plasmid, which may also slow growth rates. We can counter this by using a plasmid with a different replication origin, so that replication is tightly controlled at only one or two copies per cell (see Chapter 5). However the level of expression achieved with a low copy vector will be less than that achievable with a multicopy plasmid, other things being equal. We can adopt a similar strategy to that described above for programming gene expression, using a so-called *runaway plasmid*. If the control of plasmid copy number is temperature sensitive, then growing the culture initially at say 30°C will produce cells with only a few copies of the plasmid. Then, once sufficient growth has been achieved, the culture can be shifted to a higher temperature, say 37°C; control of plasmid replication is lost and the copy number increases dramatically, until it represents perhaps 50 per cent of the DNA of the cell. If we switch on gene expression at the same time, we will get a very substantial amount of product. Eventually the cells will die, but by that time we have enough of our product.

An alternative way of achieving a similar effect is by providing the vector with two origins of replication: one that results in many copies of the plasmid and a second that will produce only one or two copies per cell. If we can control which of the two origins is used, we can again switch conditions at an appropriate stage, so that the culture starts off with only a few copies of the plasmid per cell and then when we want to switch on gene expression we can, as well as inducing the promoter, switch to the other replication origin and increase the copy number of the plasmid.

11.2.3 Expression of lethal genes

Some genes code for products that are very damaging or even lethal to a bacterium such as *E. coli*. Expression of such genes in *E. coli* poses obvious problems. You might consider that the above approach would be equally applicable in such a situation, but it is not quite that simple. The regulation of many promoters is not tight enough. The *lac* promoter for example, although only showing full activity in the presence of an inducing agent, is nevertheless active at a lower level even in the absence of induction. If the gene product is *very* damaging, the cell will not be able to tolerate even this low level of expression. Other promoters with tighter control, such as the T7 promoter referred to earlier (see also Chapter 5), have to be used.

Even in the absence of a known promoter, there may be enough transcription to cause a lethal effect, since a wide variety of DNA sequences have an artefactual low level of promoter activity. This is especially a problem with multi-copy vectors because of the gene dosage effect, and can be reduced by using tightly controlled low copy vectors. Alternatively, obtaining high levels of production of a protein that is highly damaging to *E. coli* may require the use of an alternative host.

11.2.4 Translational fusions

If we want to provide our gene with translational signals (ribosome binding site and start codon) as well as a promoter, we can use a similar approach. For this purpose we need a different type of expression vector: one that will give rise to a *translational fusion.* In this case, part of the translation product (the protein or polypeptide) is derived from the insert and part from the vector (Figure 11.5). When using a translational fusion vector, we have to be much more careful in the design of our construct.

With a transcriptional fusion vector, as described above, we simply have to make sure the insert is the right way round. Within reason, it does not matter too much where it is – some untranslated leader mRNA can be tolerated, so it does not have to be precisely located with respect to the promoter. However, with a translational fusion, the location is all important. Because translation starts at the initiation codon in the vector sequence and the ribosomes then read the sequence in triplets, you have to make sure that the

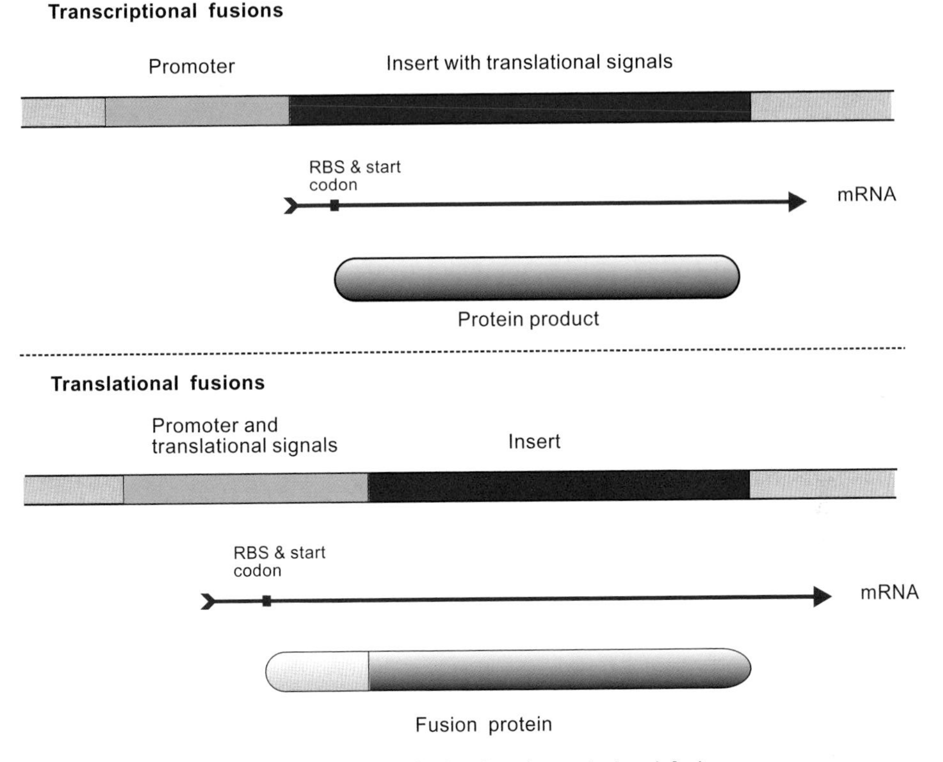

Figure 11.5 Transcriptional and translational fusions

correct reading frame is maintained at the junction. One or two bases out in either direction, and your insert will be read in the wrong frame, giving rise to a completely different amino acid sequence (and probably resulting in premature termination as the ribosomes will soon come across a stop codon in this frame) – see Figure 11.6.

You can only achieve this if you know the sequence of your insert, and the sequence of the vector, or at least of that region of the vector between the cloning site and the start codon. (Contrast this with a transcriptional fusion vector, where it is quite possible to insert an uncharacterized DNA fragment into the vector and be successful in obtaining expression.) Furthermore, you have to work out exactly what will happen at the cloning site when you cut the insert and the vector and then join them together. If you are relying on a restriction site that is naturally present in the insert, there is only a 1 in 3 chance that it will be in the right frame when joined to the vector, even assuming the insert goes in the right way round. You might need to alter your choice of vector, or to modify either the vector or the insert, or both (e.g. by using linkers or adaptors – see Chapter 4). Alternatively, you can use PCR to generate the insert fragment, in which case you must design the primer to

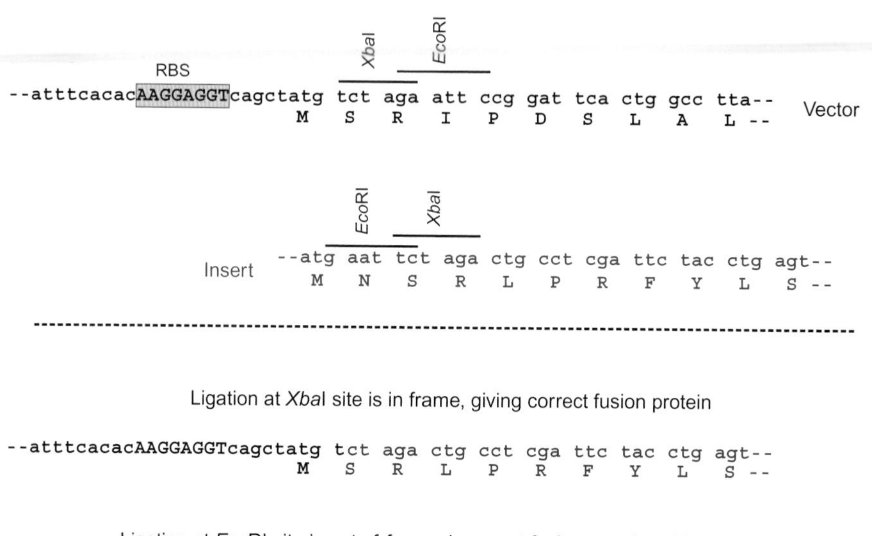

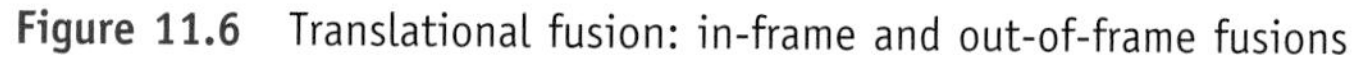

Figure 11.6 Translational fusion: in-frame and out-of-frame fusions

contain a restriction site that will produce the correct reading frame (see Chapter 8).

Finally, having done all of this, it is essential to check, by sequencing the recombinant product, that ligation has indeed resulted in the construct that you have designed, and that the reading frame is actually correct. Loss of a single base in the ligated product will result in an incorrect reading frame.

In theory, it is possible to use a vector which just has the ribosome binding site and no start codon, and to use the natural start codon of the cloned gene. However, in *E. coli* the optimum distance between the Shine–Dalgarno sequence and the start codon is only five to seven bases, so there is not much room for manoeuvre. More commonly, the vector will have the start codon as well, and the insert will not have a start codon. In this case, the translational fusion product will have a few amino acids derived from the vector sequence, and will lack some of the N-terminal amino acids that are normally found in this polypeptide. For many purposes, this is of little significance, as proteins can often tolerate a considerable amount of variation in the N-terminal sequence. In some cases it may be important to have a product that is precisely the same as the naturally occurring one – for example if you are making a protein for human therapeutic purposes. You will then need to carry out more precise manipulations rather than using off the shelf expression vectors; for example you can ensure that the vector-derived amino acids replace exactly those that are lost from the insert.

Sometimes, it is useful to be able to add a short sequence of amino acids to the N-terminus of your product. We will consider the applications of such *tagged proteins* later in this chapter.

11.3 Expression in eukaryotic host cells

Although bacteria are convenient hosts for many purposes, and *E. coli* is usually the host of choice for initial gene cloning (including the production of primary gene libraries), bacterial systems have many limitations for the *expression* of cloned genes, especially for large-scale production of proteins of eukaryotic origin. In particular, the post-translational modifications needed for conversion of the primary product of translation into the mature, folded, biologically active product are more likely to occur in a eukaryotic host. There is a wide variety of systems available; in this chapter we will consider just a few examples. Many of the concepts discussed in relation to bacterial gene expression are also relevant to expression in eukaryotic cells, even if the details are different.

11.3.1 Yeast expression systems

Yeasts have many advantages for expression of cloned genes. They grow rapidly, in simple defined media, and as unicellular organisms they are relatively easy to manipulate and enumerate. Baker's yeast, *Saccharomyces cerevisiae*, has been used as an experimental organism in microbial genetics for many years and there is now a wealth of biochemical and genetic information available to support its use for gene cloning – including the genome sequence. Powerful and versatile systems are also available for several other yeast species, notably *Pichia pastoris*.

Saccharomyces cerevisiae Vectors for cloning and expression of genes in the yeast *S. cerevisiae* were introduced in Chapter 5. In many respects, especially if using episomally replicating vectors, the concepts are similar to those involved in the design of plasmid-based expression vectors in bacteria. There is a choice between vectors carrying the origin of replication from the naturally occurring 2 μm yeast plasmid, which are maintained episomally at high copy number (up to 40 copies per cell), and centromere vectors which are maintained at low copy number (one or two copies per cell; see Chapter 5 for more detail). Both types of vectors are designed as shuttle vectors, that is, they also carry an *E. coli* replication origin, which enables the initial construction, verification and amplification of the recombinant plasmid to be carried out in *E. coli* before transferring the finished construct into yeast cells. As with bacterial expression vectors, these *S. cerevisiae* vectors are designed with a controllable promoter adjacent to the cloning site to enable expression of the cloned gene to be switched on or off. Most commonly, this involves the promoter and enhancer sequences from the *GAL1* (galactokinase) gene, which is strongly induced by the addition of galactose.

Pichia pastoris *Pichia pastoris* is able to use methanol as a carbon source, the fist step in the pathway being due to the enzyme alcohol oxidase (the product of the *AOX1* gene). This gene is tightly controlled, so that in the absence of methanol no alcohol oxidase is detectable. On addition of methanol to the culture, the *AOX1* gene is expressed at a very high level. The use of the *AOX1* promoter in the expression vectors, adjacent to the cloning site, therefore provides vectors that are capable of generating substantial levels (up to several grams per litre) of the required product.

Unlike the bacterial systems, the vectors in *P. pastoris* are not maintained episomally, but are designed to be integrated into the yeast chromosome. They do, however, contain an *E. coli* replication origin, so that they are maintained as plasmids in *E. coli*, enabling formation and verification of the recombinant plasmids in *E. coli* before transformation of *Pichia*.

Not only are the expression levels in *Pichia* usually higher than those obtainable in other systems (prokaryotic or eukaryotic), but scaling up to industrial levels of production is relatively straightforward, and the organism grows readily in simple defined media, leading to lower costs than those involved in insect or mammalian systems (as described below).

11.3.2 Expression in insect cells: baculovirus systems

Baculoviruses, such as the *Autographa californica* multiple nuclear polyhedrosis virus (AcMNPV), infect insect cells and are exploited as the basis of systems for gene expression in such cells. During normal infection, large amounts of a virus-encoded protein called polyhedrin are produced, which forms a matrix within which the virus particles are embedded. The high level of production of polyhedrin is due to a very strong promoter, which can be used to drive expression of the cloned gene, by inserting it downstream from the polyhedrin promoter. Production of polyhedrin itself is not completely essential for virus production, but in its absence the virus does not produce inclusions; the resultant plaques can be distinguished from those produced by wild-type virus. However, the viral DNA is itself too large (>100 kb) for direct manipulation to be easily carried out. The gene to be expressed is therefore first inserted into the polyhedrin gene in a smaller *transfer vector*. Once again, the transfer vector is a shuttle vector, i.e. it contains an *E. coli* origin of replication and other functions that allow manipulation in an *E. coli* host. For production of recombinant virus particles, the transfer vector must recombine with viral DNA. For this to happen, insect cells are cotransfected with the recombinant transfer vector and with viral DNA. Recombination within the insect cells leads to the production of recombinant viruses, although at a low frequency. These are separated from the wild-type, non-recombinant, viruses by picking the plaques that are characteristic of polyhedrin-deficient viruses.

More efficient versions of this system are now available. In the example shown in Figure 11.7, the transfer vector contains a multiple cloning site adjacent to (and downstream from) the polyhedrin promoter so that the cloned gene will be transcribed from that promoter. Flanking this region, on either side, are two DNA regions derived from the virus that allow homologous recombination with viral DNA. For the other component of the system, instead of intact viral DNA, a linearized DNA is used. The linearized viral DNA lacks a portion of an essential gene, thus eliminating the production of non-recombinant viruses. Recombination with the transfer vector, after cotransfection of insect cells, restores this essential gene. Additional features can be built in to this system, including the incorporation of a marker such as β-galactosidase so that blue recombinant plaques can be readily identified.

Following transfection, viral plaques can be picked and the virus characterized to verify the presence of the cloned gene. The characterized recombinant virus can then be used to infect a large-scale culture of insect cells; high levels of expressed protein are usually obtained before the cells lyse. The levels of production are generally lower than those obtainable with *Pichia* expression systems, but insect cells are claimed to provide post-translational modification that is more closely similar to that in mammalian cells. On the other hand, the yeast cells are much easier and cheaper to grow.

11.3.3 Expression in mammalian cells

There is an enormous variety of expression vectors and systems for use with mammalian cells, and a full treatment of this field is beyond the scope of this book. Although the details are more complex than the systems described so far, the general principles remain familiar.

Many vectors use the enhancer/promoters from the human cytomegalovirus (CMV), the SV40 virus, or the herpes simplex virus thymidine kinase (HS-TK) to drive transcription. These promoters give high-level, constitutive expression. As in prokaryotic systems, it is desirable to control the onset of expression. One way of achieving this is by interposing the operator sequence (*tetO*) from the bacterial tetracycline resistance operon between the promoter and the cloned gene. If the mammalian cells are cotransfected with a second plasmid containing the tetracycline repressor gene (*tetR*), also expressed using the CMV promoter, the TetR protein will bind to the *tetO* site, thus preventing transcription. When tetracycline is added to the culture medium it will bind to the TetR protein, altering its conformation and

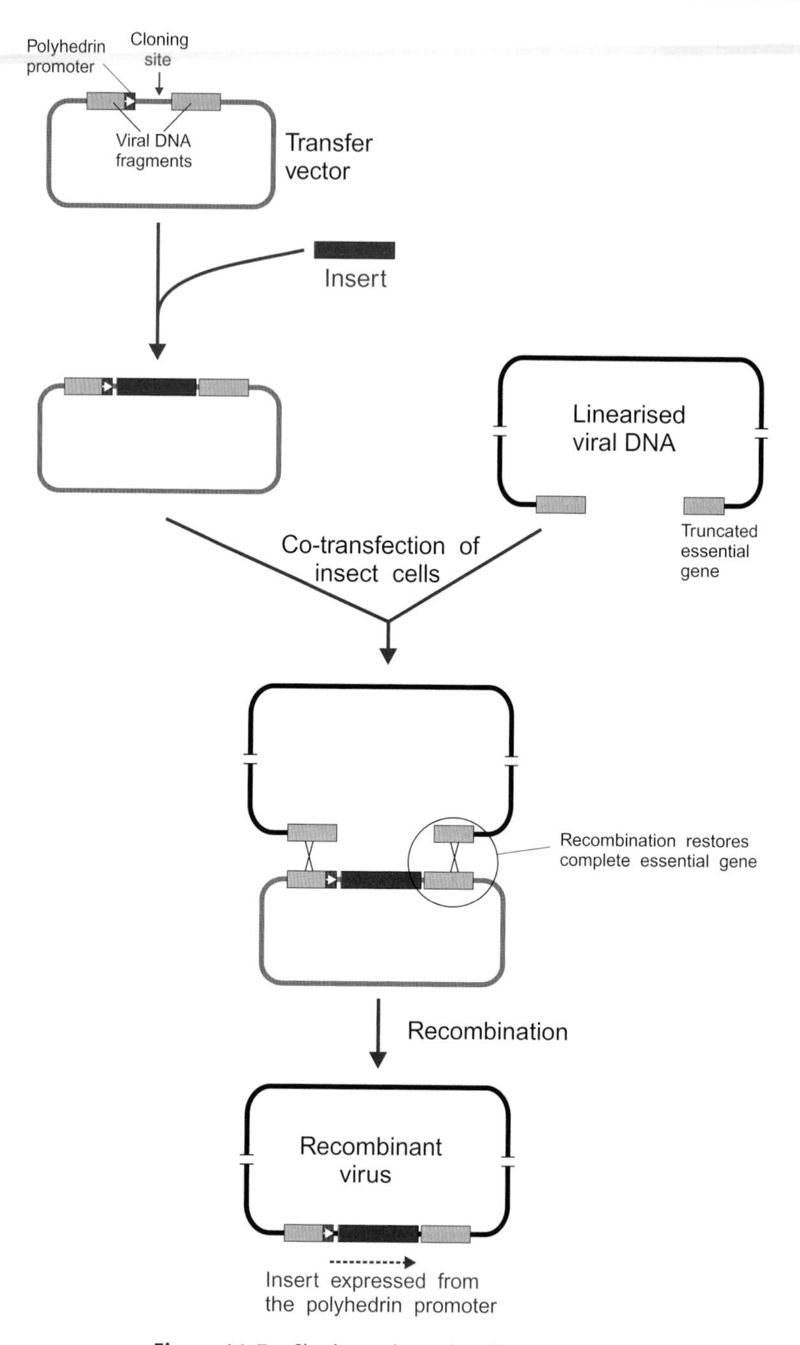

Figure 11.7 Cloning using a baculovirus vector

releasing it from the DNA, thus derepressing transcription of the cloned gene. The advantage of this is that, because this system is of prokaryotic origin, the ligand is not produced naturally in the host, and the activation it produces does not affect the induction of native mammalian genes. Similarly, an insect ecdysone-responsive element is sometimes used to induce gene expression in mammalian cells. Like the tetracycline system, ecdysone does not have any effect on native mammalian genes, and the induction is therefore specific to the genes placed under its control.

Additional sequences can be added to enable targeting of the product to specific cellular locations such as the nucleus, mitochondria, endoplasmic reticulum or cytoplasm, or secretion into the culture medium.

In contrast to bacterial cells, the introduction of DNA into mammalian cells does not depend on the independent replication of the vector; the introduced DNA can be stably integrated into the nuclear DNA. However, some expression vectors can be stably maintained at high copy extrachromosomally, such as those containing the origin of replication from the Epstein–Barr virus (EBV); with a suitable promoter system these are capable of allowing high levels of protein expression. The use of retroviral vectors for integrating foreign genes in mammalian genomes, and for obtaining gene expression, was considered in Chapter 5; for the application of these techniques in whole animals, see Chapter 15.

The advantage of using mammalian cells for expression of eukaryotic genes, especially those from mammalian sources, lies in the greater likelihood of a functional product being obtained. This is especially relevant for studies of structure–function relationships and the physiological effect of the protein on cell function. However the relative difficulty, and cost, of scaling up production, compared with either *Pichia* or baculovirus systems, makes mammalian cells less attractive if the objective is the large-scale production of recombinant proteins for other uses.

11.4 Adding tags and signals

11.4.1 Tagged proteins

The fact that a translational fusion vector (see above) adds a short stretch of amino acids to the N-terminal end of your polypeptide product can be turned to advantage. Purification of the recombinant product by conventional means can often be tedious and inefficient, and would need to be devised and optimized again for each new product. However, if the vector contains not just a start codon but also a short sequence coding for a few amino acids (known as a *tag*), the resulting fusion protein will carry this tag at the N-terminus. For example, the tag may constitute a recognition site (an *epitope*) that can be recognized by a monoclonal antibody (see Chapter 7, Box 7.2). You can

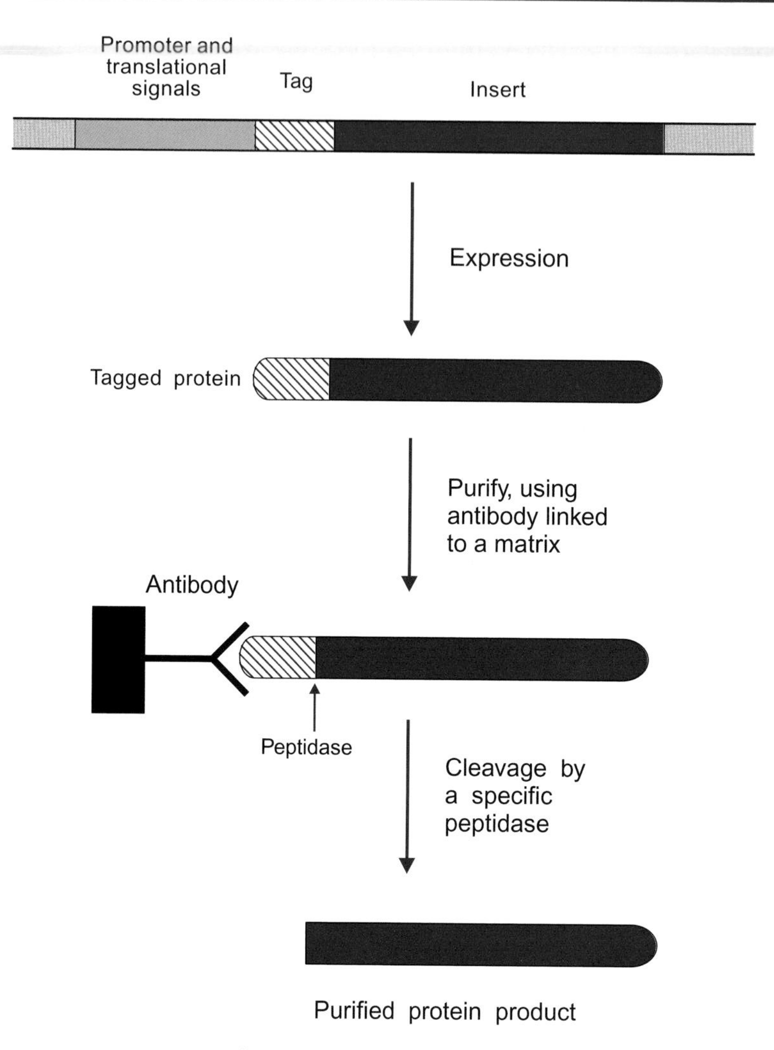

Figure 11.8 Tagged proteins

then recover the protein from the cell extract in a single step by affinity purification, using the ability of the monoclonal antibody to bind to the epitope tag (Figure 11.8). A further application of this type of vector is that you can use this specific monoclonal antibody to detect the expression of your product, if you have no other reasonably convenient assay. There are a variety of such tags (and corresponding monoclonal antibodies) available, and they can be added to either end of your insert (N- or C-termini), using modified PCR primers, rather than by incorporation into the vector.

An alternative to epitope tagging is provided by designing the expression vector (or manipulating the insert) so that a sequence coding for a number of histidine residues is added to one end of the sequence to be expressed.

This will produce a His-tagged protein which can be purified using a Ni^{2+} resin. The affinity of the histidine residues for nickel will result in the tagged protein being retained by the resin, while other proteins will be washed through. The His-tag can also be used as an epitope tag, since there are antibodies available that will specifically recognize this sequence of histidine residues, which facilitates detection of the tagged product.

Of course you may not want to have a tag permanently attached to your product. For this reason, vectors can be designed to contain not only the tag but also a site at which the product can be cleaved by highly specific peptidases. If the peptidase is sufficiently specific it will cleave the product only at this site and not elsewhere. Therefore, you can affinity purify your product, cleave it with the peptidase, and separate the protein from the cleaved tag (see Figure 11.8).

11.4.2 Secretion signals

In a similar way to the incorporation of tags into the product, we can add secretion signals. *E. coli* does not secrete many proteins into the culture supernatant – most of the proteins secreted across the cytoplasmic membrane remain in the periplasm, trapped by the outer membrane. In some other bacteria, particularly Gram-positive bacteria such as *Bacillus subtilis*, many enzymes are secreted into the culture supernatant. This can be advantageous for several reasons. Firstly, if a protein is synthesized at high levels and accumulated within the cytoplasm, it will result in a very high concentration of that protein. This can cause aggregation of the protein into insoluble *inclusion bodies*. These may be damaging to the cell, and are often very difficult to resolubilize. Secretion of the protein into the culture medium will prevent this, since the volume of the supernatant is very much greater than that of the total volume of the cytoplasm of all the cells. Even with a very dense culture, the space occupied by the cells is only a small proportion of the culture volume. Furthermore, although *Bacillus* secretes a number of enzymes, the culture supernatant will contain a much simpler mixture of proteins than the cell cytoplasm. The task of purifying the required product is therefore much simpler.

Secretion of proteins by bacteria depends normally on the presence of a *signal peptide* at the N-terminus. This labels it as a secreted protein, so it is recognized by the secretion machinery and transported across the cytoplasmic membrane. Therefore, if we incorporate a sequence coding for a signal peptide into our vector (or into the insert), the final product may be secreted. It is not inevitable, as it also depends on the overall structure of the protein. If the protein is naturally secreted in its original host, then we may be successful. If it is normally a cytoplasmic protein, the chances of getting it secreted successfully are very much lower.

As described above, in mammalian systems, signals can be added to direct the product to specific cellular locations, or to obtain a secreted product.

11.5 *In vitro* mutagenesis

In the early part of this chapter we described how codon usage may affect optimization of gene expression in *E. coli* – and in other hosts may prevent expression altogether. What can we do about it? If you are dissecting the molecular function of a protein, can you make specific modifications, such as mimicking constitutive phosphorylation by exchanging a specific amino acid residue?

If we can identify a limited number of rarely used codons in our cloned gene, which are likely to cause a reduction in gene expression, then we can use site-directed mutagenesis (see below) to alter the sequence at these specific positions, so as to change a rare codon into a more appropriate one – still maintaining the same coding properties so we are not altering the nature of the final product.

If there are a large number of such rare codons, then changing them one by one would be extremely laborious. Similarly, if we decide that the base composition of the sequence is causing problems, it would be excessively tedious to use site-directed mutagenesis to change the whole thing. In these situations we could use synthetic techniques to remake the complete gene – either by direct DNA synthesis or by PCR-based methods such as assembly PCR. All these techniques are described in the subsequent sections. Their potential applications extend far beyond adjusting codon usage, into a field known as *protein engineering*.

11.5.1 Site-directed mutagenesis

This refers to the specific alteration of either a single base or of a short sequence of bases in a cloned gene. If there is a unique restriction site within your cloned DNA fragment, at or near the point where you want to make a change, then there are several simple possibilities, including the use of exonucleases to make a short deletion, and/or the insertion of a short synthetic oligonucleotide (Figure 11.9). Although these methods are limited in their scope, they can be useful, for example in the removal or addition of restriction sites within a DNA fragment. In addition, the deletion procedure will remove various lengths of DNA, which enables you to identify the limits of regions that have specific functions, such as the ability to bind regulatory proteins (see Chapter 10). Those deletions which extend into the regulatory region will produce clones showing altered regulation of the gene concerned.

The fundamental limitation is the requirement for a unique restriction site at the appropriate place. More generally applicable techniques involve the

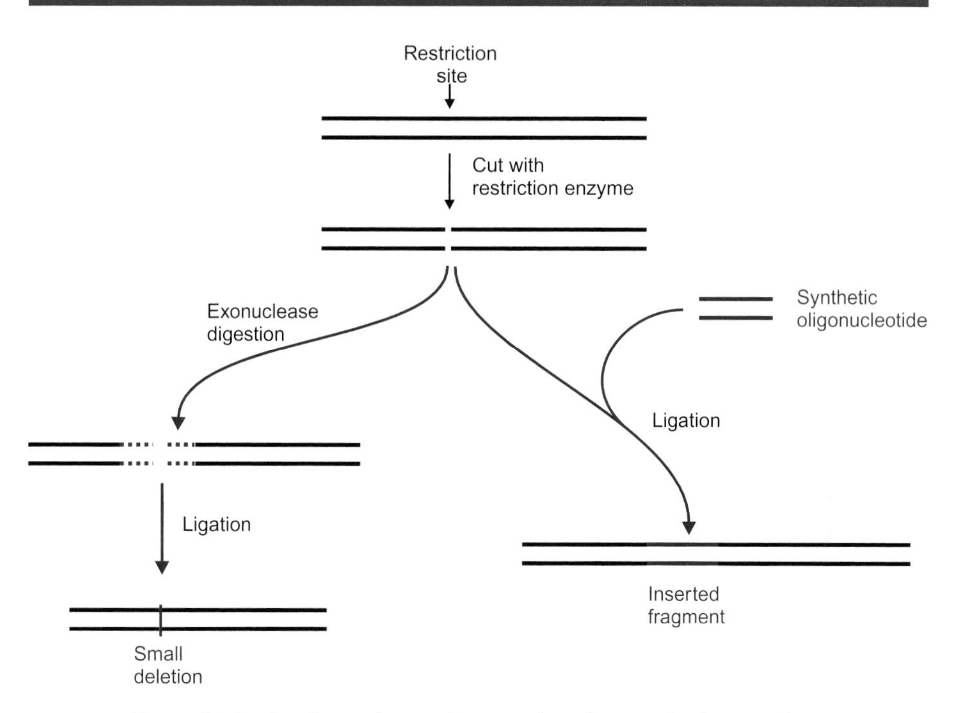

Figure 11.9 *In vitro* mutagenesis: examples of some simple procedures

use of oligonucleotide primers to introduce changes in the DNA sequence. There are a number of variations in the technique, so we will just describe the principle of the method.

Fundamentally, it relies on the ability of single-stranded DNA molecules to anneal to one another even if their sequences do not match perfectly. Therefore, if we design an oligonucleotide that contains the sequence we require, which is only slightly different from the natural sequence, it will anneal to the complementary strand at the corresponding position. This provides a primer that can be used by DNA polymerase. If we start with a single-stranded circular DNA molecule (such as can be readily obtained using an M13 vector, see Chapter 5), then DNA polymerase will synthesize the complementary strand right round the circle, producing a double-stranded molecule (Figure 11.10). DNA ligase can then seal the nick to produce a completely closed, double-stranded, circular molecule in which one strand contains the natural sequence and the second strand carries the required mutation (a heteroduplex DNA). When you transform *E. coli* with this, subsequent replication will produce some double-stranded circular DNA molecules with the wild-type sequence, and some with the mutation. If this is done with an M13 vector, the result will be a mixture of mutant and non-mutant phages.

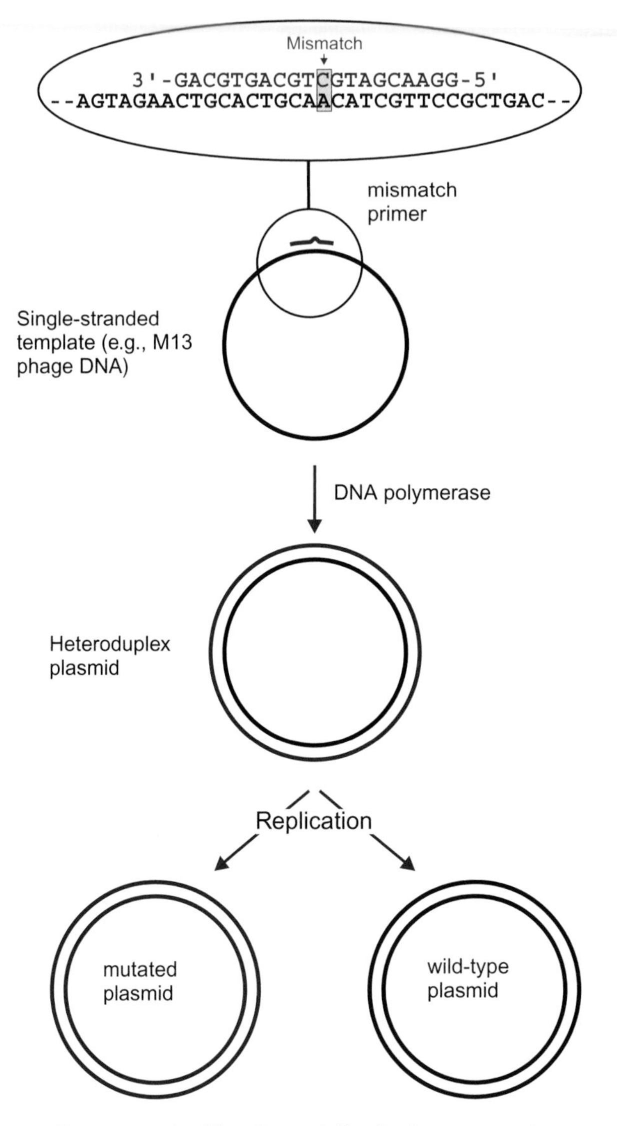

Figure 11.10 Site-directed (*in vitro*) mutagenesis

This is not a very efficient system. Even if everything goes perfectly, the best you can expect is that 50 per cent of the clones will carry the mutation. In practice the proportion of mutant clones is usually considerably lower than this, for a variety of reasons. Predominantly, we have to take account of the mismatch repair system in the host cells. The repair of mismatched DNA occurs preferentially on the new (*in vitro* generated) strand, which is unmethylated, rather than on the old, *in vivo* generated strand, which is methylated. There are ways around some of these problems, including the use of host

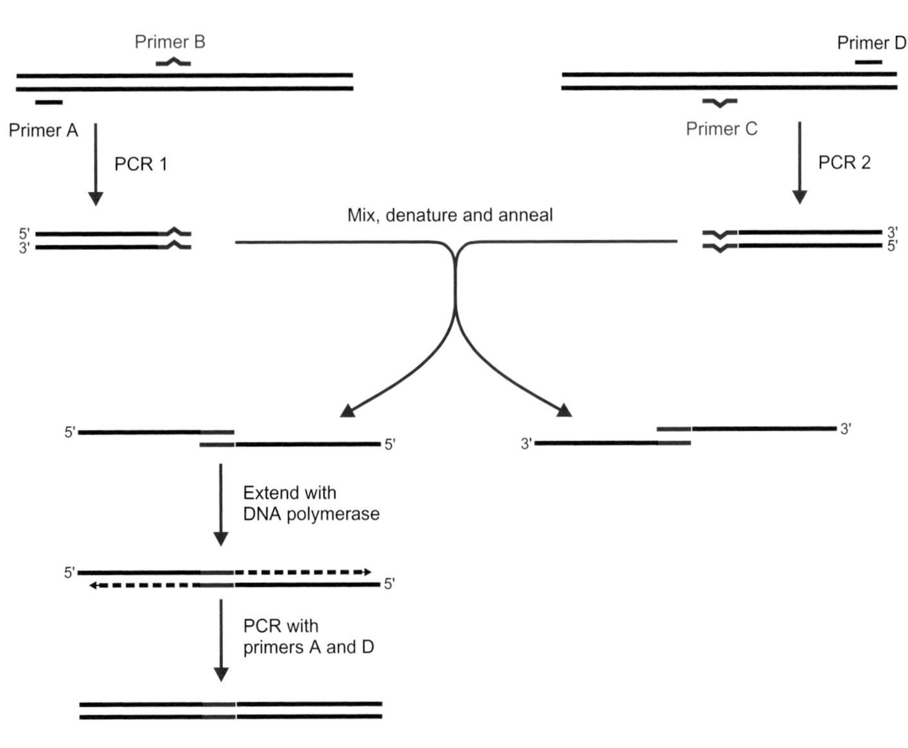

Figure 11.11 PCR-based mutagenesis

strains that are defective in mismatch repair, and other procedures that selectively eliminate the parental DNA strand.

There are also a variety of PCR-based methods, which are more efficient. A simple PCR reaction can be used if the site of the desired mutation is near enough to the end of the gene so that a primer containing the mismatch can be used. Usually, however, the site for the mutation will be too far from the end for this simple approach to be useful. In that case, you have to use a more complex method. The simplest, conceptually if not practically, is to start with two PCR reactions that produce products which overlap in the region of the mutation; in Figure 11.11, primer B, used in PCR reaction 1, is complementary to primer C used in the second PCR. Both primers contain the required alteration in the sequence (but on opposite strands). If you mix the products of the two reactions, denature and re-anneal, some of the single strands from reaction 1 will anneal to strands from reaction 2 in the region where they overlap, corresponding to the sequences of primers B and C. (Of course, there will also be re-annealing of the products of PCR 1, and similarly of the products of PCR 2; these are not shown in the figure, and will not be amplified by the subsequent PCR.) One of the two possible hybrid molecules contains 3′ ends which can act as a primer for extension by DNA polymerase to produce a complete double-stranded molecule containing the mutation (on both strands). The other type of hybrid DNA cannot be so extended, as the overlap region has 5′ ends. A subsequent PCR, using the outermost primers (A and D) will amplify only the full-length product as shown. This can then be cloned in order to test the consequences of the mutation.

The early applications of PCR for site-directed mutagenesis suffered from the risk of introducing unwanted mutations into the product, due to the absence of proof-reading ability of *Taq* polymerase. The introduction of alternative heat-stable polymerases which do have proof-reading ability, and hence show greater fidelity, has greatly enhanced the application of PCR for this purpose. Nevertheless, it is important to sequence the product to ensure firstly that the required mutation has indeed been achieved, and secondly that no other alterations have been introduced into the sequence.

11.5.2 Synthetic genes

As indicated previously, although it is possible to use site-directed mutagenesis to introduce a number of changes into the sequence of a gene, it becomes a rather laborious procedure if we want to make extensive changes – such as altering the codon usage throughout. One possibility then is to abandon the natural gene and simply synthesize a fresh one from scratch. Taking the amino acid sequence of the protein, we can use a computer to *back translate* it into a nucleic acid sequence with optimized codon usage (and/or optimized base composition). It is actually rather more complicated than that, as you would want to check that you are not introducing other undesirable features such as secondary structures that might interfere with transcription; you also would want to consider the introduction (or omission) of restriction endonuclease sites at strategic points.

In principle, you would program a DNA synthesizer to make the DNA sequence which could then be cloned. Here again, it is not quite that simple. DNA synthesizers work by adding bases sequentially, one at a time, to the growing oligonucleotide through a series of chemical reactions. With each base added, there is a very small but finite possibility of the failure of the reaction, the main consequence of which is a gradual reduction in the yield of the product. Synthesis of oligonucleotides of 30–50 bases long is very reliable, and longer sequences can be made, but the practical limit is usually not much more than 100 bases. One way round this is to make the gene in shorter fragments and ligate them, bearing in mind that the synthesizer produces single-stranded DNA, so you need to make two complementary strands for each fragment. This would clearly involve a lot of work, but if you really need to do it, it is quite possible.

11.5.3 Assembly PCR

An alternative procedure, which is a lot simpler in practice than that described above, is simply to mix together a complete set of synthetic fragments (each of which overlaps, and is complementary to, its neighbours on either side). As before, these fragments are designed to optimize codon usage and base composition. This complex mixture is then subjected to a number of rounds of PCR, which results in, amongst other products, some full-length DNA. This is further amplified, specifically, by addition of primers directed at the ends of the full-length gene, and further rounds of PCR. Although the synthesis of a complete gene of say 1000 base pairs involves a highly complex mixture of starting oligonucleotides, which might be expected to produce an impossible number of different products, the procedure really does work.

11.5.4 Protein engineering

The above discussion started off by considering site-directed mutagenesis and synthetic DNA as a means of altering the base composition and/or codon usage of a naturally occurring gene, as an aid to expressing that gene. The product is still the same as the naturally occurring one, but there is no reason why we should limit these techniques to the production of naturally occurring proteins. We can just as easily synthesize a gene which codes for an altered protein. We could introduce specific changes into the sequence and test their effects on for example the substrate specificity of the enzyme, or we could introduce cysteine residues into the sequence at strategic points, so that the protein produced would contain additional disulfide bridges. This would be expected to increase the thermal stability of the enzyme, which could be advantageous. (Unfortunately such a change often results in the loss of enzyme activity as well.)

This concept, often referred to as *protein engineering*, would in principle culminate in the production of totally novel enzymes, tailor-made to carry out specific enzyme reactions. The limiting factor now is not the techniques for producing such engineered proteins, but the inadequacy of our knowledge of how a specific sequence of amino acids will fold into a three-dimensional structure, and how we can predict the enzymic activity of such a novel protein.

11.6 Vaccines

One of the best examples of the use of genetic manipulation for product formation is the development of novel vaccines, as an alternative to the conventional use of killed or live (attenuated) pathogens, or toxoids (toxins treated to render them harmless). We will consider two approaches here: the production of an individual antigen from the pathogen (subunit vaccine), and the use of DNA itself as a vaccine. The production of live, genetically modified, vaccines will be dealt with in Chapter 15.

11.6.1 Subunit vaccines

The concept behind the use of genetic manipulation to produce a recombinant subunit vaccine is simple. If you can identify the key (protein) antigen that is needed for a protective effect, you can clone the gene responsible using an expression vector, and produce large quantities of the protein. This has many advantages over producing the same component from the pathogen itself: it is cheaper and more effective, since you may get much higher levels of expression; it is easier to produce a safe vaccine, as there is no problem with contamination with any other damaging components of the vaccine; and the production process is safer, since you do not have to grow large-scale cultures of a dangerous pathogen. The best example, illustrating the power of this approach (as well as some of the problems), is the development of a vaccine against the hepatitis B virus. Unfortunately, this virus cannot be grown in cell culture, which makes the development of a conventional vaccine next to impossible. However, we know that one virus protein, the surface antigen HbSAg (see Figure 11.12), is able to confer protection, so the expression of this gene in a suitable host should yield a product that can be used as a vaccine. However, when this gene was expressed in *E. coli*, it was found that the product did not give effective protection. The problem here is that the product from *E. coli*, although having the authentic amino acid sequence, does not adopt the correct conformation. However, it was found that the expression of this gene in the yeast *Saccharomyces cerevisiae* did give rise to HbSAg with a natural conformation, and this product was immunogenic and protective – and this is the basis of the currently used hepatitis B vaccine.

We can also use genetic manipulation as an alternative to chemical treatment of toxins as a way of producing toxoids for immunization. For example,

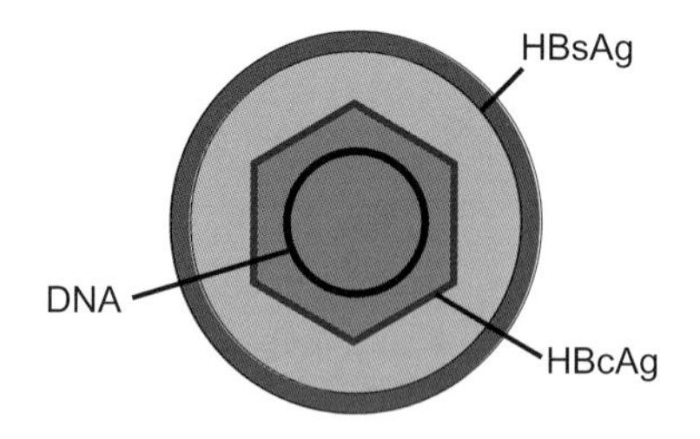

Figure 11.12 Hepatitis B virus: diagrammatic structure. HBsAg, surface antigen; HBcAg, core antigen

the cholera toxin consists of two types of subunit: the A component is responsible for the toxic effects after it gets into the target cells, while the B component is needed for attachment to cell surface receptors. The B component by itself is non-toxic. Expression of the B component alone will give a product that can cause an immune response without the toxicity of the intact toxin.

11.6.2 DNA vaccines

The most radical of the new approaches stems from the surprising observation that injection of DNA, usually as a recombinant plasmid containing the relevant genes, may stimulate an immune response to the product of those genes, in some cases leading to protective immunity. This presumably occurs because the plasmid is taken up by some of the cells in the body, resulting in expression of the genes carried by the plasmid. One advantage of this approach is the ease of administering the vaccine. The plasmid DNA can be coated onto tiny inert particles which are projected at high velocity into the skin, penetrating a very short distance below the surface of the skin (see also the discussion of *biolistics* in Chapter 15). The particles are small enough, and the velocity is high enough, for the particles to penetrate into cells within the skin. The plasmid is constructed so that the genes to be expressed are located downstream from a strong constitutive promoter (such as the CMV promoter, see Chapter 5). Although the plasmid is not replicated within these cells, there is sufficient transient expression of the cloned genes to generate an immune response. DNA vaccines represent an exciting new development in vaccine technology. However, we have to recognize that there are many questions to be answered before it is accepted for widespread use in humans.

At the same time, this technique has achieved extensive use as a means of raising antibodies to recombinant proteins in experimental animals. Instead of the time-consuming procedure of purifying the protein and immunizing animals in the conventional way (see Box 7.2), a recombinant plasmid, carrying the cloned gene together with appropriate expression signals can be used directly as an immunogen.

12 Genomic Analysis

Up to this point, we have been looking at individual genes (or small groups of genes) – how to clone and characterize them, and how to study their specific expression. In keeping with the title of the book, our theme now moves to methods that are applicable to the study of the whole genome. This includes not only the more recent developments in genome sequencing and the genome-wide study of gene expression, but also longer established techniques that are invaluable for connecting the data obtained in this way to the properties of the organism as a whole.

12.1 Genome sequencing

12.1.1 Overview

In 1975, the first complete DNA genome was sequenced – that of the small bacteriophage ØX174 (5 kb). Other complete sequences followed gradually, including the SV40 virus (5 kb, 1977), human mitochondrial DNA (16 kb, 1981), bacteriophage lambda (49 kb, 1982) and the Epstein–Barr virus, containing 170 kb of DNA (1984). A new dimension was opened up in 1995 by the first sequence of the genomes of independently living organisms, those of the bacteria *Haemophilus influenzae* and *Mycoplasma genitalium*. Although these are fairly small genomes (1.83 and 0.58 Mb, respectively), they are much larger than the previous examples. Only six years later, in February 2001, two competing entities, the publicly funded Human Genome Project and the private company Celera, published the human genome sequence of no less than 3 billion base pairs. The human genome sequencing projects provided a major driving force behind the development of the technology and of the capacity to undertake large-scale genome sequencing

From Genes to Genomes, Second Edition, Jeremy W. Dale and Malcolm von Schantz

projects. This has been put to good use. Sequencing bacterial genomes is now a relatively routine operation – to the extent that for many important organisms the complete genome sequence of a number of strains is available. Many genome sequences are also available for higher organisms – mammals, plants, insects and many others. These genome sequences, freely available, provide an invaluable tool for fundamental studies of these organisms, including their phylogenetic relationships, as well as for applied studies such as the nature of inherited disease and drug development. From a scientific point of view, these projects are obviously revolutionary.

Some examples are shown in Box 12.1. At the time of writing, there were 470 published genome sequences (29 Archaea; 397 Bacteria; 44 Eukaryotes). Two features are worth a comment. Some of the bacterial and archaeal genomes are very small – in extreme cases no bigger than a moderately large virus. Many of these organisms with extremely small genomes are endosymbionts, which rely on the host cell to carry out functions that would be necessary for an independent organism – much as a virus does. Secondly, for prokaryotic genomes there is consistently approximately one predicted ORF per kilobase of DNA. This relationship does not hold for the eukaryotic genomes, where there is much more non-coding DNA (including introns).

In addition to the finished genomes, there is a very large number of ongoing sequencing projects. Some of the genome projects have a policy of releasing all sequence data as it becomes available, on a daily basis. This is raw sequence with no annotation (i.e. nothing to tell you what it codes for, or any other information), but it can be searched for the occurrence of specific sequences, which is very useful to a research worker.

12.1.2 Genome sequencing techniques

The advances in genome sequencing have been achieved using essentially the same underlying techniques (Sanger, or dideoxy, sequencing) as described in Chapter 9, apart from the scale of the process, which demands a high degree of automation of the processes of sample preparation, setting up the sequencing reactions, and loading the products into the sequencer, coupled with extensive computer power.

An alternative technology that is beginning to be used for genome sequencing is known as *pyrosequencing*. The principle of this, as illustrated in Figure 12.1, relies on the release of pyrophosphate when a nucleotide is incorporated into a growing DNA strand. The pyrophosphate released can be detected by a coupled enzymic reaction in which ATP sulfurylase converts adenosine 5′ phosphosulfate (APS) to ATP, which enables the second enzyme, luciferase, to react with its substrate (luciferin), generating visible light, which is measured with a detector. Pyrophosphate will only be generated if the DNA synthesis reaction is supplied with the correct dNTP.

Box 12.1 A selection of sequenced genomes

Organism	Size (kb)	ORFs (predicted)
Archaea		
Nanoarchaeum equitans	500	536
Sulfolobus acidocaldarius	2225	2292
Bacteria		
Carsonella ruddii	160	182
Buchnera aphidicola	420	357
Mycoplasma pneumoniae	816	677
Tropheryma whipplei	925	817
Chlamydia pneumoniae	1229	1052
Borrelia burgdorferi	1230	1256
Helicobacter pylori	1605	1544
Coxiella burnetii	2100	2095
Neisseria meningitidis	2272	2158
Staphylococcus aureus	2813	2594
Legionella pneumophila	3344	2932
Mycobacterium tuberculosis	4411	4402
Escherichia coli	4639	4289
Pseudomonas aeruginosa	6264	5570
Streptomyces coelicolor	8667	7825
Fungi		
Saccharomyces cerevisiae	12069	6294
Candida glabrata	12280	5272
Neurospora crassa	43000	10082
Protozoa		
Plasmodium falciparum	22900	5268
Trypanosoma brucei	26000	9068
Leishmania major	32800	8272
Arthropods		
Drosophila melanogaster (fruit fly)	137000	14100
Anopheles gambiae (mosquito)	278000	14000
Plants		
Arabidopsis thaliana	115428	25498
Oryza sativa (rice)	420000	50000
Mammals		
Mus musculus(mouse)	6013257	24174
Rattus norvegicus (rat)	6823166	21166
Homo sapiens (human)	7325631	26966

The table lists only a small selection of finished genome sequences. Data are from www.genomesonline.org, from where fuller up-to-date information can be obtained.

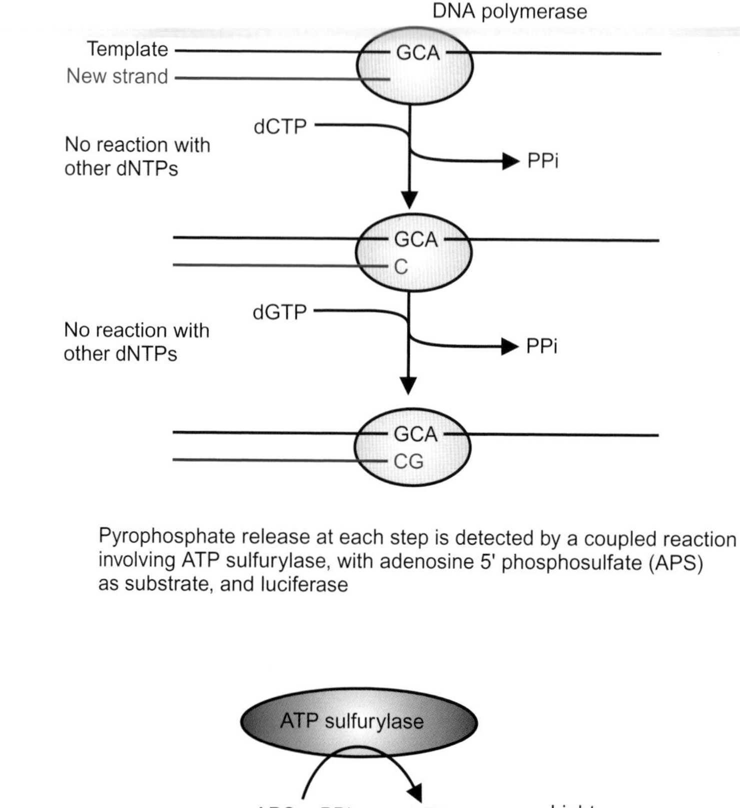

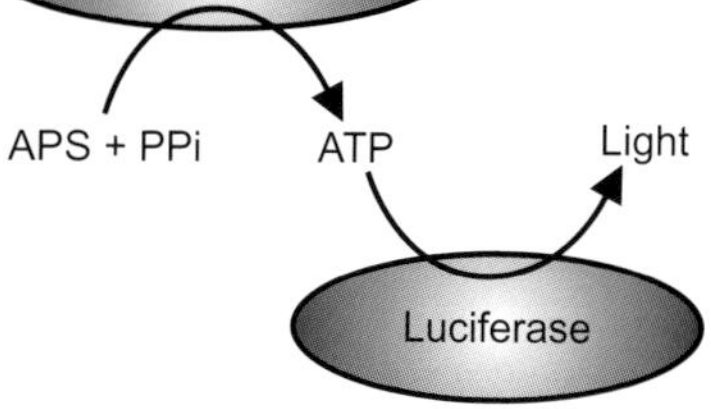

Figure 12.1 Pyrosequencing

Therefore, in the example shown, there will be a reaction in the first step if dCTP is supplied, but not if dGTP, dATP or dTTP is added. At the second step, light is only emitted if dGTP is added – so the sequence of the new strand, in the 5′ to 3′ direction, is CG.

The power of this technique comes from the ability to automate the sequencing of a very large number of DNA fragments simultaneously (*parallel sequencing*), as shown in Figure 12.2. The genomic DNA is sheared into small fragments, and an adapter is ligated to the fragments. Each fragment is then attached to a small bead, and the beads are captured in an emulsion of PCR mix and oil. In this emulsion, each bead (with its attached DNA fragment) is contained in a droplet of water, with all the reagents necessary for a PCR reaction. This enables each DNA fragment to be amplified, within

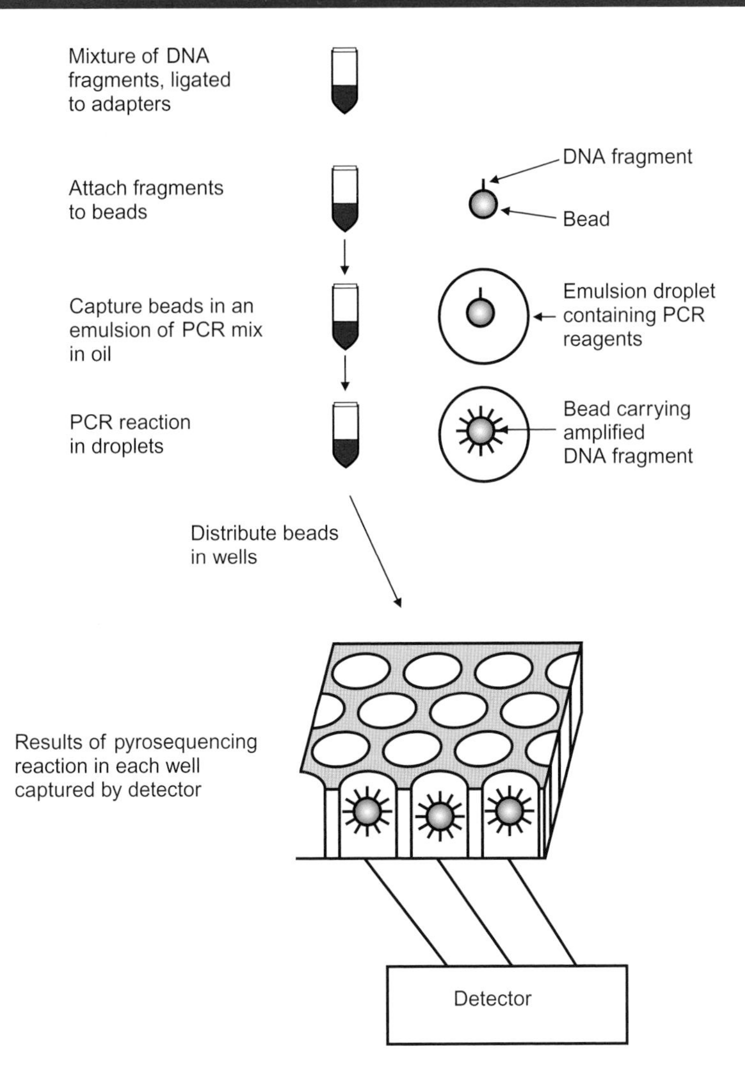

Figure 12.2 Automated pyrosequencing

the droplet, without having to purify individual fragments, resulting in a collection of beads each of which now carries millions of copies of the original fragment.

This mixture is applied to the reaction slide which contains tiny reaction wells, less than 100 pl in volume (1 pl is 10^{-9} ml), so that over 1 million wells are contained on a $60 \times 60\,\text{mm}^2$ slide. The size of the beads ensures that each well will contain only one bead, and therefore only one specific DNA fragment. The wells are also loaded with (smaller) beads carrying the pyrosequencing enzymes. The base of the slide is in optical contact with a fibre-optic bundle linked to a detector that enables the measurement of light emitted from each individual well. The reaction slide is then washed in turn with each dNTP, so that the detector senses which well reacts with each dNTP, and hence determines the sequence of the growing DNA strand.

The length of the sequence that can be read from each fragment is smaller (<100 bases) than that for the Sanger method. However, this is compensated for by the ability to sequence very large numbers of fragments simultaneously – so that the sequence of over 20 million bases can be obtained in one 4 h run, or a few days from start to finish. This is sufficient for a first draft sequence of organisms with small genomes, such as bacteria, although the small size of the sequenced fragments makes it difficult to carry out the final joining of the contigs.

Although this method has tremendous potential, it has not yet been widely used for genome sequencing, and therefore the subsequent discussion in this chapter will assume we are referring to the more established Sanger sequencing method.

As the sequence of individual fragments is obtained, they are assembled into *contigs* (see Chapter 9), which are then sequentially joined together until a stage is reached when a reasonably confident decision can be made as to where all the bits go. It can then be labelled as a *draft sequence*. Note that at this stage there are likely to be gaps in the sequence, although informed predictions can be made as to the position and size of those gaps. There will also be some uncertainties and errors in the sequence. The next stage is the finishing process, which involves filling in the gaps by more targeted methods and correcting the more obvious errors and uncertainties. A *finished* sequence will contain no known gaps, and will be accurate to a defined level. The final stage is *annotation*, when protein-coding sequences are identified, predictions made as to the nature of the products, and a wide variety of other features are identified. Aspects of the strategies available are discussed further in the next section, and annotation and further analyses are considered later in this chapter.

This description of a genome sequencing project makes it sound like a production line, and indeed the analogy is not inappropriate. In a large project, the various stages – cloning, sequencing, assembly, finishing, annotation – are carried out by separate teams, and the product from one team (clones or sequence data) is passed on to the next stage. However, you should not take the analogy too literally – information flows both ways, and one stage does not have to be completed before the next stage starts, e.g. assembly takes place concurrently with the generation of sequence data.

At the end of the process we have a finished (and annotated) sequence, and we would say that we have determined the complete genome sequence of the target organism We need to consider what we mean when we say this. For the bacterial genomes that have been sequenced, this is literally true – we know (within the chosen limits of accuracy) which base is present at every single position in the genome. However, for eukaryotic genomes, the definition of 'finished' is set lower, for example to mean more than 95 per cent of the euchromatic regions (which contain most of the genes) sequenced, to the set standard of accuracy, and with no gap more than 150 kb (this definition will vary from one project to another). This is probably not a serious limitation. For example, many of the gaps are due to the occurrence of repetitive DNA, which can be difficult to sequence (and probably does not encode any genes). Most, if not all, of the significant coding sequences will be present in the sequence that has been determined – and therefore for practical purposes we 'know' the sequence of the human genome.

12.1.3 Strategies

Various strategies, or combinations of strategies, can be used for determining sequences of DNA as long as a complete genome (or a complete chromosome). One is to extend the concept of shotgun sequencing as described in Chapter 9, that is simply to make a library of small random fragments of the whole genome and sequence them (Figure 12.3). This requires the isolation and sequencing of a vast number of clones, which is made possible by the use of robotic methods for all stages of the process – picking the clones, extracting the DNA and setting up the sequencing reactions. Furthermore, the contig assembly of such a large number of individual random sequences requires a very substantial amount of computer power. However, the vastly increased computer power that is now readily available, and the sophistica-

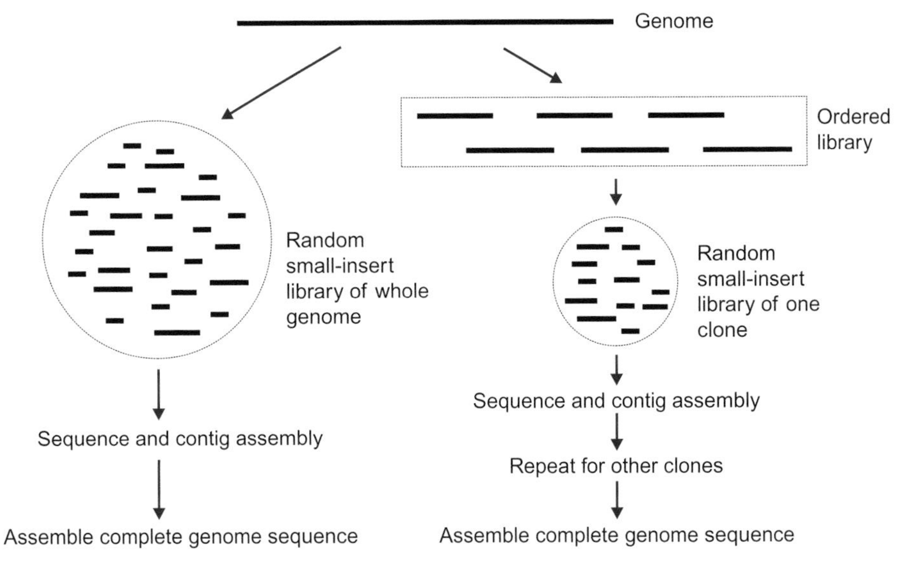

Figure 12.3 Genome sequencing strategies

tion of the robotic methods required, have made this the backbone of genome sequencing strategies. However, there are problems with it, especially those associated with repetitive sequences, and sequences that are difficult to clone and are therefore under-represented in the shotgun library. These problems are considered in the next section.

The main alternative is to split the problem up into defined units first. For a eukaryotic organism, you might want to separate the chromosomes to start with, and treat each chromosome as a separate project. You can make a library representing the whole genome or chromosome, using a vector that can carry appropriate sizes of insert (e.g. a cosmid). You would then determine the sequence of the insert in individual cosmids, and subsequently assemble the cosmid sequences into a complete genomic sequence. Some of the sequencing projects have started with random cosmids, while others have attempted to arrange the cosmids into an ordered library first (i.e. one in which the relative position of all of the cosmids has been defined – see Chapter 6).

One advantage of this 'clone by clone' strategy is that very many laboratories are capable of determining a sequence of the size of an individual cosmid insert, so it is possible to distribute the cosmids to different laboratories for sequencing. In contrast, the total shotgun approach is highly centralized and only possible in specialized genome sequencing centres (although the number of centres capable of doing this is increasing all the time). A further advantage of the 'clone by clone' approach is that useful information is generated at intermediate stages in the project – the sequence of an individual cosmid is a coherent piece of information – while the shotgun approach consists of large numbers of essentially meaningless sequences of small fragments until sufficiently large contigs have been assembled. On the other hand, the shotgun method does start generating sequence data almost from the beginning, without having to produce cosmid libraries.

It is of course not necessary to be dogmatic about which approach to use. Many genomes have been sequenced by starting with a shotgun approach, and then using individual clones (or PCR products) to deal with difficult regions.

12.1.4 Metagenomics

Since shotgun sequencing starts with a large number of random clones of DNA from the organism being sequenced, it is not a big jump to consider what happens if these random clones come from a *mixture* of bacteria. Why should we want to do this? The natural environment (e.g., soil or water, or even the human intestine), contains a very wide range of organisms. Most of these species have never been grown in the laboratory, and cannot be grown (using current methods). It is therefore not possible to isolate individual organisms and sequence the genome from a pure culture.

The use of molecular techniques to identify these unknown organisms started with PCR amplification of 16sRNA genes. These are sufficiently conserved amongst all prokaryotes that it is possible to design PCR primers that will amplify them, whatever the source. At the same time, there is enough variation to ensure that the sequence information will enable the source to be placed on an evolutionary tree in relation to known organisms.

Although this data gives an insight onto the diversity of organisms in the environment, it does not tell us much else about them. Extending the concept to genome sequencing, using shotgun cloning of DNA from the whole sample, enables the prediction of other characteristics of the organisms. This can in turn lead to the development of ways of growing individual species in the laboratory, and so confirming and extending those predictions.

These techniques are known variously as *metagenomics, community genomics* or *population genomics*. It is not necessary to obtain a complete, finished genome sequence of each organism. The random sequences are informative in relation to the variety of genes that are present, and the products they encode. And as long as you can obtain long enough contigs, sets of genes can be traced back to their common source organism.

12.1.5 Repetitive elements and gaps

Two problems encountered in genome sequencing are worth considering, repetitive elements, and gaps.

Most genomes contain *repetitive elements*: identical sequences which occur more than once (often many times) in the genome. The amount of repetitive DNA varies substantially between organisms, ranging from a few per cent to at least 50 per cent in the human genome. There are various classes of repeat sequences. *Dispersed* (or *interspersed*) *repeats* are mainly mobile elements such as insertion sequences and transposons, which occur at different sites distributed around the genome, while *tandem repeats* of shorter sequences (even as short as one to three nucleotides) may occur many times in succession at one locus. There may also be segmental duplications, where a block of perhaps several hundred kilobases of DNA has been copied to a different region of the genome. Any repetitive element has the potential to cause problems in contig assembly. Two fragments of sequence that end with part of a repetitive element may be identified by the computer as overlapping, when in fact they are derived from quite different parts of the genome (see Figure 12.4). This is more of a problem with the total shotgun approach than when sequencing individual cosmids, because the separate cosmids are much less likely to carry more than one copy of an insertion sequence. Tandem repeats cause a rather different problem: the overlapping fragments are genuinely adjacent, but the number of copies of the tandem repeat may be miscounted.

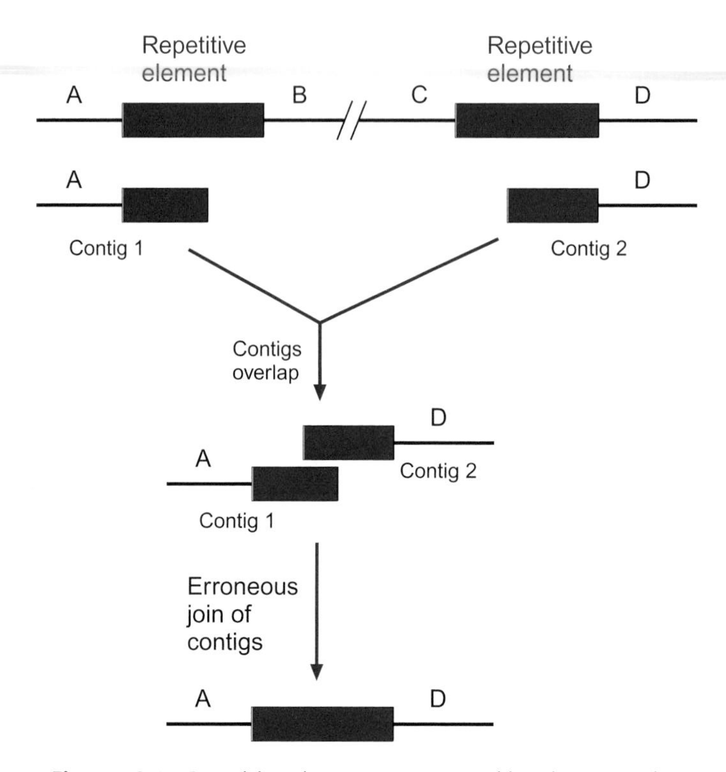

Figure 12.4 Repetitive elements cause a problem in sequencing

An additional factor has also to be borne in mind. Many of these repetitive elements are inherently hypervariable: mobile elements can transpose to different sites, and tandem repeats can vary in their copy number. Care has to be taken that the starting material is homogeneous, and the possibility of alterations during the various cloning steps has to be considered.

In the discussion of smaller-scale sequencing projects (Chapter 9), we considered the problem of *gaps* remaining after contig assembly. These may be caused by difficulties in cloning certain fragments of DNA, or by problems in the sequencing reactions, especially with GC-rich regions which may cause termination of the polymerization reaction. Primer walking, as described in Chapter 9, is less likely to be useful in this context because many of the gaps will be too big. PCR can be used to amplify a fragment to bridge the gap, but only if the sequences either side can be aligned and if the gap is relatively short. If there are a number of gaps in the sequence, it is likely to be impossible to guess the order of the sequenced fragments.

The gap problem can be addressed, at least in part, by the use of different types of clones, and in particular by the use of supervectors (especially bacterial artificial chromosomes, BACs) described in Chapter 5, which are capable of accommodating inserts of hundreds or even thousands of kilobases. These large inserts can bridge substantial gaps in the assembled genome. Hybridization can be used to identify a recombinant carrying sequences overlapping with both sides of the gap.

Gaps may arise from the lethal nature of certain sequences in multi-copy plasmids. The stringently controlled low copy number of vectors such as BACs tends to reduce this problem. If the gap arises from a sequence that is too lethal even for a low copy number plasmid, then lambda vectors can be used, although the insert size is then limited to about 23 kb. Alternatively, if other approaches have eliminated most of the gaps, the remaining gaps can be bridged by using PCR amplification of the relevant region of the genome, thus avoiding the need for any cloning step.

Once the genome sequence has been finished, to meet the criteria of accuracy and completeness laid down in the project specifications, the final, important, step is *annotation* – identifying the coding regions, the predicted products, and other features of the sequence. However, despite all the excitement that is generated by the publication of a genome sequence, knowing the sequence is only a means to an end, not an end in itself. The important factor is what you can do with that vast amount of information.

12.2 Analysis and annotation

Genome sequence data is annotated in a similar way to that described in Chapter 9, but is often, for convenience, included in the databanks in sections which may correspond to actual clones (cosmids for example). The fragments of the genome sequence may also be 'virtual clones' (i.e. arbitrarily divided sections of the genome). Whole chromosomes or whole genomes may also be included as a single sequence, and are useful for some forms of analysis – but downloading a whole chromosome sequence and analysing it may stretch the capability of your computer or your software. However you can usually perform the analysis on line, using the capacity of the servers at the databank centres. See Box 9.1 for a selection of on-line resources.

12.2.1 Identification of ORFs

Initially, the identification of ORFs follows the methods outlined in Chapter 9, that is we find all start and stop codons in each of the six possible reading frames (three in each direction). However, whereas with a specific clone we probably know the size of the protein, and may even have a partial amino acid sequence, in a genome sequencing project we want to identify all the possible ORFs, including ones that bear no relationship to any known protein. Yet we cannot simply label all these as potential protein-coding sequences. We have firstly to make a judgment as to how long an open reading frame must be before being labelled as an ORF. Unless we put a minimum size on it, the concept is meaningless – every reading frame between two stop codons is 'open' in the sense that it has no stop codons. On the other hand, by excluding ORFs below an arbitrary size, we would risk excluding genuine ORFs that do actually code for small polypeptides. For example, if we were to consider as significant only those ORFs of more than 300 bases (coding for 100 amino acids, which would make a polypeptide of about 10 kDa), then we would miss any polypeptides that are less than 10 kDa. However, if we make the size limit too small, we run the opposite risk, of predicting expression of a small protein when no such protein exists.

One extra clue is codon usage. The genetic code has many examples of different codons that are synonymous, i.e. they code for the same amino acid. Generally, these synonymous codons are not all used to the same extent – there is a preferred codon usage, which in some cases can be very marked. Some codons may be used so rarely that we could use their presence in our queried ORF as an indication that it is not really translated into a polypeptide. However, we do have to be careful about arguments like this, especially as there is a danger of circularity. If our knowledge of the codon usage of the target organism is based on the sequence of a limited number of genes, our predicted codon usage may not be accurate. If we then exclude ORFs that do not conform to this codon usage, we will reinforce our incorrect assignment of codon usage.

These factors can be exemplified by the data shown in Figure 12.5. This is actually a portion (about 7 kb) of one of the cosmids used in the sequencing of the *Mycobacterium tuberculosis* genome. It is quite straightforward to locate all the stop codons on all six reading frames (three in each direction). In this case, the sequence (as a simple text file) has been read by a program known as Artemis (freely available from the Sanger Institute) to provide this display; there are many other programs available that will perform such an analysis. In the figure, the top three bands show the position of stop codons in the three reading frames that are read from left to right, while the bottom three bands show the complementary strand (read from right to left, 5′ to 3′).

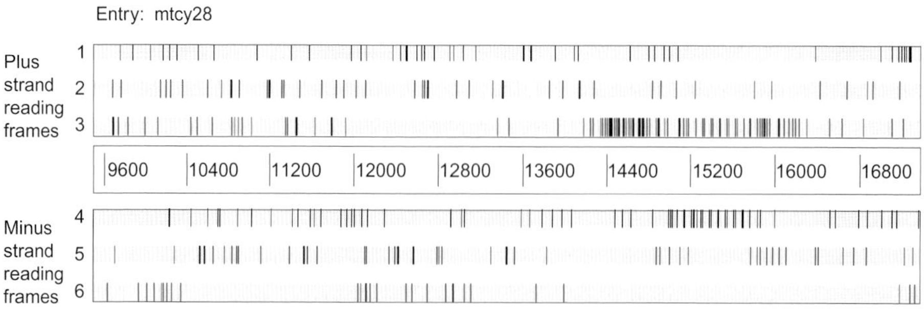

Figure 12.5 Open reading frames: computer mapping of stop codons; edited display from analysis of a DNA sequence using Artemis

You will see that the distribution of stop codons is far from uniform. If we consider that only the largest ORFs are likely to encode 'real' proteins, and therefore confine ourselves to regions of, say over 450 bases (coding for a sequence of 150 amino acids), there are three or more such sequences in each of the six reading frames. In some places, four of the six reading frames have no stop codons within such a distance, and it is highly unlikely that all of these are actually used to produce proteins (in some organisms, especially bacteriophages, overlapping genes do exist, but this rarely happens in larger genomes). So we then take start codons into account, and this reduces the number considerably, but still leaves more potential ORFs (including some overlapping ones) than are actually thought to exist.

With a eukaryotic genome sequence, we also have to identify the introns (if any) and exons before the ORF can be identified. In Chapter 9 we discussed how this can be done for individual clones, by using consensus splice sites to recognize intron/exon boundaries. For a whole genome sequence, this approach is likely to be overwhelmed by the volume of data, making the predictions less reliable. More sophisticated approaches are needed.

Computerized identification of ORFs in a genome sequence uses these (and other) factors through a statistical tool known as a hidden Markov model (HMM). A Markov model, or Markov chain, describes the probability of an event at one position (in this case, the occurrence of A, G, C or T) where that probability is influenced by events at one or more previous positions. For example, if you consider the words in this paragraph, the probability of finding the letter h is much higher if the previous letter is a t. In a zero-order Markov model, the event at one position is independent of previous events, so you have a random DNA sequence, which may be true of non-coding DNA (although even non-coding regions are often not random in this sense). In a second-order model, the probability of a specific base at one position is influenced by the nature of the previous two bases, which is of course relevant to the situation with triplet codons. Technically, an HMM considers a number of chains, only one of which consists of the observed states, the other chains being 'hidden', but the details are beyond the scope of this book. HMMs can be 'trained' to recognize the parameters associated with coding sequences, using DNA regions (from the organism being investigated) in which the genes have already been identified. HMMs also provide a way of overcoming the problem of the prediction of intron/exon boundaries. Note that we have previously encountered another use of HMMs, in the identification of protein domain families (Chapter 9).

Another clue that we can use is a comparison of ORFs with predicted proteins from other organisms. (We do not have to rely on the identified ORFs from other sequences; we can ask the computer to compare the product of our suspected ORF with all the possible products that would arise from translation of other complete sequences in all six reading frames.) If the

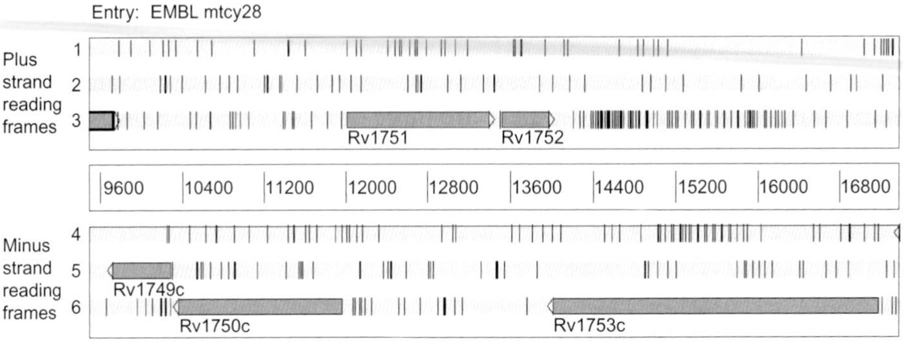

Entry: EMBL mtcy28

Plus strand reading frames 1 2 3

Rv1751 Rv1752

9600 10400 11200 12000 12800 13600 14400 15200 16000 16800

Minus strand reading frames 4 5 6

Rv1749c Rv1750c Rv1753c

CDS	9674	10231	Rv1749c	Possible membrane protein
CDS	10315	11913	Rv1750c	Possible CoA ligase
CDS	11967	13349	Rv1751	Possible hydroxylase
misc_feature	13224	13250		Dihydrofolate reductase signature
RBS	13464	13468		Possible ribosome binding site upstream of Rv1752
CDS	13476	13925	Rv1752	Unknown
CDS	13960	17121	Rv1753C	PPE family of Gly-, Asn-rich proteins

Figure 12.6 Open reading frames: display of coding sequences; edited display from analysis of a DNA sequence and databank annotations using Artemis

predicted polypeptide from our query sequence is similar to polypeptides that might be made by other organisms, this suggests that it is a real coding sequence.

As we saw in Chapter 9, databank entries contain a substantial amount of *annotation*, in addition to the sequence itself. This annotation shows the location of significant features of the sequence, and, in particular, the assignment of ORFs. Artemis is capable of reading the full EMBL entry and displaying these features. In Figure 12.6, we can see the output from Artemis when it is supplied with this extra information, for the same sequence as in Figure 12.5. This shows which of the potential ORFs in this region were considered (by those doing the annotation) to represent 'real' coding sequences, as well as (in some cases) the likely function of the product.

We have to remember that the assignment of ORFs, and the possible function of the corresponding proteins, is only a prediction, and is subject to a degree of uncertainty in the absence of direct evidence as to the existence and properties of the encoded protein. The necessary assumptions about the likely minimum size of a protein means that statistics on the number of predicted proteins have to be treated with care as they may miss an unknown number of small proteins.

It must be remembered that the genome also encodes important RNA transcripts that do not encode proteins. The most well-known of these are ribosomal and transfer RNA (rRNA and tRNA). A more recent discovery is microRNA (miRNA) in eukaryotes, which regulates expression of other genes (see Section 12.5.4 below).

12.2.2 Identification of the function of genes and their products

The methods described above will lead to a provisional assignment of ORFs, or coding sequences, in our genome sequence. The next question is the nature and function of those proteins. When we were dealing with the sequence of a specific clone (Chapter 9), we started with a strong hypothesis as to what the protein actually does.

However, if we are dealing with a whole genome sequence, we would be starting with no preconceived idea as to what each putative ORF actually codes for. In this case, we have to resort to comparing each predicted sequence against all the known protein sequences held in the databanks. The methods for doing this were discussed in Chapter 9; as we saw then, it is possible to carry out such comparisons with DNA sequences as well as with predicted protein sequences, but proteins usually give better results. There are a number of possible outcomes to such a search. If we are lucky, our protein may have a high level of similarity to a well-characterized protein from another source (or even better, to a number of such well-characterized proteins). In this case, we are on reasonably safe ground in attaching that label to our protein as well.

There is a snag. Many proteins in the databanks, especially those predicted from gene sequences, have a functional label attached to them not because of an established function, but because of their similarity to another protein with an established function. Unless we look carefully at the evidence for the identity of these proteins, there is a risk of building up a chain of such similarities, becoming less and less reliable. In other words, gene A in one genome is identified as coding for enzyme X. Then gene B, in a second genome, is similar to gene A in the first genome, so it is also labelled as coding for enzyme X. In a third genome we find a gene that is similar to B, so this is also labelled as coding for enzyme X, and so on. We may end up with a gene that is only remotely similar to the originally identified gene A.

We may be less lucky. Our predicted gene product may not have a high degree of similarity to any known protein, so we cannot be certain as to its function. However, it may show some features which are characteristic of certain classes of protein, which would enable us to provisionally label it as, for example, 'probable membrane protein', or 'possible oxidoreductase'. These could be identified by the methods described in Chapter 9, using ProSite or Pfam for example. Alternatively, it may show no discernible resemblance to any sequence in the database, in which case it has to be labelled as 'unknown function'. This applies to a substantial proportion of the predicted proteins in a typical genome sequence. We may actually find that, although it does have similarity to other proteins (or predicted proteins) in the database, these proteins are already labelled as 'unknown function'; you can see one example (identified as Rv1752) in Figure 12.6.

However good these predictions look, they are still only predictions, which need direct experimental evidence to prove them. This includes not only verification of the biochemical functions of the gene product, but also testing of its role in the physiology of the whole organism – including its potential role in causing a specific disease. The approaches outlined in Chapter 9, for testing a specific cloned gene, are not so easy to apply to a whole genome. Later in this chapter we will consider some ways of surveying the whole genome for specific gene functions.

In the course of these sequence comparisons, we may also come across *pseudogenes*. These are DNA sequences which have significant sequence similarity to 'real' genes, i.e. to DNA sequences that are known or believed to code for proteins, but the pseudogenes contain changes in the sequence that make it unlikely to be functional. The simplest type of change is one that puts a stop codon (or several stop codons) within what should be the coding sequence. Some pseudogenes may be transcribed into mRNA (*transcribed pseudogenes*), but these mRNA molecules cannot be translated into functional proteins.

12.2.3 Other features of nucleic acid sequences

The analysis of our DNA sequence does not end with the identification of protein binding sites. One of the simplest, and yet very informative, analyses is the base composition of the DNA, i.e. the ratio between G + C and A + T bases. Overall, this ratio is characteristic of a particular species, and tends to be similar between different species within a genus, while varying more widely between less related organisms (sometimes as low as 30 per cent G + C or as high as 70 per cent). Base composition can therefore be used as an aid to establishing the taxonomic relationship between different species.

Within bacterial genomes, the base composition tends to be reasonably uniform from one region to another, but there are notable exceptions. Figure 12.7 shows an example from the genome of *E. coli*, where the region indicated

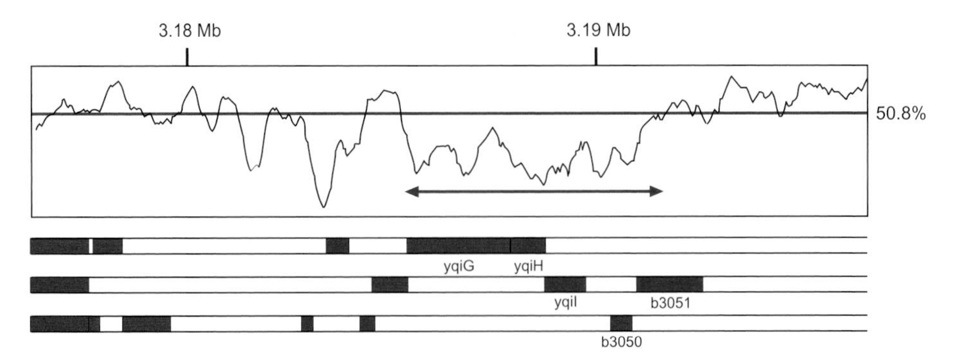

Figure 12.7 Variation in genomic GC content; the figure shows edited Artemis output displaying part of the genomic sequence of *E. coli*

has a G + C content of 40.9 per cent, compared with the average overall composition of 50.8 per cent G + C. Such a region is known as an *island*, and is usually (as it is here) marked by a sudden change in base composition at each end. Many of the examples that have been studied concern groups of genes that are connected with bacterial pathogenicity, and are hence referred to as *pathogenicity islands*, but the phenomenon is not restricted to virulence determinants. What is the significance of these islands?

Processes such as DNA replication and transcription, and particularly the regulation of these processes, are to some extent sensitive to the base composition of the DNA. Therefore over an extended evolutionary period, the enzymes involved in these processes, and the composition of the DNA, have evolved together to produce a well-balanced system. Furthermore, the codon usage of the genes is also related to the base composition – so as the codon usage and the specificity of the available tRNAs co-evolves, this will also be reflected in the base composition of the DNA. The inference from this is that these *islands*, with a different base composition, are relatively recent arrivals. They represent DNA that has been acquired by the bacterium by horizontal gene transfer from a different species. This inference has been substantiated in some cases by direct evidence. For example, the genome of *Vibrio cholerae* contains an island in which the genes for the cholera toxin are found. It has been established that this island is in fact phage DNA that has been integrated into the bacterial chromosome. Many other islands have been shown to be integrated bacteriophages, either by direct evidence or by comparison of the sequence with that of known bacteriophages. In the example shown in Figure 12.7, the genes indicated are predicted to be associated with the production of fimbriae (pili), and may have been introduced by a bacteriophage.

Mammalian DNA also contains ‘islands’ with a different base composition from that of the remainder of the genome, but these include regions with a different significance. The dinucleotide CG (usually written as CpG to emphasize that we are referring to consecutive bases on one strand) occurs much less commonly than would be expected from a random distribution of bases. Yet in some regions, the frequency of this doublet is very much higher, forming regions (CpG-rich islands) up to 2 kb in length, with a much higher G + C content than the whole genome. There are many thousands of such regions in the genome. The frequent association of CpG islands with promoter regions means that the identification of such an island can be taken as suggesting a transcriptional regulatory region.

Amongst the many other features of DNA that are amenable to computer analysis, we can single out the occurrence of inverted repeat sequences. It is important to be clear about the meaning of an ‘inverted repeat’. Because DNA strands have a direction to them, and the two strands are in opposite directions, an inverted repeat of, say, CAT is *not* TAC but ATG (see Figure 12.8). A pair of inverted repeats, in close succession, can anneal together, so

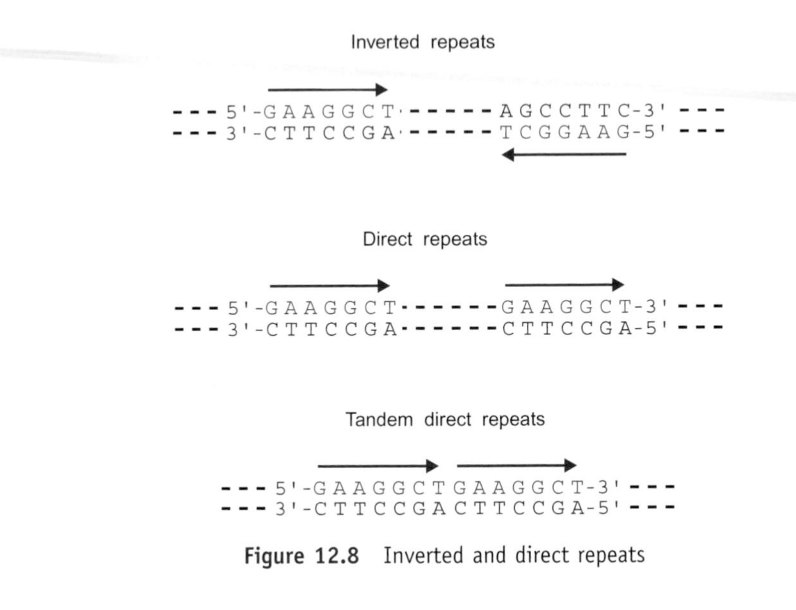

Figure 12.8 Inverted and direct repeats

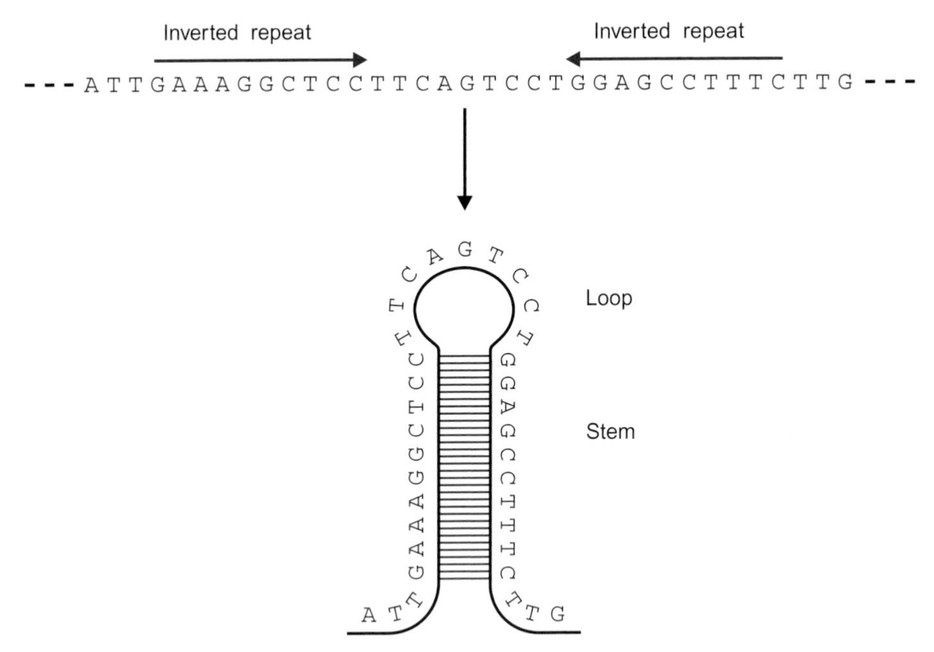

Figure 12.9 Formation of stem–loop structure

that a single strand containing such a sequence will give rise to a *hairpin* structure, or if they are separated by a few bases, a *stem–loop* structure (Figure 12.9). A well-known example of such a structure is the tRNA molecule. Ribosomal RNA also exists in a highly folded conformation, and mRNA transcripts will also form folded structures. Although such structures are most common in single-stranded nucleic acids (mainly RNA), they can also give rise to localized destabilization of double-stranded DNA, most notably when the double helix is unwound during replication. One of the significant aspects of such structures is the role that they play in transcriptional termination; the formation of a stable stem–loop structure in the mRNA favours the dissociation of the nascent mRNA strand from the template DNA, allowing the two DNA strands to reanneal. This causes the RNA polymerase to pause, and ultimately stop transcription.

Regulatory proteins often bind to inverted repeats. Many such proteins are dimeric, so one subunit can bind to one copy of the inverted repeat, and the other subunit to the second copy. The requirement for two binding sites gives a higher degree of specificity to the interaction, as well as establishing a strong interaction.

Repeated sequences are also a frequent cause of variation (see Chapter 13). Inverted repeats are typically found at the ends of mobile genetic elements such as transposons and insertion sequences. In addition, recombination between inverted repeat sequences will lead to inversion of the region between them. In contrast, recombination between two direct repeats will cause deletion of the region between them. These sources of variation are not only an interesting natural phenomenon; they can be a nuisance in gene cloning as the occurrence of repeated sequences within your insert can cause instability of the construct. Tandem direct repeats (that is, direct repeats with no intervening region) also cause variation in other way – by *replication slippage*. When the replication apparatus encounters a tandem direct repeat, it may (very occasionally) jump forwards or backwards, causing a loss or gain of additional copies of the repeated sequence. One example of this is found in Huntington's disease, where a CAG repeat in a gene called huntingtin has a tendency to slip and extend itself. Above a certain length, the encoded poly-glutamate stretch in the resulting protein will cause the disease.

Because the number of copies of a direct repeat at a specific site may therefore vary from one individual to another, it forms the basis of one method of molecular typing (see Chapter 13).

12.3 Comparing genomes

12.3.1 BLAST

In principle, the comparison of two genomes is not fundamentally different from the searches described in Chapter 9, using BLAST. However in this case, the query sequence is not an individual gene or fragment, but a whole genome. The sheer size of the query sequence presents some problems both with the analysis and the presentation of the results. Modified versions of BLAST are therefore used for this purpose. The problems of course become more severe if you want to compare the whole of one genome with the complete database.

12.3.2 Synteny

Approaches such as BLAST will identify regions of two genomes that match one another, within the degree of matching required. However, it does not tell us whether the genes are arranged in the same order. Generally speaking, the sequence of individual genes is more highly conserved than is their arrangement in the genome – which is interesting since it indicates that genes get shuffled around in the course of evolution. The matching of the gene order between organisms is known as *synteny*. A useful tool for displaying the degree of synteny, and the presence of insertions and deletions (*indels*), between two organisms is the Artemis Comparison Tool (ACT), from the Sanger Institute (see Box 9.1). An example of the comparison of part of two genomes (*Shigella sonnei* and *E. coli*) is shown in Figure 12.10. The blue blocks (red in the original) show regions that match, in the same orientation, while the grey blocks indicate regions that are in the opposite orientation. While there is a very large amount of agreement between the two genomes, there is one large block of DNA (and many smaller ones) that is inverted, and a number of smaller regions that are found in different places in the two genomes. In Figure 12.10b, we have zoomed in on a small part of the genome, which shows a small inverted region. Checking the annotations, we find that this contains the insertion sequence IS2. The mobility of insertion sequences can make a substantial contribution to the plasticity of bacterial genomes.

Another useful tool for genome sequence comparison is gMAP (see Box 9.1); an example of the use of this tool is shown in Figure 12.11. In the first step, a comparison of the entire genome sequence, based on pairwise BLAST comparisons, is shown for three strains of the bacterial pathogen *Streptococcus pyogenes*. As in the previous example, it can be seen that the genome sequences are made up of blocks that are broadly similar, but that these blocks vary in order and orientation between the strains. Clicking on region 3 produces a comparison of this part of the genome (Figure 12.11b), which confirms the overall similarity of this region (despite the different position and orientation in the genome), but also identifies a small region that is absent in one of the strains. This is labelled 3 in Figure 12.11b – the regions identified are renumbered each time, so this should not be confused with region 3 in Figure 12.11a. Clicking on this region produces the full comparison shown in Figure 12.11c, with individual genes identified by additional coloured bars. Each of these has a link to the sequence annotation so the genes involved can be identified. Three of them are labelled in the figure – identified as a phage tail protein, a phage repressor, and a phage integrase,

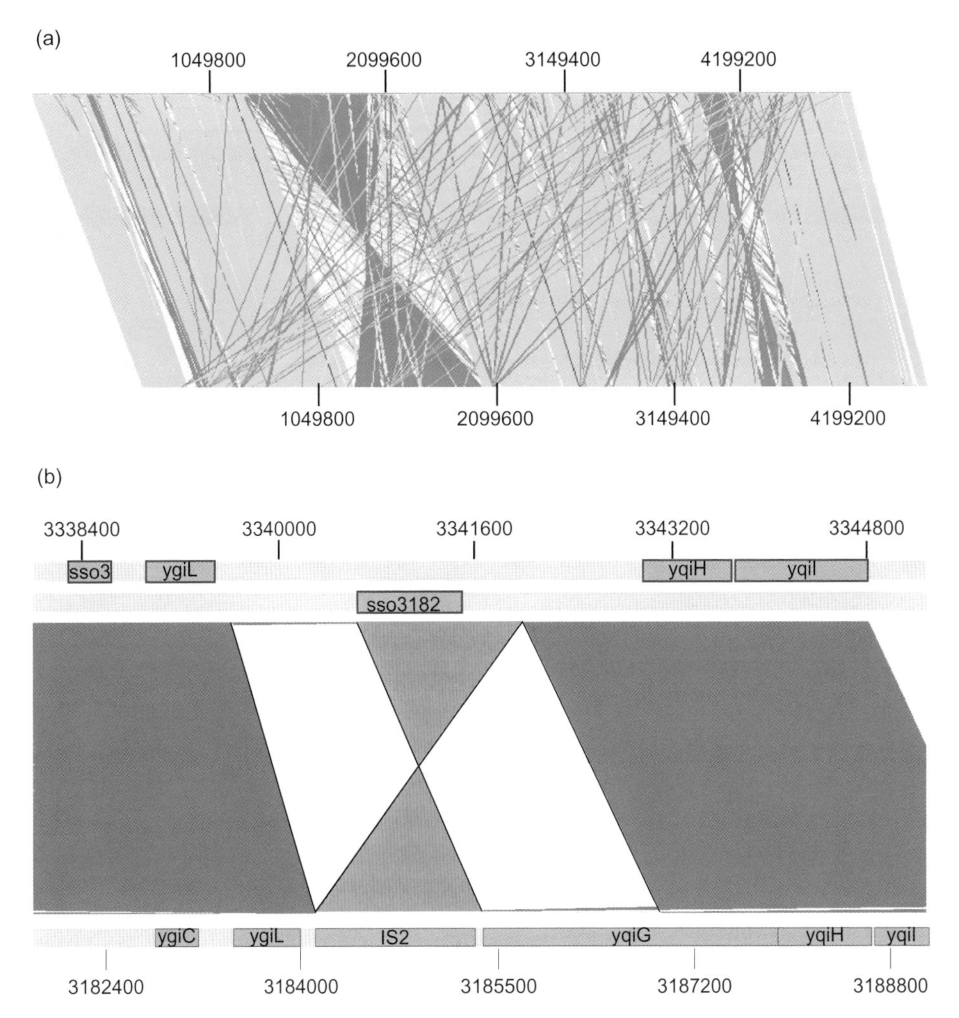

Figure 12.10 Genome sequence comparisons using ACT. (a) Comparison of the genome sequences of *Shigella sonnei* (top) and *E. coli* (bottom). Blue = matched sequences in the same orientation; grey = matched sequences in the reverse orientation. (b) Detailed comparison of a shorted region of the same sequences

on the basis of similarity to known bacteriophage genes. This suggests strongly that this region consists of an integrated bacteriophage, which is absent from strain MGAS8232.

12.4 Genome browsers

So far in this chapter, we have described some of the ways in which we can make sense of genome sequence data. Fortunately, we have access to some immensely useful tools for viewing longer or shorter chromosome segments

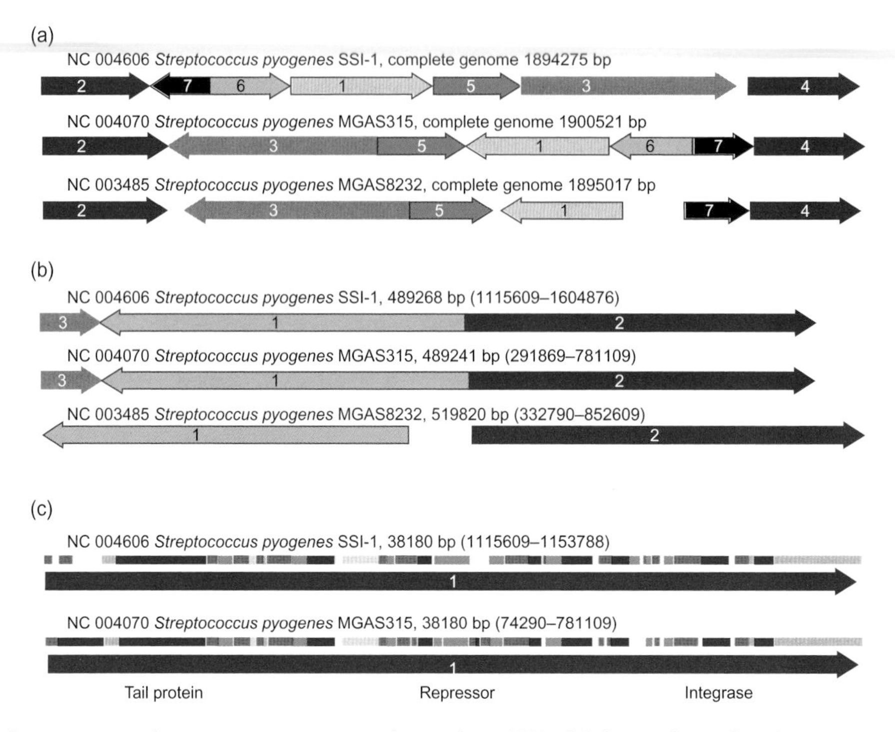

Figure 12.11 Genome structure comparison using gMAP. (a) Comparison of entire genomes; (b) comparison of selection region 3, from (a); (c) detailed comparison of selected region 3, from (b)

and their annotations using web-based *genome browsers*. These are basically viewing tools that all access the very same information from the databases from the latest version of each finished genome sequence. Because the data that they all access are the same, you have the choice of using whichever genome browser you feel most comfortable with. The original genome browser is hosted by the University of California at Santa Cruz (genome.ucsc.edu), while the most advanced and versatile one is called ENSEMBL (www.ensembl.org). A third, simpler browser, the NCBI Map Viewer (www.ncbi.nlm.nih.gov/Genomes) is directly connected to GenBank. All genome browsers display your choice amongst the currently available sequenced animal genomes, with many other eukaryotes thrown in for good measure.

Basically, all of these browsers allow you to zoom back and forth between the chromosome view and the sequence view, with a number of different degrees of resolution in between. Through the browser's default annotation tracks, and through a number of optional ones, the sequence view is decorated with annotation information retrieved from the databases, including the position of genes, exons and coding regions, but also with optional features such as known polymorphisms, alternative splice variants, regions that are conserved in selected related organisms, etc. Apart from browsing, you are able to search for a specific gene (or a specific chromosomal region) and see it displayed. You are able to access links which produce graphic displays of predicted transcripts and proteins, and domains in these. The browsers also allow you to define and export a smaller or larger part of a chromosome, and to compare it with other species.

12.5 Relating genes and functions: genetic and physical maps

We have already, in Chapter 9, discussed how we can use the primary DNA sequence of a gene to make predictions about the structure and function of its protein products. However, this approach has limitations. A substantial proportion of the genes identified in any genome sequence are not related to any gene with a known function, although there may be apparent ORFs coding for similar hypothetical proteins in other genome sequences. Always there is also the cautionary note that bioinformatics can make predictions about function, but does not provide definitive answers. In Chapter 9, we also considered some of the ways in which we can address the relationship between a cloned gene and its function. We now want to look at some of the ways that are applicable on a genome-wide scale.

Another way of looking at the problem is to contrast the data obtained by classical and molecular genetics. Classical genetics works primarily with phenotypes, and produces a *genetic map*, which shows the position on the chromosome of genes associated with specific phenotypes. However, the genetic map does not tell us the structure of those genes, nor does it directly tell us about the biochemical function of those genes.

On the other hand, molecular techniques are, in the first place, concerned with the structure of genes and their sequence. This could start with cloning and sequencing a fragment of DNA; you could then use hybridization techniques to find the position of that sequence on the chromosome, producing a *physical map* of the chromosome. Physical maps can take other forms as well, including maps of restriction sites. Ultimately, as we have seen, you can determine the complete sequence of the genome, which is the definitive physical map. It tells you exactly the DNA sequence at any position on the genome, but, taken in isolation, tells you nothing about the nature or function of the genes or their products, and still less about the phenotype associated with them.

Although we can make inferences on the basis of sequence comparisons, to advance our understanding further we have to be able to relate the genetic and physical maps.

To some extent we can tackle the problem from either end. For example, we could start with the classical approach, i.e. isolating specific variants and mapping the genes concerned. We would then follow that with the techniques described in the previous chapters to isolate and clone the DNA region that is different in the mutant and wild-type organisms – in this way linking the genetic map to the structure of a specific gene. The techniques described earlier in this book for individual genes are extended by other methods, discussed later in this chapter, which are suitable for genome-wide studies.

Alternatively, we can start from the other end. If we know the sequence of a piece of DNA, or the entire genome, we can (as described in Chapter 9) infer the likely nature of the enzyme or other product coded for by each gene, by comparison with the sequence of known genes from other organisms. Thus we can work backwards from the physical map towards the genetic one. However, this approach has limitations. Firstly, a substantial proportion of the predicted genes identified in any genome sequence are either not related to any known gene, or are related to another gene of unknown function. Secondly, even where there is a good degree of similarity with another gene that is labelled as coding for an identified enzyme, that identification is only as good as the identification of the gene we are comparing with. As discussed earlier in this chapter, there is a risk of setting up a chain of increasingly unreliable similarities. Furthermore, we cannot be sure in all cases that enzymes with a similar structure actually carry out the same biochemical reaction. The enzyme β-lactamase, responsible for ampicillin resistance in many bacteria, is similar in many respects to a serine protease, but it is not a proteolytic enzyme.

There is a final limitation to this approach that is more fundamental in nature. We may have correctly identified the biochemical reaction carried out by the enzyme for which our gene is responsible, but this does not necessarily tell us what role that gene plays in the characteristics of the cell. For bacteria, and other unicellular organisms, it may be relatively straightforward to understand the role of enzymes that are components of a simple metabolic pathway, such as synthesis of an amino acid – but, even at this simplest of levels, such understanding is not always completely straightforward. The organism may have more than one gene coding for enzymes that carry out the same reaction, so we would have to ask under what conditions each of those genes is used. With more subtle processes, it may be very difficult to ascertain the role of specific proteins – and if we consider complex processes such as the regulation of cell division, it is likely to be impossible to determine the role of individual proteins just by examining their structure. If we then move on to consider a multicellular organism such as an animal, there is an even bigger jump from knowing the biochemical function of the protein to understanding its role in the whole animal. A multicellular organism may also utilize the gene product for more than one function. For example, in many vertebrates the gene encoding the enzyme alpha-enolase is also expressed at high levels in the lens of the eye, where it has no catalytic function but contributes to the transparent and refractive properties of the lens.

We can now look at some of the techniques that are available for constructing more direct links between genetic and physical maps, i.e. for establishing (or confirming) more directly the actual function in the cell of specific genes.

12.5.1 Linkage analysis

Linkage analysis is a classic technique for establishing how close two genes are on the chromosome. Thus if our genetic mapping data tells us that the gene we are interested in is closely linked to another marker that has been characterized, we can narrow down the search for the gene of interest to a much smaller region of the chromosome. However this would require the mapping a very large number of genes if we are to be sure that there will be a mapped gene very close to our unknown gene. The distances separating known linked genetic markers in mammals commonly run to thousands of kilobases (1 per cent recombination corresponds to about 1 Mb of DNA), but the second marker does not have to be a functional gene; it can be a polymorphic marker such as the microsatellites described in Chapter 13. If the unknown gene is often co-inherited with such a polymorphism, linkage analysis can identify its position to a comparatively short region, which can then be cloned and characterized to identify the nature of the mutation that is responsible for the observed variation. This technique, known as *positional cloning*, has been used for the identification of important human genes such as *BRCA1*, the presence of which predisposes to breast cancer. The identification of genes associated with human diseases is covered further in Chapter 13.

12.5.2 Ordered libraries and chromosome walking

In Chapter 6, we described the construction of a special type of gene library known as an *ordered library*, which consists of a set of overlapping clones so that the position of each clone is known with respect to the clones on either side on the genome. This provides, in essence, a form of physical map of the genome (and ordered libraries have played a significant role in some of the genome sequencing projects). Any gene that has been cloned can easily be located to one of these clones by hybridization (of course if the genome sequence is known this is not necessary). We can then use that as a starting point for locating other genes that are known to be linked to the first marker.

Construction of ordered libraries, especially of large genomes, is a laborious undertaking. A more generally applicable version of the technique is that known as *chromosome walking* (see Figure 12.12). Again, this requires as a starting point a marker that is known to be linked to the gene in question. This marker is used to identify a clone from a gene library, by hybridization. That clone is then used to screen the gene library in order to identify overlapping clones; one (or more) of these clones is then in turn used as a probe

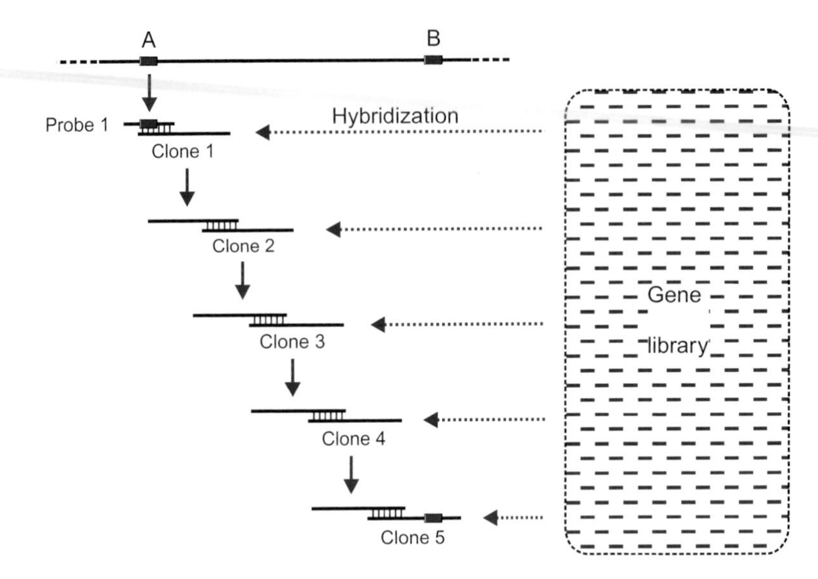

Figure 12.12 Chromosome walking

to identify other clones that overlap with it. These steps are repeated until the required sequence is reached.

12.6 Transposon mutagenesis and other screening techniques

12.6.1 Transposition in bacteria

Transposons are DNA sequences that have the ability to move from one DNA site to another. Part of the DNA of a transposon codes for an enzyme (transposase) that is capable of carrying out a special form of recombination, involving inverted repeat sequences at each end of the transposon, which results in insertion of the transposon at a new position, either on the same DNA molecule or on a different one. Transposons can thus move from one site to another on the chromosome, or they can move from a plasmid to the chromosome, or from one plasmid to another. The details of the process can vary quite considerably from one transposon to another, but that does not need to concern us, apart from noting that some transposons can insert more or less at random while others have varying degrees of specificity. One of the most commonly used transposons is Tn5 (or derivatives thereof), which is not very specific in its insertion site requirements, and hence can insert at a large number of positions.

A further feature of transposons that is relevant here is that they generally carry antibiotic resistance genes. Indeed they play, together with plasmids, a major role in the spread of antibiotic resistance genes amongst pathogenic bacteria. However, transposition not only moves genes between different sites. Insertion of a transposon within a coding sequence will usually inactivate that gene, thus producing a mutation. The site of that mutation is now marked by the presence of the resistance gene, which makes it relatively easy to clone, and thus to identify, the affected portion of the DNA.

The procedure in practice (illustrated in Figure 12.13) is to use a plasmid, carrying the transposon, which is unable to replicate in the host species being

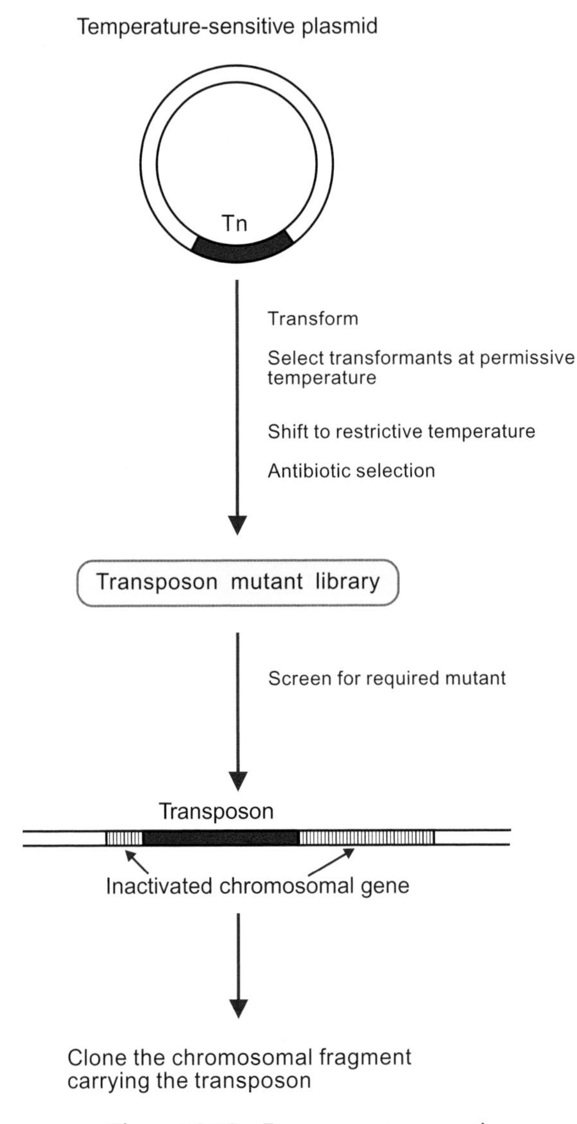

Figure 12.13 Transposon mutagenesis

investigated; this is known as a *suicide* plasmid. Even better is to use a plasmid that is temperature-sensitive for replication, so you can establish the plasmid at a low temperature (e.g. 30°C, the *permissive* temperature) and subsequently prevent its replication by shifting the incubation temperature to, say, 42°C (the *restrictive* temperature). Inside the bacterial cells, the plasmid is unable to replicate at the restrictive temperature. Thus, if we plate the transformed bacteria on a medium containing the relevant antibiotic, only those cells in which the transposon has hopped onto the chromosome will be able to survive and grow to form colonies. If this happens, the transposon will be replicated as part of the bacterial chromosome.

Of course we do not know where the transposon will have jumped to, but there is a large number of possibilities. We can store this collection of cells as a *transposon mutagenesis library*.

We can now use this library to identify all the genes that are related to a specific phenotype, by screening the library for mutants that are altered in this characteristic. We can even, using more powerful methods described later in this chapter, screen the library for inserts in essential genes (including genes necessary for virulence).

Once we have identified the mutant clones of interest, the next step is straightforward. We can extract genomic DNA from those cells, digest it with a restriction enzyme, and ligate these fragments with a suitable vector. In effect, we create a genomic library. However, we do not need the complete library. We are only interested in those fragments that carry the transposon. We can identify these quite easily because they will contain the antibiotic resistance gene that is part of the transposon. Therefore, we just need to plate the library onto agar containing the relevant antibiotic, and only those clones that carry the transposon will be able to grow.

These clones will contain not only the transposon but also a portion of the DNA either side of the insertion side. Determining the sequence of this flanking DNA will therefore enable us to identify the gene into which the transposon has inserted, and we thus have a direct link between the sequence and the phenotype, i.e. we know (subject to certain limitations that are discussed below) that inactivation of that gene gives rise to that phenotype, and hence we can infer the function of that gene in the normal life of the cell.

As is so often the case, PCR provides us with an alternative to cloning for identifying the insertion site of the transposon. We cannot do a straightforward PCR, because that would require knowledge of the flanking sequence, for designing the primers, and that is exactly what we do not know. One strategy that we can adopt in such circumstance is known as *inverse PCR* (see Figure 12.14). If we cut the DNA with a restriction enzyme (using one that does not cut the transposon itself), then, instead of ligating these fragments with a vector, we can carry out a ligation in the absence of a vector and under conditions that promote self-ligation (intramolecular rather than inter-

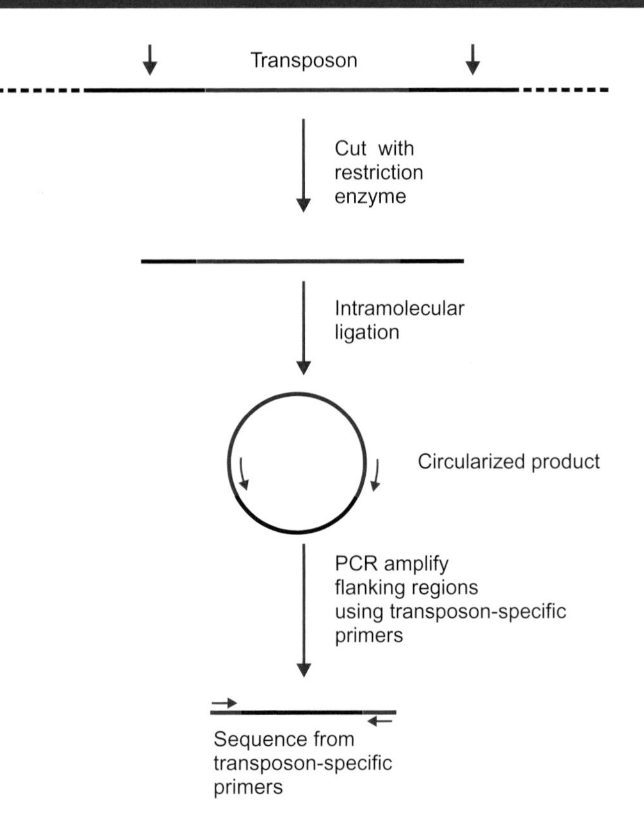

Figure 12.14 Locating an integrated transposon by inverse PCR

molecular ligation). Reference to Chapter 4 will show this requires ligation at *low* DNA concentrations, whereas usually we do ligation at high concentrations of DNA to promote ligation of DNA fragments with the vector DNA. The consequence of self-ligation is that, amongst many other fragments, we have circular molecules containing the transposon and the flanking sequences. Although this will be only one amongst thousands of products, the flanking sequences can be amplified by PCR, using primers derived from the known sequence of the transposon and directed outwards from the transposon.

In transgenic multicellular organisms (Chapter 15), similar situations are sometimes encountered when an introduced transgene happens to lodge itself inside another gene. Because only a small fraction of eukaryotic genomes actually code for gene products, this is a much less likely occurrence than in bacteria. However, although useless from the point of view of the intended experiment, these strains are sometimes useful for gene mapping and/or as disease models.

Further applications of transposons, coupled with reporter genes for detecting genes that are expressed under different conditions, and for identifying secreted proteins, are discussed in Chapter 14.

12.6.2 Transposition in *Drosophila*

The discussion of transposons and transposition has so far focussed on bacteria. However, transposable elements of one sort or another are common in all types of organisms. The family of transposable elements known as P elements, which occur in the fruit fly *Drosophila melanogaster*, are especially important – both in providing vectors for the integration of foreign genes into the *Drosophila* genome and in providing a system for transposon mutagenesis of *Drosophila*.

Transposition of P elements, as with bacterial transposons, requires the action of a transposase, acting on inverted repeat sequences at the ends of the element. In a P strain, which carries multiple copies of the P element dispersed throughout the genome, the transposase is repressed and so no further transposition occurs. However, if sperm from a P strain fertilizes an egg from a strain that does not contain a P element, the temporary absence of the repressor causes extensive transposition, resulting in a high rate of mutation.

The P element is also able to transpose into the genome from an injected piece of DNA. Therefore, if we insert a piece of foreign DNA into a P element contained on a plasmid vector, and then inject that construct into a fruit fly embryo, the P element will transpose into the genome, carrying our inserted DNA fragment with it. However, it is not easy to insert DNA into a P element without disrupting the transposase gene. Figure 12.15 shows how we can get round this problem. The transposase can act *in trans*, i.e. it can be expressed from a different piece of DNA. In the example shown, the foreign DNA fragment has replaced most of the P element genes, leaving the inverted repeat ends intact. The transposase is expressed from a second copy of the P element, and it will recognize the inverted repeats flanking the foreign DNA, resulting in transposition of the insert into the chromosome

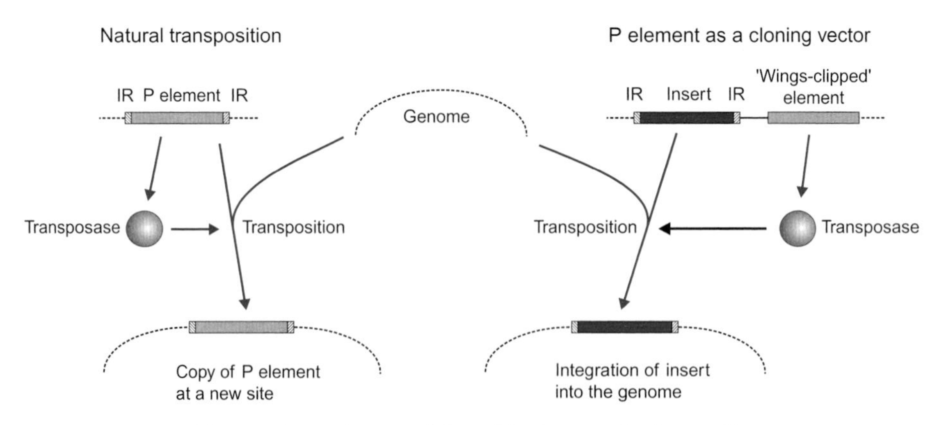

Figure 12.15 Transposition of P elements in *Drosophila*

(with the IR ends). At the same time, we do not want the element with the intact transposase to be inserted as well, as it would cause additional mutations, so we remove the inverted repeat ends from the P element that has the transposase, rendering it non-mobile; this is referred to as a 'wings-clipped' element.

The main applications of this approach lie not in the expression of genes from other organisms in *Drosophila*, but in identifying or confirming the relationship between specific genes and identified phenotypes. As a simple example, Rosy$^-$ flies have brown eyes rather than red ones. Insertion of a DNA fragment coding for the enzyme xanthine dehydrogenase will restore the wild-type eye colour, thus confirming the function of the *rosy* gene; this is an example of *complementation* (see Chapter 9). Fruit flies have been used extensively as a model research system for a multitude of simple and advanced functions, especially the differentiation and development of multicellular organisms. The ability to link phenotypes with specific DNA sequences in this way has been an important component of these advances.

The applications of P elements do not end there. Insertion of a P element into the chromosome can cause a mutation, and since the affected gene is tagged with the transposon, it is readily identified, as in bacterial transposon mutagenesis described above. Another application involves a P element containing a reporter gene (such as the β-galactosidase gene, *lacZ*) with a weak promoter. Random insertion of this element into the genome will occasionally result in integration adjacent to an enhancer element, resulting in activation of expression of the reporter gene. This technique, known as *enhancer trapping*, enables the identification of enhancers and their specific activity in certain cell types; related ways of using reporters to identify regulatory sequences, such as bacterial promoter-probe vectors, were described in Chapter 10.

These applications to fruit flies represent an example of *transgenics*, in that they include the manipulation not just of individual cells but of the whole organism. Further examples of transgenics, as applied to higher animals and plants, are discussed in Chapter 15.

12.6.3 Transposition in other organisms

Transposon mutagenesis can be used in a similar way with many other organisms. In yeast (*S. cerevisiae*) the endogenous *Ty* element provides a way of obtaining libraries of mutants. This is a retrotransposon, i.e. it transposes via an RNA intermediate, produced by a reverse transcriptase encoded by the element – and is thus related to retroviruses. However insertion is not random, as it tends to insert preferentially within genes that are transcribed by RNA polymerase III (mainly tRNA genes), rather than protein-coding genes (which are transcribed by RNA polymerase II), which is a limitation. An alternative strategy is to use bacterial transposons such as those based on Tn*3*. These do not transpose in yeast, but transposition can be carried out in *E. coli* containing cloned yeast DNA. The mutagenized DNA is then re-introduced into yeast cells where it combines with the genome, replacing the native sequence.

Retroviruses and genetically engineered retrotransposons have also been used for gene tagging in mice and other vertebrates, primarily with cell lines, although techniques are becoming available for application to whole animals.

In plants, T-DNA from *Agrobacterium tumefaciens*, which is also used as a vector for introducing foreign genes (see Chapter 15), can be used for insertional mutagenesis. Alternatively, endogenous transposons such as the maize transposon *Activator* (*Ac*) have been used to generate libraries of mutants in various plants including *Arabidopsis*.

12.6.4 Signature-tagged mutagenesis

A form of transposon mutagenesis can also be used to identify genes that are necessary for the virulence of pathogenic bacteria. Insertion of the transposon into such a gene will attenuate the organism; i.e. it will destroy (or reduce) its virulence. This will be manifested by a reduced ability to grow or survive following administration to an experimental animal, or in some cases by a reduction in its ability to survive attack by macrophages in culture. It is not possible to select directly such a mutation, and testing the thousands of mutants in a transposon library is impractical.

However, we can modify the transposon by incorporating a highly variable sequence tag so that each copy of the transposon is uniquely identifiable (Figure 12.16). We then produce a transposon mutant library, with these tagged transposons, and infect mice with a pool of transposon mutants. Those clones in which the transposon has inserted into a gene that is essential for virulence will be unable to replicate in the mice and will therefore be absent when we recover the bacteria from the infected mice. We then use PCR to amplify all the tags that are present in the recovered bacteria, and label the collection of PCR products for use as a probe. Identification of the tags that are absent in this mixture is carried out by probing a membrane that contains a gridded array of each of the clones from the original transposon mutant library. Absence of hybridization means that the clone concerned was not present in the material recovered from the mice – and hence identifies this as a mutant in which an essential virulence gene has been inactivated by the transposon. The gene can then be recovered and identified as described above. This technique, known as *signature tagged mutagenesis*, has proved to be an extremely powerful tool for the identification of virulence genes – or in principle for the identification of any gene that is essential for the growth of the bacteria under defined conditions.

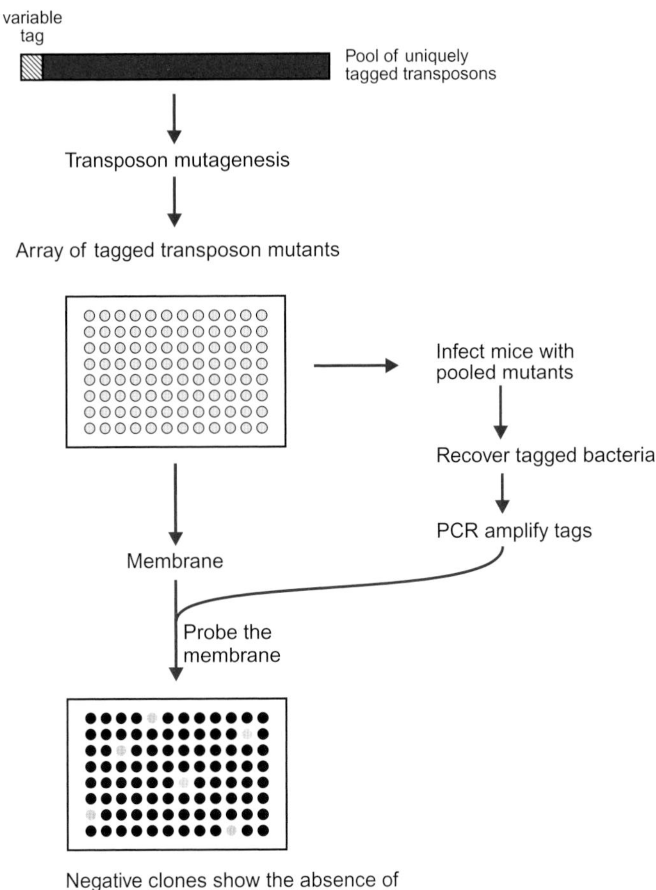

Figure 12.16 Signature tagged mutagenesis

12.6.5 Gene knockouts, gene knockdowns and gene silencing

In Chapter 9, we described the use of gene knockouts, or allelic replacement, for studying the function of a specific gene. Although it might seem that this procedure is not suitable for genome-wide analysis, as it has to be done on a gene-by-gene basis, it is still possible to attempt it – for example, a large-scale collaborative project succeeded in knocking out 95 per cent of the 6200 predicted protein-coding genes in the yeast *Saccharomyces cerevisiae*. A project is underway to knockout every predicted mouse gene (NIH knockout mouse project or KOMP).

However, the discovery of RNA interference (RNAi) has made an alternative, and much more convenient, approach available. RNAi, or *gene silencing*, was discovered by finding that injecting double-stranded RNA into the

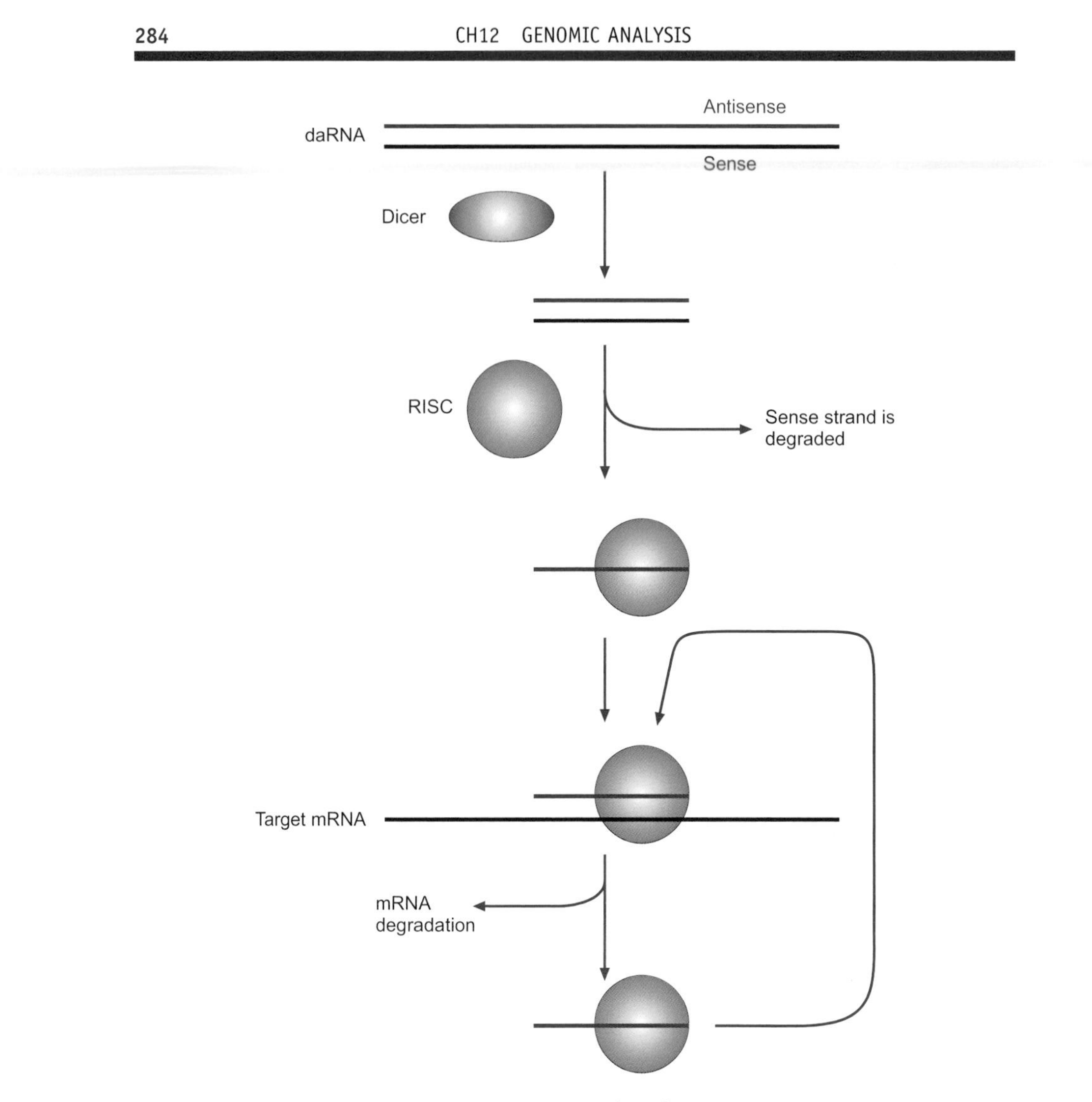

Figure 12.17 RNA interference

nematode *Caenorhabditis elegans* resulted in the corresponding gene being switched off. Andrew Fire and Craig Mello were awarded the Nobel Prize in 2006 for this discovery. The mechanism responsible (Figure 12.17) is that the dsRNA is cleaved by an enzyme, called Dicer, into short fragments (short interfering RNA, or siRNA) which are 19 bases long with an extra two unpaired bases at each end. These are recognized by an ribonucleoprotein complex called RISC (RNA-induced silencing complex), which specifically degrades the sense strand of the siRNA. The antisense strand then binds specifically to the complementary region of the target RNA, and guides the protein complex to it, leading to specific degradation of that message. This is a very widespread phenomenon, in organisms ranging from protozoa to mammals, indicating that it is an evolutionarily ancient mechanism for genome defence against viruses, and may also act in controlling mobile genetic elements such as transposons.

The relevance for this chapter is that, instead of the laborious process of knocking out genes one by one to determine their function, it is possible, with *C. elegans*, to inject the worms with dsRNA, or simply to put them in a solution containing dsRNA. Alternatively, you can feed them with bacterial clones expressing dsRNA targeted to specific genes. This is simple enough to do one gene at a time, or you can use mixtures (libraries) of dsRNA to screen for genes associated with complex properties of the organism. Human cells in culture can also be screened by transfection with libraries of siRNAs; such libraries are commercially available. Use of an expression vector designed to express short hairpin RNAs (shRNAs), which are also cleaved by Dicer to siRNAs, enables long-term studies of the effects of silencing specific genes.

It is worth noting that a related phenomenon plays a major role in the regulation of natural gene expression in a wide range of organisms, including probably all multicellular organisms. The genomes of these organisms encode a large number (as many as 1000) different *microRNAs* (miRNAs). The initial transcription product is a relatively long, hairpin-containing RNA which is cleaved within the nucleus to precursor miRNA (pre-miRNA) about 70 bases long, with a hairpin structure. These are exported to the cytoplasm where they are cleaved by Dicer to 22-base dsRNA molecules, one strand of which participates in RNA interference through binding to mRNA. The principal differences from siRNA are that the miRNAs bind to mRNA molecules that are similar but not completely complementary, and that the mechanism (in animals) does not usually involve degradation of the mRNA but probably an inhibition of translation of the message. However in plants miRNAs show a high degree of complementarity to their targets, and usually result in RNA cleavage.

RNAi also has a role in transgenesis and, potentially, in gene therapy, and we will return to those subjects in Chapter 15.

12.7 Conclusion

In this chapter, we have moved from a consideration of genome sequencing and analysis of gene structure to linking the genome sequence with genome-wide studies of gene function. Further evidence of gene function comes from genome-wide studies of transcription and translation, producing lists of all transcribed genes (*transcriptome*) and all translated products (*proteome*). These are covered in Chapter 14. However, before that, in Chapter 13, we want to look further at how we can study the variation between strains and species at the genome sequence level.

13 Analysis of Genetic Variation

The basic structure and organization of genes, in eukaryotes and prokaryotes, was considered in Chapter 2. It is a central dogma of molecular biology that the inherited characteristics of an organism are a reflection of the structure and organization of its genes. However, it is important to realize that this includes an extremely complex set of interactions between different genes, and their products, as well as environmental factors. The gene(s) directly responsible for the observed characteristic may be absolutely identical but the effects may be different because of variation in other genes that affect their expression, or because of alteration in other cellular components that affect the activity of the proteins encoded by those genes. Environmental influences will therefore have a major role in determining the observed characteristics of the organism. Excessive reductionism – ascribing every change to one single gene – is an ever-present pitfall in the analysis of variation, as many traits are much more complex than that.

We can use the study of genetic variation to examine differences between members of the same species, ranging from the study of bacterial characteristics (such as antibiotic resistance) to investigation of human genetic diseases, or to differentiate between individuals (for example in forensic analysis). Or we can compare the genetic composition of members of different species – even over wide taxonomic ranges – which can throw invaluable light on the processes of evolution as well as helping to define the taxonomic relationship between species. Many of the concepts involved are similar, but of course the differences are (usually) much greater when we are comparing widely different species. Some methods of comparing genomes were considered in Chapter 12; we now want to look further at some methods that can be applied to the analysis of variation within a species.

From Genes to Genomes, Second Edition, Jeremy W. Dale and Malcolm von Schantz

13.1 Single nucleotide polymorphisms

The simplest form of genetic variation consists of a change in the sequence of bases at a single point. Such differences are called *single nucleotide polymorphisms* (SNPs), although in human genetics the term is usually reserved for differences that are relatively common in the population. It is worth reflecting briefly on the consequences of such a simple change. The first possibility is that the polymorphism is not even within a gene; indeed in eukaryotes that is the most common occurrence (this is both because most eukaryotic DNA is not a part of any gene, and because a non-coding sequence is where a polymorphism is least likely to have a negative effect). Secondly, if the change is within a gene, the chances are (again in eukaryotes) that it is in an intron and (unless it affects a splice site) does not affect the resulting gene products. Even if it is within an exon, there may be no change at all in the sequence of the protein encoded by that gene. A single base change may alter for example a leucine codon such as UUA into CUA, which also codes for leucine (Figure 13.1). This is referred to as a *synonymous* substitution. This would be expected to have little or no effect on the organism (and hence is also referred to as a silent mutation) – although the change in codon usage could have some effect on the translation of the mRNA. Other changes may also be silent, although not synonymous. A change from UUA to GUA would replace the leucine with valine at that position in the protein, which may have little consequence for protein structure and function (depending on how critical that specific position is), since leucine and valine are similar amino acids. Other changes, such as UUA to UCA (serine), are more likely to have an effect, although such a change (and even more radical changes) can be tolerated at some positions. These non-synonymous (or missense) substitutions therefore may or may not give rise to changes in the characteristics of the cell.

Most commonly these changes will either have no significant effect or will simply impair or destroy the function of the product of the gene. Very rarely, they will lead to a protein that can function in a slightly altered way – for

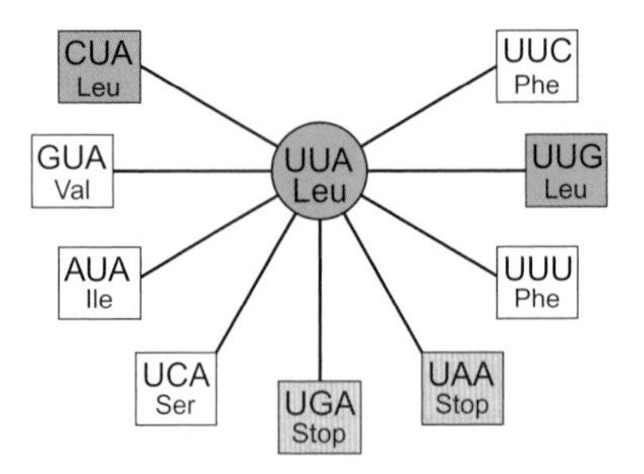

Figure 13.1 Codons arising by single base substitutions from UUA

example, an enzyme that may be able to use a different substrate. An accumulation of such mutations may eventually give rise to an enzyme which carries out quite a different reaction. This is of course the basis of the gradual process of evolution. It is also important to keep in mind the differences that *diploidy* makes in species such as our own. Clearly, having two copies of each gene allows more leeway for evolutionary drift in one allele, as long as the other one functions normally. This is why carriers of a recessive disease gene generally are not affected by it. When analysing genetic variation in diploid organisms, it is obviously paramount to keep in mind that the organism has two copies of each gene (with some exceptions such as the X chromosome genes in mammals), and that the individual can be either homozygous or heterozygous for any given genetic variation.

However, it is much easier to obtain a mutation that results in the loss of function than one which changes the characteristics of the product, so although we can study the products of changes that have accumulated over a long time scale, we rarely see these changes in action. The main exception to this is where there is a strong selective pressure in favour of such a change – the best example being some types of antibiotic resistance in bacteria, such as resistance to rifampicin (see below).

A further possibility is exemplified by the change from UUA to UGA, in Figure 13.1. This is a *stop codon* (in the standard genetic code), i.e. it causes termination of protein synthesis (because of the absence of a tRNA that could recognize it). Such a change will therefore cause a truncated protein, which will usually have lost activity.

If instead of changing a single base we envisage a situation where a base is removed, then we have a *frameshift mutation*. Once the ribosomes pass that point in the mRNA, they are reading the message in the wrong frame – and the translation product will be totally different from that point onwards. Usually they will encounter a stop codon quite quickly, as the unused reading frames are generally well supplied with stop codons.

A point mutation may also occur within an exon but outside of the coding region in the 5′ or 3′ untranslated regions (UTR). One might assume that such mutations would not cause any discernable differences in the phenotype, but they may have an effect, because the UTRs have a function, too, although a less easily predicted one than the coding region. They serve as a type of 'bar code' specifying parameters such as sorting and stability of the RNA, which is accomplished by the action of RNA-binding proteins. An alteration to the mRNA in this region may therefore affect its recognition by such a protein, which may have a significant effect on the translatability of the mRNA.

We can analyse the occurrence of sequence polymorphisms in different strains, or different individuals, either by direct sequencing of PCR products, or by using allele-specific probes.

13.1.1 Direct sequencing

The principle here is to amplify the gene, or a suitable fragment of it, and to sequence the PCR product. For studying a human genetic disease, you would then compare this sequence in individuals with or without the condition. However things are not necessarily that simple. For example, you may find a change at one position in all the patients compared with all the controls. That does not necessarily mean that you have identified the cause of the disease. Assuming this is an inherited condition, that change may be genetically linked to another elsewhere on the same gene or on a neighbouring gene. On the other hand, you may find that not all the patients have the same mutation. What presents as a single condition may in fact be due to any one of several mutations. For example, the human eye disease retinitis pigmentosa can be caused by no less than 90 different mutations in the rhodopsin gene, but it could also be caused by any one of hundreds of other mutations in dozens of other genes.

In a diploid organism you have to bear in mind that the mutation may affect the binding site for one of the PCR primers. In that case, in a heterozygote, the mutant allele will fail to amplify. On the other hand, the wild-type allele will amplify perfectly, and give you the incorrect impression that you are dealing with a homozygous wild-type. In these cases, RFLP analysis remains a valuable tool to ensure that there are no major deletions or other rearrangements in your gene of interest.

A similar approach can be applied to studying the causes of variation in bacteria. For example, we know that the antibiotic rifampicin acts by inhibiting RNA polymerase. Can bacteria become resistant by means of mutations that alter the structure of this enzyme so that it is no longer affected by rifampicin? If we take a number of strains (in this example, of *Mycobacterium tuberculosis*), including both sensitive and resistant ones, we can use PCR to amplify the RNA polymerase gene, or more specifically the *rpoB* gene (coding for the beta subunit of RNA polymerase), so that we can test the hypothesis by sequence analysis. This will show that most of the resistant isolates carry mutations that result in a single amino acid change at one of a limited number of positions (Figure 13.2). These changes are not found in the sensitive controls. If this remains true after sequencing a larger number of strains, you can become increasingly confident that these mutations include

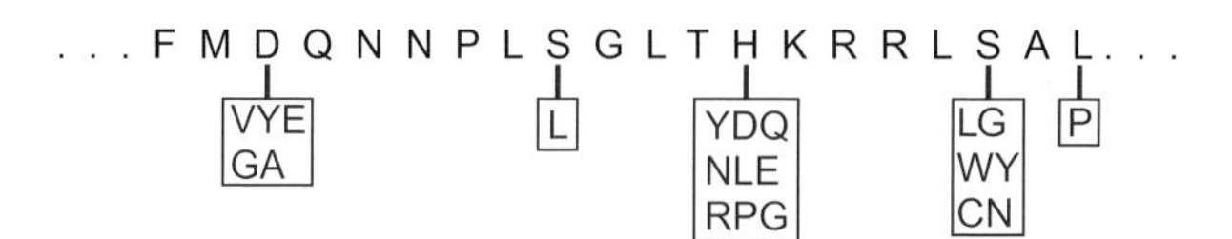

Figure 13.2 Point mutations in RNA polymerase causing rifampicin resistance in *M. tuberculosis*. The diagram shows a portion of RpoB from *M. tuberculosis*. The amino acid changes shown account, between them, for over 90 per cent of rifampicin resistance

those responsible for rifampicin resistance. In this case you have the advantage of being able to test the hypothesis experimentally, by deliberately introducing each of those mutations (by allelic replacement, see Chapter 9) into a rifampicin-sensitive strain and seeing if it does become resistant. There is still the limitation that you are likely to find some strains that are resistant to rifampicin but do not have any of these mutations; this indicates that there is another mechanism that you have not identified.

13.1.2 Allele-specific oligonucleotide probes

If you are confident that you have identified all the likely mutations at a specific site, then using sequencing to identify them is rather unnecessary. In this case it is possible to make a pair of short oligonucleotides, one of which will hybridize (under stringent conditions) only to the mutated form, and the other only to the corresponding wild-type. The pair of probes can then be used to test the PCR-amplified product for the presence of that mutation. This is normally done using yet another variation of filter hybridization – *dot* or *slot blot*, where you pipette all your PCR amplicons from the individuals you are testing in a row on one membrane for each probe that you are using. You can either do this by hand, each pipetted drop forming a dot, or you can use a dot or slot blot manifold, through which your sample will be transferred onto the membrane by vacuum suction in a geometrically defined pattern. Duplicate filters are then hybridized with the mutant and wild-type probes. Since a defined amount is applied to the membrane, it is possible to obtain a quantitative result, using densitometry to assess the strength of the signal obtained after hybridization with the probe. If the number of known mutations is too large to make this approach feasible, you can produce an *array* (see later in this chapter), with oligonucleotides corresponding to each mutation immobilized on a membrane or a glass slide, which is then hybridized to your PCR-amplified sample. In this case, the PCR product would be labelled, rather than the oligonucleotides.

Quicker and more sensitive results can be obtained by combining this concept with that of real-time PCR, in which specific fluorescent probes are used to monitor the progress of the PCR reaction as it happens. If these probes are designed to detect a specific mutation, you have a sensitive and rapid test for that specific lesion. Real-time PCR was described more fully in Chapter 8.

13.2 Larger-scale variations

In addition to these *point mutations*, there is a wide range of larger scale variations in the structure of the genome. These include *deletions*, *duplications*, *insertions*, *transpositions*, *inversions* and other rearrangements. Of particular importance in the context of this chapter is the activity of mobile genetic elements, including insertion sequences and transposons. Genome sequencing projects have disclosed large numbers of such elements in most species, where they play an important role in the generation of variation by inactivation of the genes into which they are inserted; they also provide a convenient tool for the differentiation of distinct strains and species. Strains may also differ in their genome structure through the insertion of viruses. Examples of phage insertion and insertion sequences were shown in Chapter 12. Extrachromosomal agents such as plasmids are also a significant component of variation. In eukaryotic cells, we also have to recognize the contribution of the prokaryote-derived mitochondria and chloroplasts to the overall genetic makeup of the cell.

13.2.2 Microarrays and indels

The most convenient way of detecting differences between closely related organisms (such as two strains of the same species) through insertions and deletions is to use a *microarray*. This is a glass slide containing a large number of tiny spots of a series of PCR products (or synthetic oligonucleotides) which correspond to all the predicted genes from a specific organism. Alternatively, for analysis of small genomes, you could use sequences spaced at regular intervals over the whole genome. To provide reasonable coverage of the whole genome you would need thousands of such fragments for a bacterial genome, or millions for an organism with a larger genome. Whether we use PCR products or synthetic oligonucleotides, we need to know the genome sequence of the standard strain. A robot is used to deposit each spot in a closely defined position on the slide. Once the DNA fragments are made, the machine can produce many identical copies of the same microarray. Hybridization to a labelled DNA sample can then detect which of the spots contain a DNA fragment that is present in the sample (Figure 13.3). Further details of various types of arrays are considered in Chapter 14.

For genomic analysis, we would normally want to compare the standard strain (the one used to produce the microarray) with another strain. Genomic DNA fragments from the standard strain are labelled with a fluorescent dye, and fragments from the test strain are labelled with a second dye. The two DNA samples are then mixed and hybridized to the microarray. A microarray reader is used to compare the fluorescence arising from the two dyes on each spot. If both hybridize equally, the reader will display a yellow spot. If a gene is absent in the test strain, only the standard probe will hybridize, and you will see a red spot. If the converse were to happen – only the test DNA hybridizing – you would get a green spot, but in this example this would not be expected to happen, as the standard probe should hybridize to all the spots. This is therefore a very effective way of detecting regions of DNA that are present in only one of the two strains. As it is not easy to determine

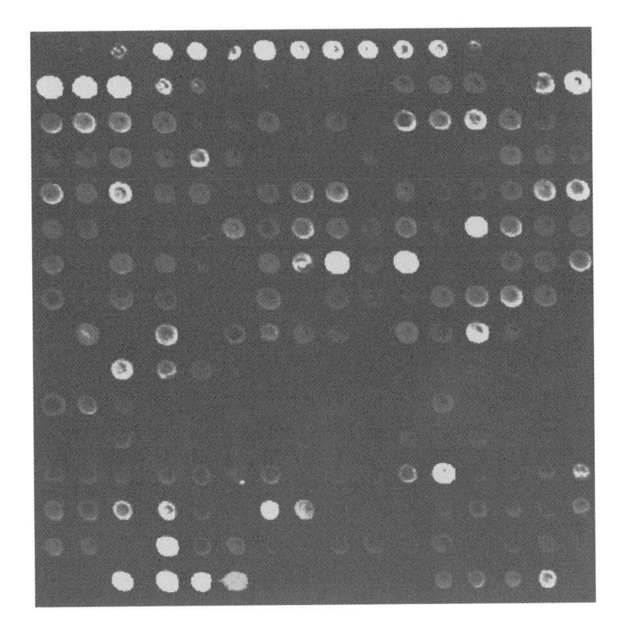

Figure 13.3 Use of a microarray. The illustration shows a representation of a part of a microarray, hybridized with a fluorescent-labelled probe. In practice, two probes with different labels would normally be used and the results displayed in red, green or yellow, depending on which probe hybridized to the greater extent

whether this represents an insertion in one strain or a deletion in the other, the term *indel*, meaning insertion or deletion, is often used.

An alternative to this global approach is to use an array of oligonucleotide sequences representing all the known mutations or polymorphisms in a gene, allowing a quick screen in cases where there are many potential differences such as the range of rhodopsin mutations as described above.

In Chapter 14, we will look at another important use of microarrays, for genome-wide analysis of gene expression (*transcriptomics*).

13.3 Other methods for studying variation

Ultimately, comparison of complete genome sequences is the only way of detecting all the possible types of variation that may occur. This may seem to be an extravagant and unrealistic suggestion, but it is a measure of the speed with which genome sequence technology is advancing that complete sequences are already available for more than one strain of several bacterial species (to say nothing of two human genome sequences). Global comparisons of genome sequences between strains, species and genera are playing a major role in our understanding of the evolution of the genome as a whole. Thus we are not far from a position when a serious analysis of the differences between, for example, two closely related bacterial strains would rely on a comparison of the complete genome sequences – although we should remember that the genome sequence by itself does not directly explain which of the differences is responsible for the observed variation in the characteristics of the organism. Indeed, if you had enough spare money, there are no technical obstacles to comparing your own complete genome sequence with that of your best friend. Again, you would no doubt find many interesting differences but none that would be likely to explain why you like different kinds of music.

However, we are not quite at the point yet where genome sequencing would be an effective *routine* method of identifying variations between two organisms. Furthermore, although genome sequencing is escalating rapidly, there is still an important role for methods that do not rely on such knowledge.

13.3.1 Genomic Southern blot analysis: restriction fragment length polymorphisms

One of the mainstays of the analysis of genetic variation is the Southern blot, which enables us to detect which band on an agarose gel is able to hybridize to a specific probe. The technique was described in Chapter 7 as a way of verifying the nature of the insert in a recombinant clone. In that case, there were only a few bands on the gel. If we digest a total DNA preparation from an organism, we will usually get so many fragments that they will appear as a smear on the gel rather than as discrete bands. (We will consider later on, under *Pulsed-field gel electrophoresis*, what happens if you use an enzyme that cuts so rarely that you get only a few very large fragments.) In this situation, Southern blots not only enable us to identify specific gene fragments, but the method also enables us to compare the DNA samples according to the pattern of hybridizing bands.

If you use a probe that hybridizes to a DNA sequence that is present as a single copy in the genome, then you will detect a single band on the Southern blot. (This assumes that the enzyme used to digest the genomic DNA does not cut within the region that the probe hybridizes to; if it does, you will get more bands.) The size of that band (or bands) will be determined by the position of the restriction sites flanking the detected sequence. If the structure and location of that gene is the same in each organism tested, then the band will be in the same position. On the other hand, if there is variation in the distance separating the two restriction sites, there will be a difference in the size of the fragment, and you will see a difference in the position of the detected band. This is referred to as a *restriction fragment length polymorphism* or *RFLP*.

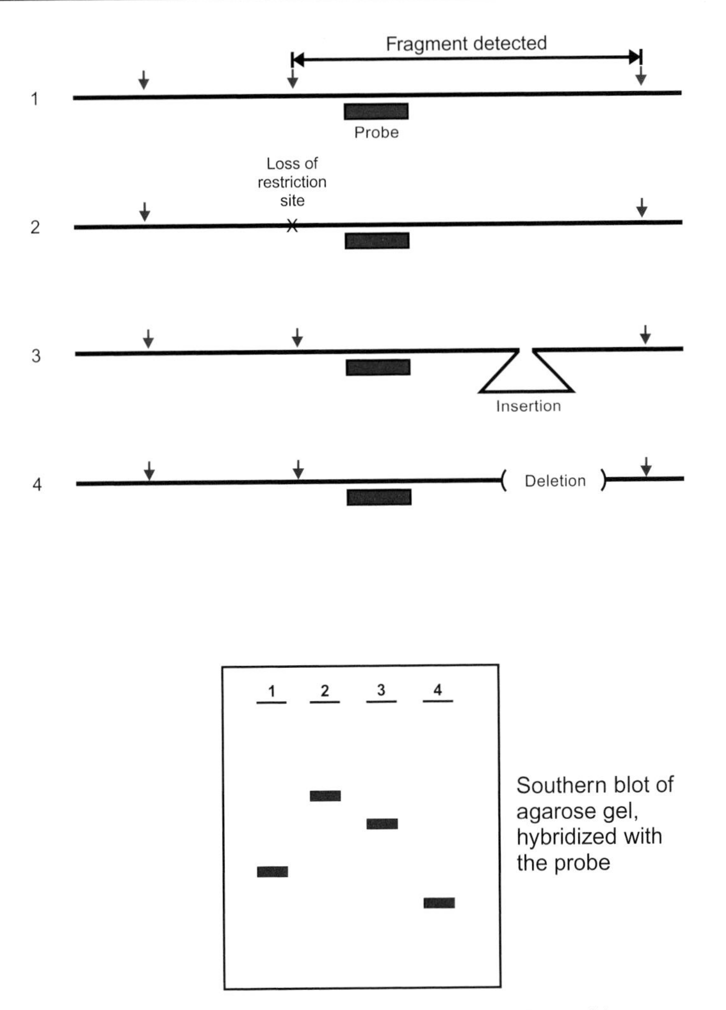

Figure 13.4 Restriction fragment length polymorphism

There are a variety of changes that can give rise to such a polymorphism (see Figure 13.4). The most obvious one – a point mutation causing loss (or gain) of a specific restriction site – will usually happen at a very low frequency. Insertions and deletions will make the fragment larger or smaller, respectively, although if the insert contains a cleavage site for the enzyme used (or if the deletion removes a restriction site) the opposite effect may occur. Such events, in a specific sequence, are also likely to be rare, although in some regions of the genome, deletions may be more frequent, usually arising from recombination between two homologous, or partly homologous, sequences.

For RFLPs to be useful for routine comparison of individuals or strains, we need to look at events that happen more frequently. The most useful

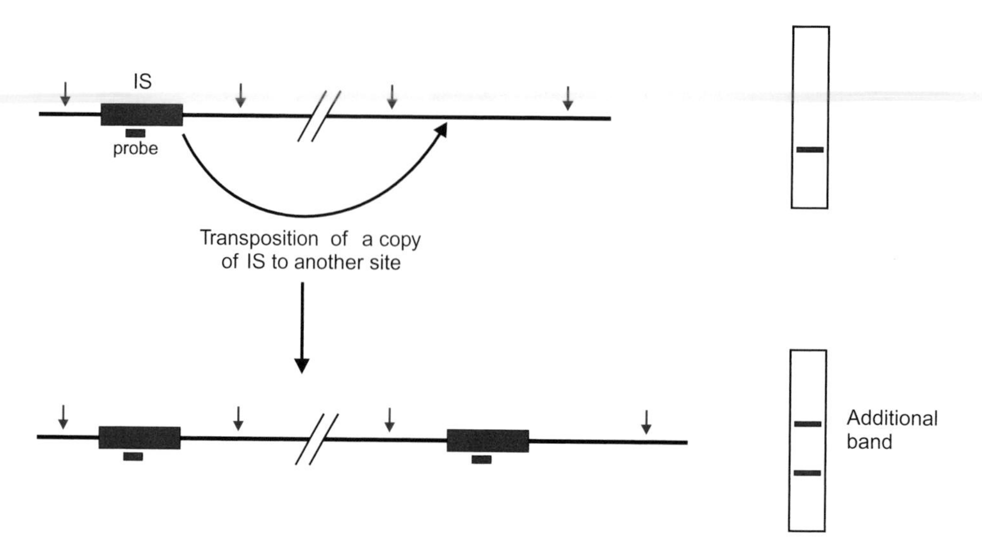

Figure 13.5 Restriction fragment length polymorphism arising from transposition of an insertion sequence

polymorphisms arise through the duplication or transposition of repetitive sequences. Duplication is often associated with the occurrence of tandem repeated sequences, i.e. a short sequence of bases occurs twice (or more) in succession. When this region is replicated, mistakes can occur: the replication machinery may slip back from the second copy to the first, giving rise to an extra copy of the repeat; or it may (less commonly) slip in the other direction, causing a reduction in the number of copies. The repeat sequence can be quite short – for example just a pair of bases, or even a run of the same base – in which case the change in the length of the restriction fragment will be small. PCR-based techniques are then more useful in detecting these changes (see the section below on *VNTR and microsatellites*).

If we use a probe that hybridizes to a mobile DNA element such as a transposon or insertion sequence, then when a copy of that element moves to a different site on the chromosome we will pick up a new band on the Southern blot (Figure 13.5). This is a widely used technique for the differentiation of bacterial strains (*typing* or *fingerprinting*) for epidemiological purposes. For example the standard method for typing strains of *Mycobacterium tuberculosis*, the bacterium that causes tuberculosis (TB), is Southern blotting of digested DNA using a probe that hybridizes to an insertion sequence known as IS*6110*. If two patients have caught TB from the same source, the IS*6110* patterns will be identical (see the arrowed tracks in Figure 13.6). If they are different, we can conclude that they are not part of the same outbreak. It should be noted that in this case we are mainly looking at changes arising from the location of the sequence detected by the probe, rather than

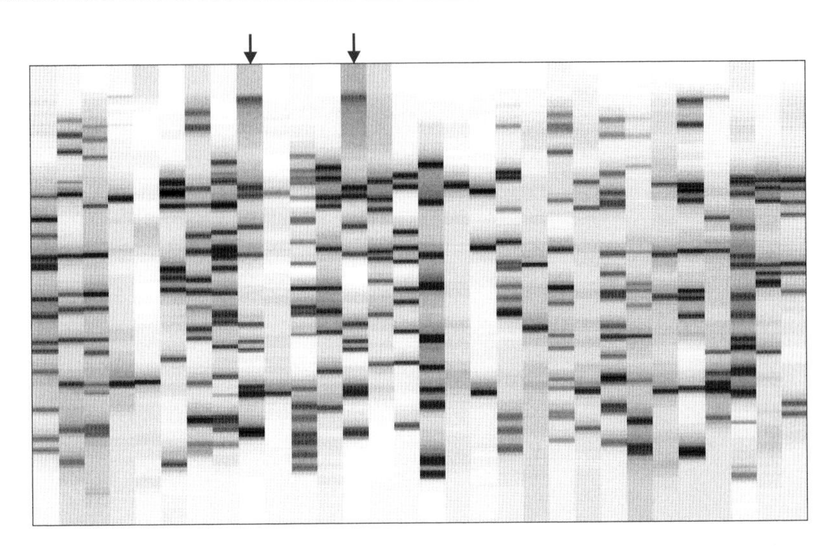

Figure 13.6 Fingerprinting of *M. tuberculosis* using IS*6110*. Each track shows a difference isolate of *M. tuberculosis*. Total DNA preparations were cut with *Pvu*II and electrophoresed through an agarose gel. Following Southern blotting, the membrane was hybridized with a labelled probe derived from the insertion sequence IS*6110*. The arrowed tracks show identical patterns, which may represent two strains from the same outbreak of tuberculosis

alterations in a specific region of the genome, although the latter can contribute to the overall polymorphism.

From the above discussion it will be evident that using Southern blots to detect RFLPs will mainly identify larger-scale changes in the structure of the genome, or a specific part of it, and are much less likely to identify minor sequence changes in a specific gene (unless that change destroys or creates a restriction site). However, Southern blots, or other hybridization-based techniques, can also be used to detect specific sequence changes. If you use a small oligonucleotide probe, and high stringency conditions, a single base change in the target sequence can cause a detectable change in the ability of the probe to hybridize to the target DNA. This has applications as diverse as the diagnosis of human genetic disease and the detection of bacterial mutations causing antibiotic resistance, as we saw earlier in this chapter. Generally, however, the probe would be used in combination with PCR rather than being applied directly to genomic DNA.

13.3.2 VNTR and microsatellites

One of the practical limitations of the conventional RFLP method described above is that it requires a relatively large amount of DNA for the signal to be detectable. For many practical applications, including forensic examination and many types of clinical diagnosis, this approach is not possible. We have to use PCR to amplify the tiny amount of DNA that is available, and we then need a different approach to identify variations.

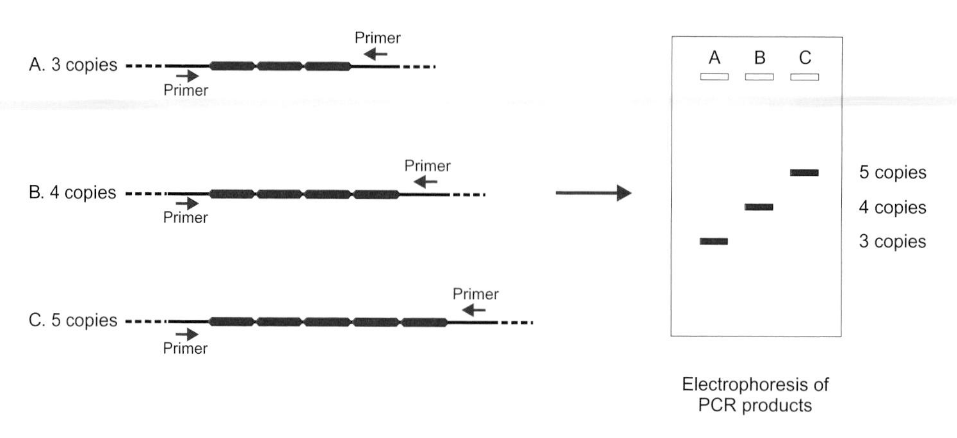

Figure 13.7 Variable number tandem repeats

One approach exploits the tandem repeats referred to previously, but instead of using a probe and a genomic Southern blot, we can amplify the region containing the tandem repeat, using primers that hybridize either side of the tandem repeats. If we run the product on an agarose or (preferably) acrylamide gel, we will see that the size of the product varies in a stepwise fashion according to the number of copies of the repeated sequence (see Figure 13.7). This gives the technique its name, VNTR, standing for variable number tandem repeats. Each strain can then be given a number corresponding to the number of that tandem repeat that is present. By identifying several loci that contain VNTRs, and determining the number of copies at each position, we arrive at a composite designation such as 3433243. This is now increasingly used in bacterial genotyping as an alternative to RFLPs.

A very similar concept that has proved extremely useful in human genotyping involves the use of shorter tandem repeats known as *microsatellites*, or *short tandem repeats* (STRs). These are curious short stretches of sequence of 200 bases or less. They look almost as if the genome got bored with randomly distributed bases and instead decided to go for a straight stretch of one or two (less commonly three or four) nucleotide bases repeated a hundred times or so (such as TTTTTTTTT . . . or GTGTGTGTGTGTG . . .). Nobody knows what these repeats do, but it cannot be particularly important, because they are extremely variable in length (no doubt because of ‘slipping’ during replication) as well as being randomly distributed throughout the genome, and it is this that makes them so useful. By performing PCR with a battery of primers flanking microsatellite markers, a genetic fingerprint can be obtained that makes it possible to draw conclusions about an individual’s identity and genetic relationship with others with great confidence.

13.3.3 Pulsed-field gel electrophoresis

Most of the methods we have looked at so far are based on the examination of variation in limited regions of the genome. For example, even with a method such as RFLP fingerprinting of bacteria using an insertion sequence, where the insertion sites giving rise to the polymorphism may be more or less randomly distributed over the whole chromosome, the actual regions involved in the comparison of two strains represent less than 1 per cent of the whole chromosome. Short of complete sequencing of all the genomes to be compared (see the discussion earlier in this chapter), how can we look at the global structure of the genome?

In principle, you might expect that, since a restriction endonuclease will cut genomic DNA at defined positions, you would get a characteristic pattern of restriction fragments from a given strain. Any deletions, insertions or transpositions (and some base substitutions) would give rise to a change in that pattern. In principle you would be correct. However, if we consider an enzyme that requires a six-base recognition sequence (a 'six-base cutter') then, assuming an equal proportion and random distribution of all four bases, we would expect this enzyme to cut on average once every 4096 bases. If you are working with a bacterial genome of, say, 4 million (4×10^6) bases, you would expect about 1000 fragments. Agarose gel electrophoresis does not have enough resolving power to separate out such a large number of fragments (although it is possible, albeit extremely difficult, to obtain characteristic patterns for typing using only a part of the range of sizes). For organisms with larger genomes, the problem is of course correspondingly greater.

There are enzymes available that will cut DNA less frequently because they have a longer recognition sequence and/or a recognition site that occurs less frequently than expected on a random distribution (see Chapter 4). Such an enzyme will produce fewer fragments – but the problem now is that the fragments are too big to be separated by conventional gel electrophoresis. These large fragments *will* actually move through the gel, but in a size-independent manner. This is most easily visualized as considering the DNA fragments as very long and very thin structures; once they eventually get lined up end-on, parallel to the direction of the electric field, they can then slide through the pores in the gel. As long as the field direction is constant, they will keep moving. The limiting factor is the time taken to orientate the molecules.

The technique of *pulsed-field gel electrophoresis* (PFGE) exploits this. The simplest version to describe is one where the polarity of the electric field is briefly reversed periodically throughout the run (Figure 13.8). Each time the field is reversed, the molecule has to be re-orientated in order to move through the pores in the opposite direction, and the longer the molecule is, the slower this process will be. Consequently, the net migration of the

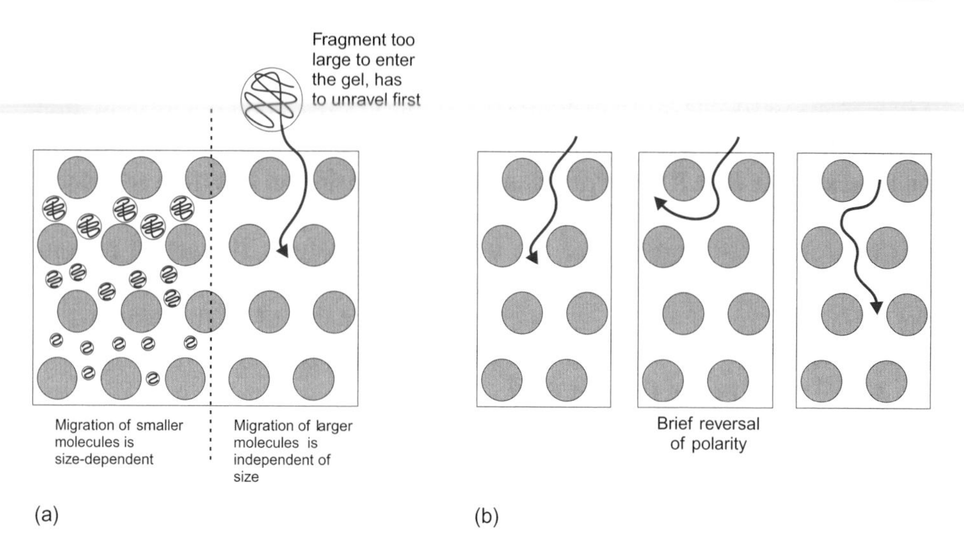

Figure 13.8 Pulsed-field gel electrophoresis. (a) Conventional electrophoresis; (b) PFGE, migration is dependent on the rate of re-orientation, which is related to fragment length

fragments through the gel will be size-dependent. (In reality, this approach leads to unacceptable distortions of the lanes, and the apparatus actually used for PFGE employs more complex systems for changing field direction in order to achieve straight and coherent lane patterns.)

Successful application of PFGE requires the isolation of genomic DNA intact or at least as extremely large fragments. This is virtually impossible by conventional means, as the DNA tends to fragment when the cells are lysed. Instead, intact cells are embedded in agarose, and both lysis and restriction digestion are carried out in a plug of agarose excised from the initial block. The agarose plug is then inserted into the gel and the DNA fragments migrate from the plug into the gel proper during electrophoresis.

PFGE has been employed for bacterial typing, but the technical demands of the process make it less popular than the other methods described. It should be noted, however, that it does provide a method (although not the only one) for the preparative separation of chromosomes from eukaryotic cells.

13.4 Human genetic diseases

13.4.1 Identifying disease genes

One of the most important applications of gene technology lies in the ability to detect genetic mutations that result in specific diseases, or in a predisposition to the development of such a disease. We have already discussed, in different chapters of this book, various methods employed for identifying mutations that cause disease (or other distinguishing traits). Here, we will take the opportunity to look back at the general strategy and integrate the different parts we have touched upon earlier.

The classical approach towards identifying a disease gene is based on the concept of linkage analysis, as introduced in Chapter 12, which enables the position of genetic markers to be mapped to specific regions of one of the chromosomes. Before the advent of molecular methods, the construction of a genetic map in model organisms (especially *Drosophila* and mice) was carried out by painstaking hard work, involving large numbers of controlled crosses between strains with different genetic markers. The extent to which two different genetic markers are inherited together (co-segregate), or in other words the degree of linkage between the markers, provides an estimate of their relative position. Genes that are close to one another will tend to be inherited together. In this way, a genetic map (or linkage map) can be gradually built up, showing the relative position of the available genetic markers. The position of specific disease-associated genes can be located by the degree of linkage to other markers. Of course linkage analysis in humans is much more difficult, as controlled crossing is not an option.

In some cases, it is possible to supplement this with another form of mapping, known as *cytogenetic mapping*. Staining of eukaryotic chromosomes with a dye such as Giemsa stain produces a characteristic banding band, as some parts of the chromosome take up the stain much more heavily than others (see Figure 13.9). If the mutation causes a change in the banding pattern, then its position can be detected microscopically. This is especially useful in organisms such as *Drosophila*, where the salivary gland cells contain bundles of sister chromatids lying side by side. This *polytene* structure makes the banding pattern much more obvious. However cytogenetic mapping is also useful in other organisms, as an initial stage in gene mapping – for example the human gene coding for the enzyme phenylalanine hydroxylase (which is defective in people with phenylketonuria) can be located to the

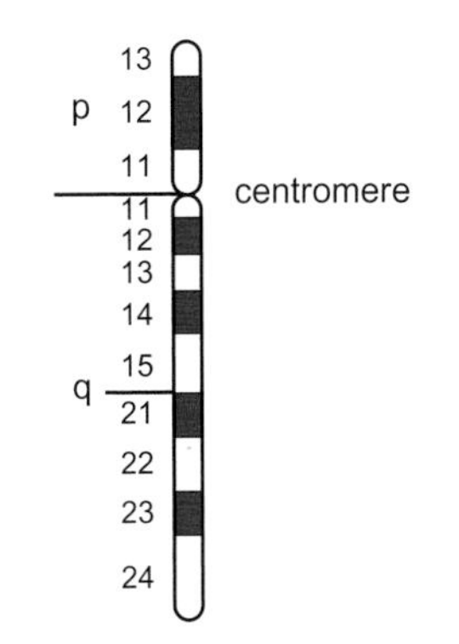

Figure 13.9 Chromosome banding patterns. This shows a simplified representation of the banding pattern of human chromosome 12

band designated 12q24 on the long arm of chromosome 12. However, this is a relatively imprecise method of mapping, with a resolution of several million base pairs. If the gene concerned has already been cloned, *in situ* hybridization can be used to determine its position on the chromosome more precisely. This usually involves fluorescently labelled probes, and hence is known as *fluorescent in situ hybridization* or *FISH*.

One limitation of genetic linkage analysis is the need for markers that have an identifiable phenotype. Since there are a limited number of such markers, even in well-studied organisms, the linkage map will be relatively crude. However, we can use molecular markers, such as RFLPs and microsatellites (see earlier in this chapter), to locate a disease-associated gene more precisely, by mapping its linkage to a specific polymorphism. This approach can be applied to human genetics by examining very large pedigrees from which DNA samples are available, rather than by making controlled crosses. The availability of the human genome sequence has lead to the identification of a large number of *single-nucleotide polymorphisms* (SNPs), which are also important as molecular markers.

These forms of gene mapping simply show you the chromosomal location of a gene that determines susceptibility to a specific disease (or some other trait). They do not identify the gene, its product, or its function. However, they do provide a route for cloning such a gene. It is important to realize that we may not have any information about the function of this gene, beyond knowing that there is a gene in which mutations result in a certain disease, so none of the approaches outlined in Chapter 7 for identifying genes in a library are available to us. However, if we have mapped its position well enough, so that we know it is close to a defined marker (whether this is another gene, a polymorphism or a microsatellite), then we can use a probe for that locus to screen a large insert gene library (using for example YAC vectors) to obtain a clone that carries the defined marker and a substantial amount of adjacent DNA, which we hope contains the disease gene. Since this form of cloning relies solely on the position of the gene, it is known as *positional cloning* (Figure 13.10).

The distance that can be covered by this approach can be extended by *chromosome walking* (see Chapter 12). In this technique, the first clone, carrying the known marker, is used to probe the gene library again, so as to identify clones carrying adjacent sequences; these are in turn used as probes to obtain the next region of the chromosome, and so on. When your 'walk' reaches the region where you expect the gene to be, analysis of the sequence by the methods described in Chapter 9 will enable the identification of genes and their putative products. This will lead to a candidate gene that may be responsible for the disease in question, and the structure of this gene can be tested in individuals with and without the disease to confirm the association.

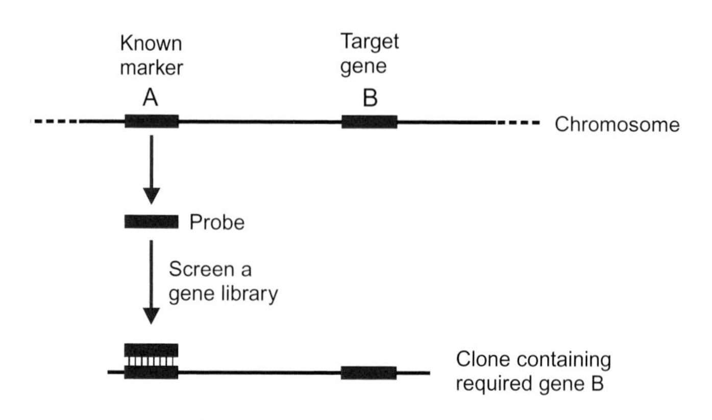

Figure 13.10 Positional cloning

The processes of positional cloning and chromosome walking are simpler to describe than to carry out. However, the availability of the sequence of the human genome, and of genomes of other organisms, makes it a lot simpler. It is now possible to screen putative genes in the area of interest in a matter of hours on a computer, rather than in months or years in a laboratory.

An alternative approach is to base the search on gene products (mRNA or proteins) rather than on the DNA itself. This is based on the assumption that expression of the relevant gene will be altered in the diseased state, or that the structure of the product will be different. Both of these assumptions can be used – either separately or in combination – to devise protocols for the identification of candidate genes. The most important of these protocols are described in Chapter 14.

The final identification of the relevant gene will usually rely on a combination of techniques. The putative genes identified by positional cloning and chromosome walking can be tested for their association with the disease state by specifically testing the expression of those genes in the diseased tissue, or by analysing the structure of the protein. In experimental animals, final confirmation will come from a combination of gene knockout and complementation experiments (Chapter 15).

The power of these techniques therefore relies on a combination of approaches. Classical genetics (often termed *forward* genetics) starts with a phenotype (whether a diseased state or yellow peas) and works towards an identification of the genes responsible. On the other hand, the molecular approach offers the possibility of starting with differences in a defined DNA sequence, and investigating the phenotypic effects of such a change. This is often called *reverse* genetics.

Sometimes one or more candidates are obvious without any further experiments being necessary. For example no reverse genetics was needed to conclude that mutations in the red and green cone photopigment genes were candidates for the cause of red/green colour blindness. Nor was any forward genetic analysis needed – the linkage of these conditions to the X chromosome is well established from observations of their pattern of inheritance. Thus, the mapping of these two *a priori* candidate genes to the X chromosome constituted strong supporting evidence for the initial hypothesis. In this way, forward and reverse genetics interact – the chromosomal location and the expression pattern are both important parts of the jigsaw puzzle.

The central piece of evidence, however, is quite naturally the sequencing of the allele from affected individuals and the confirmation that it is different from the wild-type. This is not, however, sufficient on its own. The mutation must first be analysed, using methods that we have described in Chapter 9. First of all, is it located where it could be predicted to have a biological effect? This would often involve the alteration of the encoded amino acid sequence in a subtle or more drastic way. However, even if the coding region is completely unaltered from the wild-type, mutations in the promoter region that affect the expression of the gene, mutations in splice sites or insertions that otherwise affect mRNA processing could still cause disease. However, even if you did show the cause to be one of these, the scientific community would normally demand even more evidence before your discovery could be published. Some polymorphisms, such as amino acid substitutions, occur completely normally in the population without causing any abnormality. One way to show that the mutation causes disease is to express the mutated protein and compare its biological properties to its wild-type counterpart. Another one, preferably used in combination, is of course to show that the mutation that you have discovered co-segregates with the disease to a significant degree in a genetic pedigree.

13.4.2 Database resources

Of course the availability of the human genome sequence has had an enormous impact on the ease with which genes associated with disease can be identified. There are a number of databases that can be used in this context (see Box 9.1 in Chapter 9 for a list of on-line resources with web addresses). In particular, the dbSNP database contains information about short insertion and deletion polymorphisms as well as single nucleotide substitutions, while OMIM (Online Mendelian Inheritance in Man) is a catalogue of human genes and genetic disorders.

13.4.3 Genetic diagnosis

The identification of mutations that give rise to human genetic diseases not only leads to significant insights into the nature of those diseases, but also provides a mechanism for molecular diagnosis. This is of particular benefit in prenatal diagnosis, which involves obtaining samples containing foetal cells by amniocentesis or chorionic villus sampling – the latter technique having the advantage of being able to be carried out at an earlier stage of pregnancy, which is important if abortion is considered as an option. The increasing use of *in vitro* fertilization makes other options available, since it is possible to remove a single cell (blastomere) at the eight-cell stage, by micromanipulation; this cell can then be tested (pre-implantation genetic diagnosis). The remaining seven cells continue to develop normally and can subsequently be implanted in the uterus, if the genetic tests are satisfactory. The gene(s) to be tested can be amplified by PCR and specific point mutations detected by the use of a pair of labelled oligonucleotides, one of which will hybridize to the normal sequence and the other will be specific for the mutant (see earlier in this chapter). Alternatively, real-time PCR procedures (see Chapter 8, also described earlier in this chapter) can be devised that will differentiate the two versions of the sequence directly, producing much quicker results.

A further advantage of molecular techniques for detecting mutations is the ability to detect the presence of the defective gene in heterozygous carriers. If the mutation is recessive, the heterozygotes will show no symptoms of the condition, but a proportion of the offspring of two heterozygous parents will have the disease. In the past, genetic counselling of prospective parents from families with a history of a genetic disease has relied on statistical probabilities to assess the likelihood that both parents carry the affected gene. With molecular techniques, it is possible to be certain, one way or the other.

Similar considerations apply to genes which only show their effects later in life, and to those genes that only cause a predisposition to a specific disease. For example, in some families there is a particularly high incidence of breast cancer, indicating an inherited predisposition to this disease. In many cases this has been traced to a specific gene, *BRCA1*. Individuals carrying a mutated gene have a higher risk of developing breast cancer, although the presence of the affected gene does *not* mean that it is certain that they will develop the disease.

There are substantial ethical issues raised by the availability of these techniques. Even if one accepts abortion as an option under some circumstances, its use as a method to prevent the birth of a 'defective' child raises the uncomfortable spectre of eugenics. Furthermore, the ability to test people for the presence of genes that may affect their subsequent health or lifespan has serious implications for their ability to obtain life or medical insurance, subject to legal restrictions that may limit the powers of insurers in this respect.

13.5 Molecular phylogeny

Traditionally, systematics – the classification of organisms into species, genera and higher groupings – relied on phenotypic properties, such as physiology and anatomy for higher organisms, and biochemical characteristics for bacteria. Once the role of DNA as the genetic material was established, in the middle of the last century, it became apparent that molecular methods, based on DNA similarity, could be used to supplement the phenotypic approach. Initially this was based on rather imprecise methods such as DNA hybridization, but the advent of the ability to sequence proteins and genes caused a revolution in systematics, and since these species, genera and so on have arisen by the accumulation of gradual changes by evolution, a study of the current similarities and differences between organisms can be used to infer the order and time of the divergence of the various species. This is molecular phylogeny.

Not all of the methods described in this chapter are equally suitable for this purpose. For evolutionary analysis, establishing the order in which species diverged from one another, we do need to be confident that the rate of accumulation of variations is similar for each species over time (i.e. there is a reasonably constant *molecular clock*). This is not necessarily true for methods such as RFLP analysis, especially when it is based on a mobile genetic element such as an insertion sequence. We can illustrate this by reference to Figure 13.11. Here we can see that A and B are more similar to one another than either is to C, and a computer program will generate a dendrogram to display that relationship visually. However, unless we know that the rate of variation is constant, we should not jump to the conclusion that C diverged from a common ancestor of A and B, and that the difference between A and B happened more recently. If for some reason variation has occurred more rapidly in C (and its ancestors) than in A and B, then C might have diverged more recently, but has undergone more rapid variation since – for example due to enhanced mobility of the insertion sequence. This does not matter if we are simply typing or fingerprinting the organisms, or the samples involved. We would simply need to be confident that only the same strain, or samples

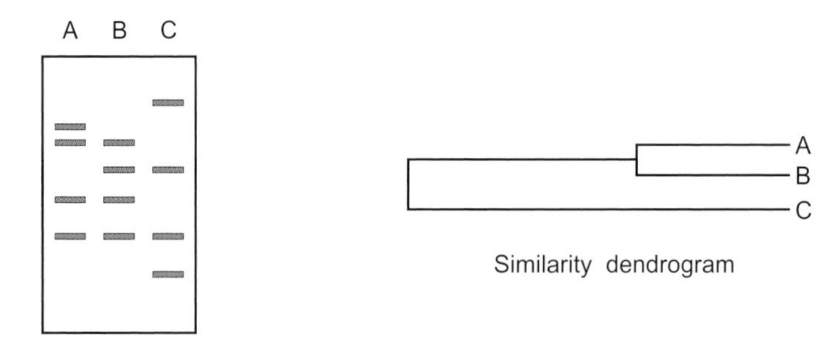

Figure 13.11 Analysis of similarity of RFLP patterns. Based on the RFLP patterns, A and B are more similar to one another than either is to C. The dendogram is a way of displaying these relationships visually. However, it does not necessarily represent the true evolutionary history of these organisms

from the same individual, would give identical patterns. We will consider the generation of phylogenetic trees in more detail in the next section.

Comparison of DNA or protein sequences provides a much more reliable method of establishing phylogenetic relationships. In Chapter 9, we looked at a simple example, based on comparison of a single protein (thymidylate synthase) from a variety of organisms. There are a number of potential problems with such an approach. If the protein is an essential one, it may be highly conserved, so you would get less variation than expected. This would not matter much if the effect was the same in all organisms, but it might not be. Some organisms could be more tolerant of variation, for example because of other differences in their physiology. More seriously, you could find that proteins with the same function have evolved from different ancestral proteins – an effect known as *convergent evolution*. You may therefore find that you get different phylogenetic trees depending on your choice of protein. A common approach to counter these problems is to use a number of proteins for the comparison. You do not need to compare them individually – you can join the sequences together to produce a single artificial protein sequence. However you are still left with the objection that by examining protein sequences you are ignoring all the non-coding DNA, and you are also ignoring any synonymous changes in the DNA sequence (i.e. changes which replace one codon with another one coding for the same amino acid).

The alternative is to base your analysis on the DNA sequence itself. This could be selected fragments of the genome, coding or non-coding, or the complete genome sequences, provided that you can identify the regions that align with one another. One advantage is that this can be applied to incomplete genome sequence data, without having to identify the coding sequences.

It is worth noting that the choice between protein and DNA sequence comparisons depends, in part, on your view of evolution. A deterministic view would regard evolution as occurring solely through natural selection, and therefore only changes in the protein sequences are relevant. The opposing view is that a substantial component of the variation between species occurs by random genetic drift, with neutral changes becoming fixed because of limitations in the population size. These variations will only be detected at the DNA level. In practice, evolution is a mixture of both effects, but the relative contribution may differ.

13.5.1 Methods for constructing trees

There are many methods available for inferring a phylogenetic tree from sequence data, none of them perfect and all quite complex, so a full treatment is beyond the scope of this book. However, it might be helpful to introduce some of the basic concepts.

Four DNA sequences for comparison

A: C A T A G A C C T G A C G C C A G C T C
B: C A T A G A C C C G C C A T G A G C T C
C: C G T A G A C T G G G C G C C A G C T C
D: C C T A G A C G T C G C G G C A G T C C

Distance matrix, showing the number of base differences in each pairwise combination

	A	B	C
B	5		
C	4	7	
D	7	10	7

A and C show the fewest base differences, so they are combined into a single operational taxonomic Unit (OTU) and the matrix is recalculated using the average number of differences (e.g., A to B = 5, C to B = 7; mean = 6).

	AC	B
B	6	
D	7	10

The fewest changes are shown between B and the combined OTU AC, so B and AC are combined into one OTU, and the matrix is recalculated:

	ACB
D	8.5

These data can now be used to construct a tree - see Figure 13-13

Figure 13.12 Phylogenetic trees based on DNA sequences; sample data and UPGMA calculation

As an example, we can start with four short sequences (Figure 13.12). The first step is to count the number of base differences between each pair of sequences, and construct a difference matrix, as shown in the figure. We immediately come up against two problems. Firstly, the sequences to be compared may not be the same length, so we should really divide each count by the length of the sequence to get a true measure of dissimilarity. In this case, the sequences are the same length, so it is clearer to use the actual number of differences. The second problem is less obvious. As the sequences diverge, an A may change to a G, for example; as they diverge further, there may be a second mutation which changes the G back to an A. A simple count of the number of differences will therefore underestimate the true evolutionary difference. A statistical correction can be applied to account for this, but in this case we have left it as it is, to keep it simple.

Clustering methods, such as UPGMA (unweighted pair group method with arithmetic means), are often used to develop trees from such a matrix. To explain this, we have to introduce the concept of operational taxonomic units (OTUs). This simply means the sequences, or groups of related sequences, that we are using at each point. Thus we start with four OTUs, and the aim is to combine these sequentially until all the branches in the tree have been defined.

We inspect the matrix and identify the pair of sequences (OTUs) that are most similar, which in this case are A and C, with only four differences. (Note that we will come back to this later on, as it is an example of how this method can give rise to an incorrect tree.) We combine A and C into a single OTU (AC), and calculate the average distance from this combined OTU to each of the others. This gives us a new matrix, now with only three OTUs. Again, we combine the two most similar OTUs (which would be B and AC) into one OTU (ACB), and calculate a further matrix, now with only two OTUs. The final step is to calculate the average distance from ACB to D, and we have a tree (Figure 13.13) which appears to show the relationships between these OTUs and the genetic distance separating them. The depth of the

Using the data in Figure 13-12, A and C are first combined into a single OTU. The 4 sequence differences are distributed equally between the two branches of the initial tree

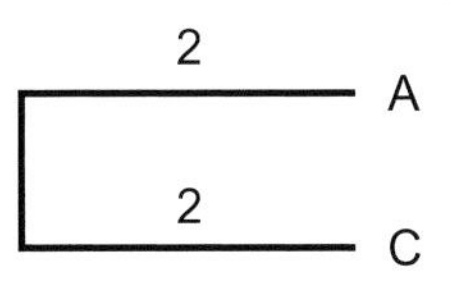

Following recalculation of the matrix, AC and B are combined into one OTU. Again, the mean number of differences (6) is divided equally between the two branches. Add the numbers together as you trace the path from A or C to B

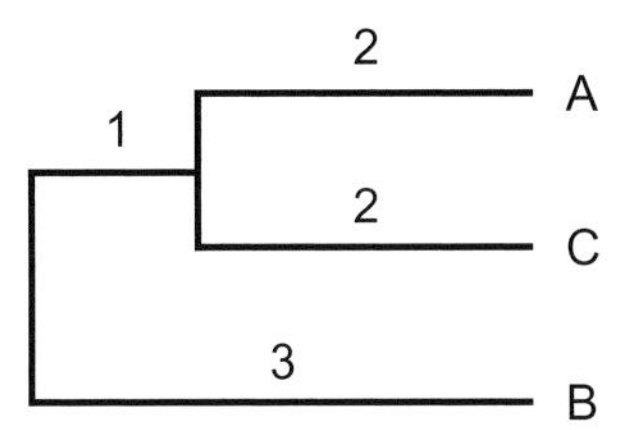

Finally, ACB is combined with D into a single OTU

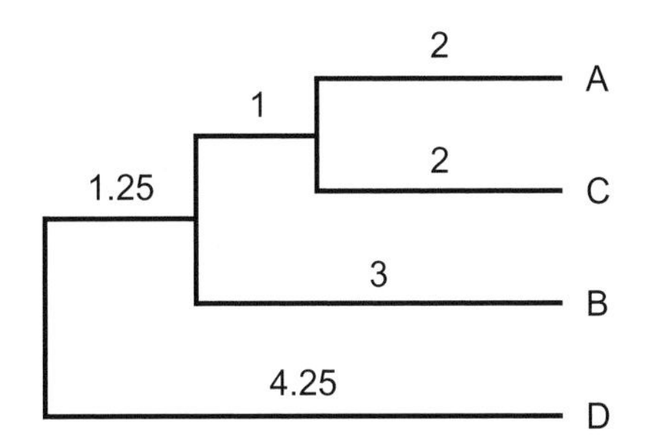

Figure 13.13 Derivation of a phylogenetic tree using UPGMA

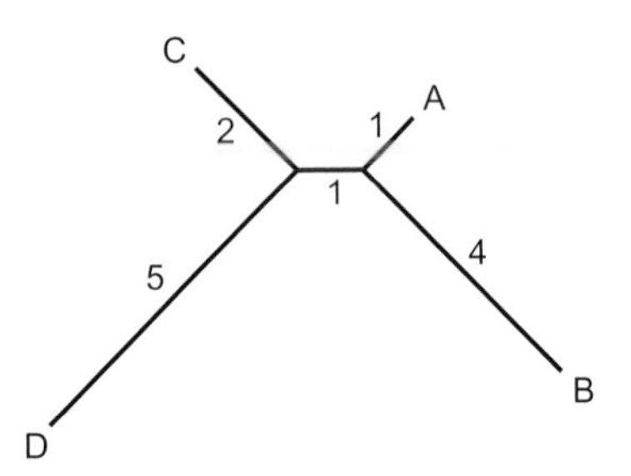

Figure 13.14 An unrooted phylogenetic tree. Produced by the neighbour joining method, from the data in Figure 13.12

divergence of each branch is shown as half the distance separating the OTUs – for example the A and C differ by four bases, so each branch is given a length of 2. This is a limitation, as it means that the individual path lengths are not accurately represented. Therefore, while the true distance from A to B is 5 changes, and that from B to C is 7, both lengths are shown on the tree as $(2 + 1 + 3) = 6$ (the average of 5 and 7).

Furthermore, this method only gives viable results if the rate of change is the same for all branches, i.e. there is a uniform linear molecular clock. As this condition is frequently not met, clustering methods can easily give erroneous trees. They were popular initially because of their relative simplicity and being less demanding on computer power than other methods. Increased computer power makes this much less of a problem. There are ways of overcoming these limitations, but the availability of superior methods means there is little reason to persist with clustering for phylogenetic inference.

One such method, which is conceptually similar to clustering methods, but allows for variation in the clock speed, is the neighbour joining (NJ) method. Neighbours are two OTUs that are connected by a single node in the tree – so if you look ahead to Figure 13.14, A and B are neighbours, as are D and C, but A and D are not. Of course we do not yet know the tree, so there a number of possible trees, and potential neighbours. The first step is to select a pair of OTUs on the basis that separating that pair from the others will give the smallest sum of all the branch lengths in the hypothetical tree. Those two OTUs are then combined to form a single OTU (or node) and a new distance matrix is computed. This process is repeated until all the branch points have been defined.

With the data in Figure 13.12, the NJ method gives the tree shown in Figure 13.14. This is an *unrooted* tree. It does not imply any direction to evolution, nor what the ancestral organism looked like. The usual way of obtaining a *rooted* tree is to include an *outgroup* in the comparison, i.e. an organism that we know (or can be reasonably confident) diverged from all the others before

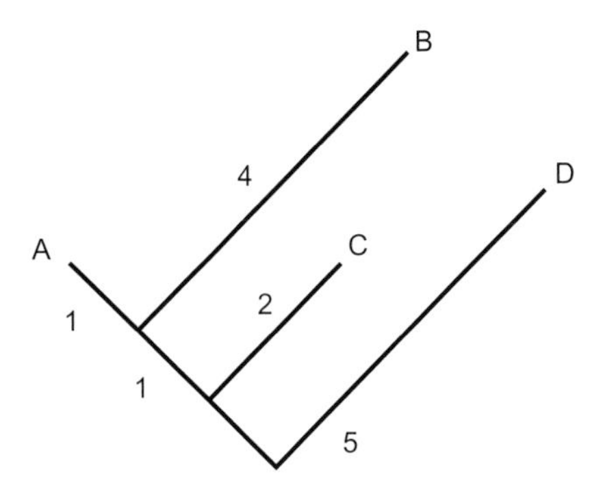

Figure 13.15 A rooted phylogenetic tree. Produced with the same data as Figure 13.14 using D as an outgroup to root the tree. Note that, although the root must be along the branch between D and ABC, its precise position cannot be determined in this analysis

they diverged from one another. This may be from independent evidence, such as fossil records, or we may infer it from the sequence data itself. In this case, we may be able to take D as an outgroup, since its sequence is so different from each of the others. Using D as an outgroup, we obtain the rooted tree shown in Figure 13.15. Note that this method does not tell us exactly where the root is, except that it is somewhere along the branch joining D to the node where C diverges from A and B.

We can now consider why the UPGMA and NJ trees are different. The reason lies in the length of the branch leading to B, presumably as a consequence of a higher rate of variation in this lineage. As a result, A appears to be more similar to C than to B, and therefore, in the clustering (UPGMA) method, A is initially clustered with C. On the other hand, the NJ method considers all the path lengths, which show for example that B is much closer to A than it is to either C or D. The NJ method therefore produces a much more reliable tree.

A real phylogenetic study would use longer sequences, and would usually include a larger number of sequences. In such a case, there are a large number of possible trees. How confident can we be that the tree we have constructed is the right one? The short answer is that we cannot be certain. We are attempting to infer evolutionary history from an examination of the current state of things, and that is inherently unreliable. However, we can obtain a measure of how reliably the tree depicts current relationships. A common procedure for doing this is known as *bootstrap* analysis. Essentially this means taking random samples of the data and repeating the construction of the tree each time. If the tree is reliable, we will get the same tree every time. The extent to which this is true gives us the measure we need of the robustness of the tree.

If we knew not only that the clock speed was uniform and constant, but also how fast it ran – i.e. how many base changes occurred in a period of time – then we would be able to estimate how long ago these species diverged. In general, there are too many uncertainties to do that reliably, but if there is other evidence to place one of the nodes at a specific point in time, then we can use the tree to put a time scale on the other changes. For example, from a comparison of Neanderthal DNA with chimpanzee and modern human DNA, and using an independent estimate of the divergence of humans and chimpanzees of 6 500 000 years ago, it has been estimated that modern humans and Neanderthals diverged about 516 000 years ago.

14 Post-genomic Analysis

In a previous chapter we considered ways in which we can study the expression of individual genes. Now, having dealt with the sequencing of complete genomes, we can move on to look at how we can analyse gene expression across the genome as a whole. These methods, collectively described as post-genomic analysis, include *transcriptomics* (the study of the complete collection of mRNA transcripts in a cell, tissue or organism at a particular time) and *proteomics* (analysis of the complete set of translated proteins).

14.1 Analysing transcription; transcriptomes

In Chapter 10, we described methods such as northern blotting and qRT-PCR for studying the expression of specific genes. There are a variety of reasons why we might want to study individual genes, including the identification of disease-associated genes when a limited number of strong candidate genes are known. On the other hand, we commonly want to identify and compare the complete spectrum of genes that are expressed under different circumstances. This would include genes that are expressed in one organism but not in another, genes that are only expressed in specific tissues, and genes that are differentially expressed under varying environmental conditions. Over the years, a number of techniques have been developed for identifying the mRNA species that are expressed in different samples. Such methods are able to answer some very important questions, or at least give us a strong clue. For example, they can tell us which gene is associated with a degenerative disease, by comparing diseased and healthy tissue. They can tell us which genes of a pathogenic bacterium are expressed during infection of a suitable host, as opposed to those that are expressed in laboratory culture. Comparisons of gene expression can also indicate which genes are involved in various

From Genes to Genomes, Second Edition, Jeremy W. Dale and Malcolm von Schantz

stages during development of an organism. This is far from an exhaustive list.

Modern methods rely mainly on the use of microarrays, but we will first describe some of the older procedures. These still have some potential applications, especially as they can be performed in the absence of genome sequence information.

14.1.1 Differential screening

One approach is to identify cloned genes that are differentially expressed by using *differential screening* of a gene library (see Figure 14.1). This is similar to the method for screening a gene library, as described in Chapter 7, except that here we are comparing the effects of screening with two different, mRNA-derived, probes. Starting with the library in the form of colonies or phage plaques on an agar plate, the procedure involves producing two identical filters containing DNA imprints of the gene library (either by repeated transfer from the same plate, or by transfer from two identical plates produced by replica plating). To obtain the probes for screening the library, we extract total mRNA from two samples (two strains, or two tissues, for example), and subject each of these complex mRNA mixtures to reverse-transcription in the presence of labelled nucleotides. One of the filters is hybridized with one of these pools of labelled cDNA, while the second pool

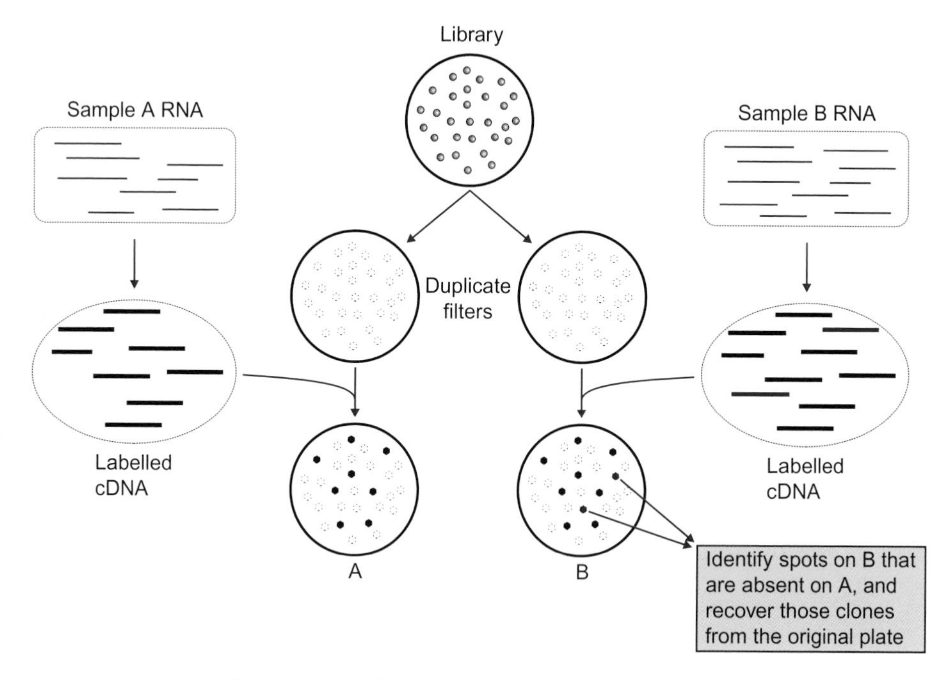

Figure 14.1 Differential screening of a gene library

is used to probe the second filter. Clones that show a positive signal with one probe but not the other are candidate clones of differentially expressed genes.

This description assumes that the specific transcript is completely absent from one sample (sample A in the figure). However, genes are not necessarily turned off that absolutely; there may still be a small amount of the specific mRNA present. This might result in a different intensity of some spots with the two probes, and we could use that for our differential screening. Yet the situation is rather better than that. If there is only a very small amount of a specific mRNA species (a very low abundance), then the amount of the corresponding cDNA in the probe will be too low to produce a visible signal. Furthermore, the synthesis of cDNA is likely to be more efficient for the high abundance mRNA, leading to even further depression of the signal from the low abundance species. So any low-abundance mRNA in sample A may give, not just a weak signal, but no signal at all. This actually causes a problem in some cases, in that mRNA species that are less abundant, but are still differentially expressed, may not be detectable in either sample. RT-PCR can be used to generate more representative cDNA probes, but the more representative the probes become, the more difficult it is to pick up differential expression where the ratio between the two sources is not high.

One advantage of this approach, especially when dealing with an organism for which the genome sequence is not available, is that we not only identify the differentially expressed genes, but also in the same step find clones carrying those genes. These can then be sequenced and characterized.

14.1.2 Subtractive hybridization

An alternative procedure is to remove the common sequences from the probe, leaving only those which are present in the test material; this is known as *subtractive hybridization*, and is illustrated in Figure 14.2. We are looking for transcripts that are present in one sample (which is called the *tester*) but are absent from the other (the *driver*). For example, the driver may be derived from diseased cells so we would like to identify genes that cannot be expressed in those cells; the tester would then come from normal cells. Alternatively, we could be interested in genes that are only expressed under certain conditions, for example bacterial genes that are switched on during anaerobic growth, in which case the driver is the 'normal' sample. Whichever way round we look at it, the procedure is designed to find mRNA that is present in the tester and not in the driver.

The details of the procedure vary, but in the example shown the first step is to produce cDNA copies of the mRNA from both samples, by reverse transcription. The cDNA from the driver is labelled with biotin (in the figure it is shown as end-labelled for clarity, whereas in practice the biotin will be

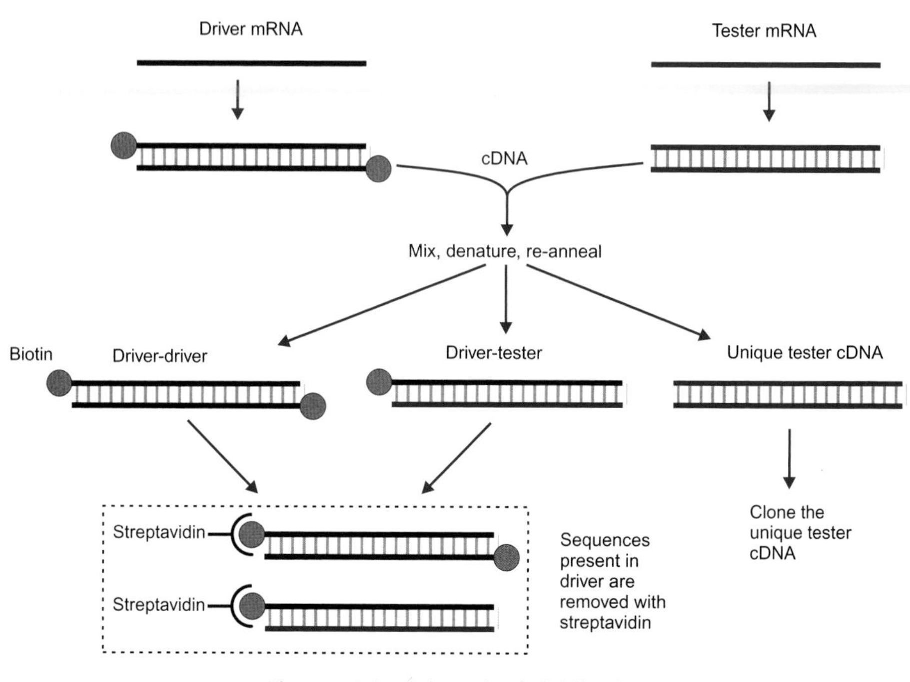

Figure 14.2 Subtractive hybridization

incorporated throughout the molecule). Biotin has strong affinity for streptavidin, so all the molecules containing biotin can be easily removed. The next step is to mix a large excess of the biotin-labelled driver cDNA with the tester cDNA, and to denature and re-anneal the mixture. Any tester cDNA that is complementary to driver cDNA will form hybrid molecules, with one tester strand and one driver strand. Since the driver strand carries biotin, the hybrid molecules can be removed by streptavidin (together with any excess driver–driver double-stranded DNA). You will then be left with a mixture that is enriched for those tester cDNA molecules that are unique, i.e. they are not present in the driver population.

The subtracted tester transcripts can be used as a probe for screening a gene library. However, that way you only have one shot at enjoying the fruits of your labour. Moreover, rare transcripts are unlikely to yield any detectable signals even though they may be the ones you are looking for. A better method is to produce a subtractive cDNA library by adding adapters (see Chapter 4), and ligating the tester cDNA into a vector. This library is then available for repeated testing, either by systematic analysis of all clones, or by comparative screening of duplicate lifts with labelled tester and driver cDNA. This method has been used successfully on a number of occasions for the identification of disease genes, by using a wild-type source as the tester and cells from the disease phenotype (lacking a specific gene product) to generate the driver.

One disadvantage of this procedure is that it requires relatively large amounts of mRNA. The use of PCR, combined with the principles of the above procedure, provides a range of more powerful methods. One such procedure, known as representational difference analysis (RDA) is illustrated in Figure 14.3. In this example, linkers are ligated to the ends of the cDNA to provide priming sites for PCR analysis. In the first part of the procedure, the tester and driver samples are treated in the same way. Then after removing the initial linkers from the PCR products, a different pair of linkers is added to the tester PCR product, but not to the driver product. When the two products are mixed and allowed to anneal (again using a large excess of driver), any DNA strands that are present in both samples will anneal to form a hybrid molecule, while the unique tester strands, which will not be able to anneal to the excess driver DNA, will form specific tester–tester molecules. These can be amplified by subsequent PCR using primers that are specific for the linkers that were added only to the tester DNA, which will thus undergo exponential amplification. Any driver–driver molecules (not shown) will fail to amplify, because they lack this primer-binding site. The tester–driver hybrids carry unpaired ends (from the linkers that were added to the tester molecules), and these can be removed by digestion with a nuclease that is specific for single-stranded DNA (such as mung bean nuclease).

14.1.3 Differential display

Differential display is a PCR-based method of comparing two cDNA populations. One primer is designed to bind to the poly-A tail, while the other is a random sequence. By running PCR at a very low annealing temperature, a subpopulation of the cDNA molecules in the sample is amplified. This subpopulation will be small enough to be resolved in a polyacrylamide gel. If you perform identical PCR reactions with two cDNA populations, you will be able to identify cDNA species that are expressed at different levels in the two samples, or even completely absent in one of them. These can then be excised from the gel, reamplified and sequenced. If novel, their full sequence can be obtained by library screening or by RACE (see Chapter 8).

The presence of these cDNA species amongst the amplified mixture is a random event governed by the degree of sequence similarity with the primers; the method itself does not favour the amplification of differentially expressed messages. Thus, a considerable number of primer combinations will have to be tried, and even so, the analysis will be far from exhaustive. Nonetheless, although nicknamed ‘differential dismay’ by some frustrated practitioners, the PCR-based nature of this method carries some specific advantages, including the use of small amounts of starting material, and the ability to

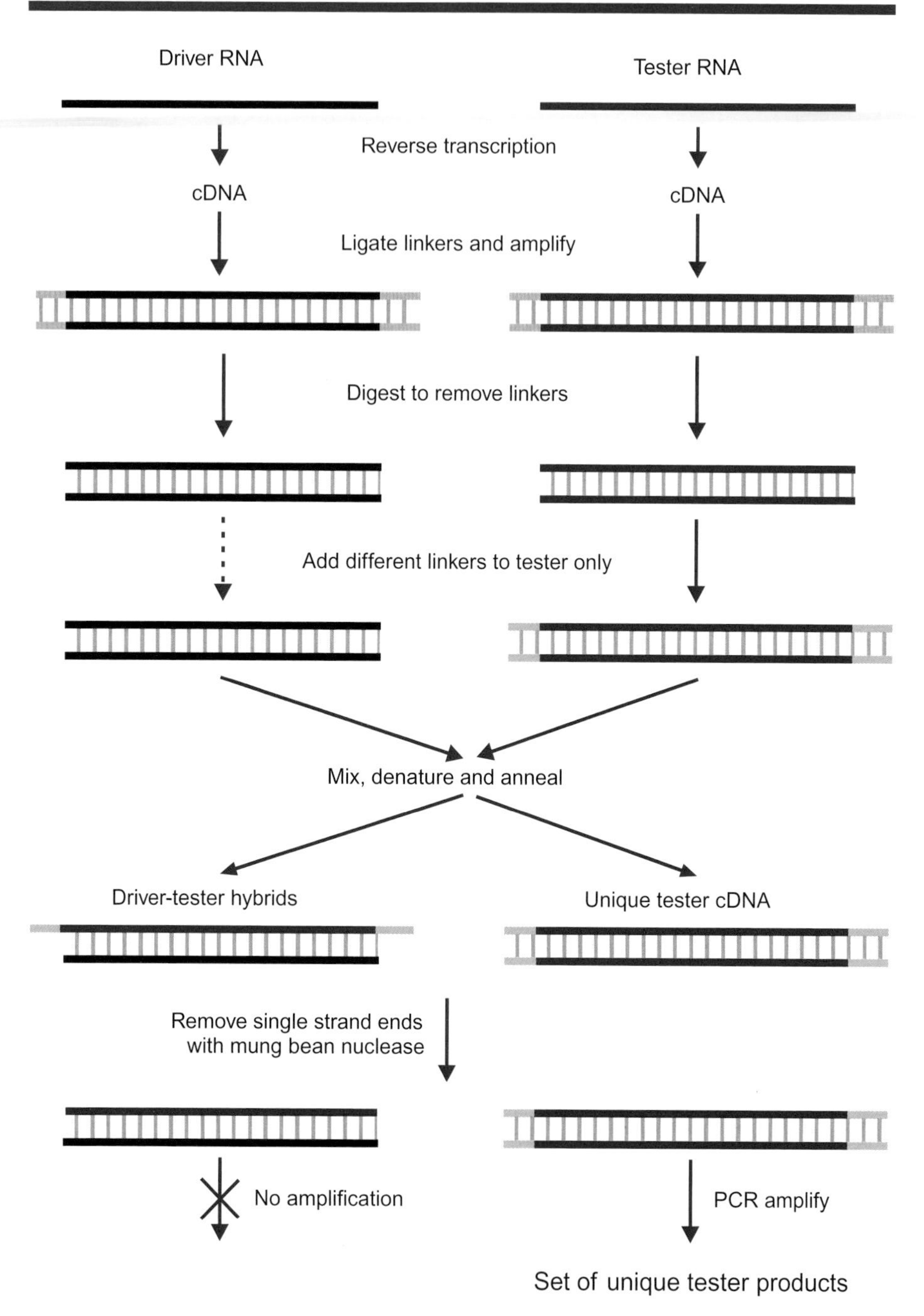

Figure 14.3 Representational difference analysis

identify differential expression even of very rare transcripts. However, this procedure has been largely supplanted by more sophisticated techniques such as representational difference analysis as described above, or by array technology (see below).

14.1.4 Other methods – transposons and reporters

Transposons are mobile genetic elements that are able to move (transpose) from one site to another in the genome, or between plasmids and chromosome. In Chapter 12, we discussed the use of transposons to produce mutations, due to inactivation of a gene into which the transposon has inserted. Since the gene into which the transposon has inserted can be readily identified, this can be employed to link a phenotype with a specific gene. We can extend the concept to the study of gene expression, by inserting a *reporter* gene (see Chapter 10) into the transposon.

This is especially valuable in bacteria for identification of the range of genes that are expressed under a specific set of conditions. For this purpose, we would first make a *transposon library* (see Chapter 12), containing a large number of clones with transposon inserts at different, random, sites. We would then expose the transposon library to the environmental conditions in question, such as anaerobic growth, and identify those clones that show expression of the reporter gene. Cloning, or PCR amplification, of the transposon and its flanking DNA then enables the insertion sites to be identified.

The principle can be extended to the identification of secreted proteins, by including an alkaline phosphatase gene as a reporter (this is referred to as a Tn*pho* transposon). This procedure depends on the fact that alkaline phosphatase activity is only exhibited if the enzyme is secreted, and not if it remains in the cytoplasm. Insertion of Tn*pho* within a gene coding for a secreted protein can give rise to a fusion protein carrying the secretion signals from the native protein joined to the alkaline phosphatase. If this product is secreted, then phosphatase activity will be detected.

14.2 Array-based methods

The development of methods such as subtractive hybridization allowed scientists to pan for differentially expressed gene products, rather than merely examining the expression of a specific gene. Often, the objective has been to search for a specific disease gene, or one or a few genes whose products fulfil a particular role. With the availability of genome sequence data, it became possible to use array technology to address the global question of identifying the complete spectrum of genes that are expressed under certain conditions and not otherwise. This is a much more powerful technique, and much simpler, than those described above.

The basis of the use of *arrays* was described in Chapter 13. However, in that case the arrays were probed with a mixture of genomic DNA fragments, to identify genes that were present in one sample and not another. Now we want to identify genes that are *expressed* differently. The only change is that we will hybridize the array with labelled cDNA, rather than genomic DNA. Those spots in the array that correspond to an expressed gene will yield a positive signal. Using machines to produce many identical copies of the array, you can test for the presence of a wide range of specific cDNA molecules derived from cells from different environments. Various types of arrays, and how they are used, are described below.

Conceptually, this is not really different from differential screening. Subtractive cDNA libraries in plasmid vectors have often been painstakingly picked clone by clone and inoculated in a grid formation on an agar plate, and duplicate colony lifts used for their analysis. The advent of robotic picking and high-density spotting technology, and of electronic image analysis, not only made it technically simpler to produce such arrays, but also largely eliminated the need to limit the array to a subpopulation of particularly suspect transcripts. Furthermore, knowledge of the complete genome sequence of an organism makes it possible to produce arrays of small DNA fragments covering the whole genome, or selected parts of it. If we want an array specifically for testing expression (as opposed to genomic comparisons), we may want to restrict the array to protein-coding regions, either by using cDNA or by identifying ORFs. There are various possible approaches for many organisms.

14.2.1 Expressed sequence tag arrays

If you want a cDNA array, the simplest method would be to plate an unamplified cDNA library, say from the pancreas, and pick the clones one by one. This has its distinctive disadvantages, however. Firstly, the use of your array will be restricted to samples from one tissue. Your array will have limited use with samples from brain tissue, for example, because the brain and the pancreas will have their specific repertoires of genes which are expressed in either tissue but not in both. Since you will want to use your array to test a variety of tissues, you want to make it as widely applicable as possible.

Secondly, when you made your library, you were wise to base it on poly-A-enriched mRNA (if you were working with a eukaryote), thus avoiding a library where 95 per cent of the material consists of endlessly repeated rRNA and tRNA sequences. However, even so you will have quite a lot of repetition. This is because the frequency of any given gene in a cDNA library (unlike a genomic library) is biased according to the levels of expression of that particular gene. Thus, you will find major components such as structural genes (e.g. tubulin), housekeeping genes (metabolic enzymes) and major tissue-specific products (e.g. insulin) greatly over-represented in your library.

A project that has simultaneously helped us towards independence of tissue-specific libraries and the problem of over-representation is the development of *expressed sequence tags* (ESTs). ESTs are cDNA clones where a laboratory equipped with a robot has done just what we outlined above, picked every clone from a library. This has been done with libraries from many different tissues, creating a sort of transcriptome fingerprint for each one of them. They have all been subjected to one sequencing run, yielding a few hundred (unconfirmed) bases from one end, numbered and archived jointly in large public facilities. The sequence tag database is available through GenBank, so that you can search it for similarity to a gene you are interested in, and obtain the clones for a nominal charge.

Systematic computer comparison has made it possible to identify which cDNA clones are unique. It is thus possible to produce cDNA arrays with non-redundant clones. By spotting these onto a membrane (macroarray) or a glass slide (microarray), resources have been created that enable simultaneous study of the majority of the genes in the human, mouse and rat genomes, independently of the sequencing of these genomes. The label of choice of macroarrays is ^{33}P (providing better resolution than ^{32}P), and the labelling is then captured by a phosphoimager. For microarrays, fluorescent labels are used, and the data is collected by specialized microarray readers (as described in Chapter 13). A virtually unlimited number of copies of each array can be produced at a negligible marginal cost. Thus, by obtaining a pair of identical arrays, and screening them with labelled tester and driver cDNA populations, you can in one experiment collect more data than many months of differential display analysis would have given you.

14.2.2 PCR product arrays

One objective of the EST programme was to try to obtain at least one clone representing each gene. Although, in the case of the human genome, this objective was not quite achieved, ESTs have been immensely helpful in the task of identifying those genes that were represented, and their open reading frames. Some genes, however, lack EST correlates. In addition, in the cases of those that do, the method of spotting bacterial clones carrying plasmids is far from ideal. Inevitably, some eukaryotic probes will have high enough sequence similarity to the bacterial sequences to bind non-specifically to the array.

In bacteria, by contrast, ESTs are not available. However, we now know the entire genome sequence for a number of bacteria (see Chapter 12),

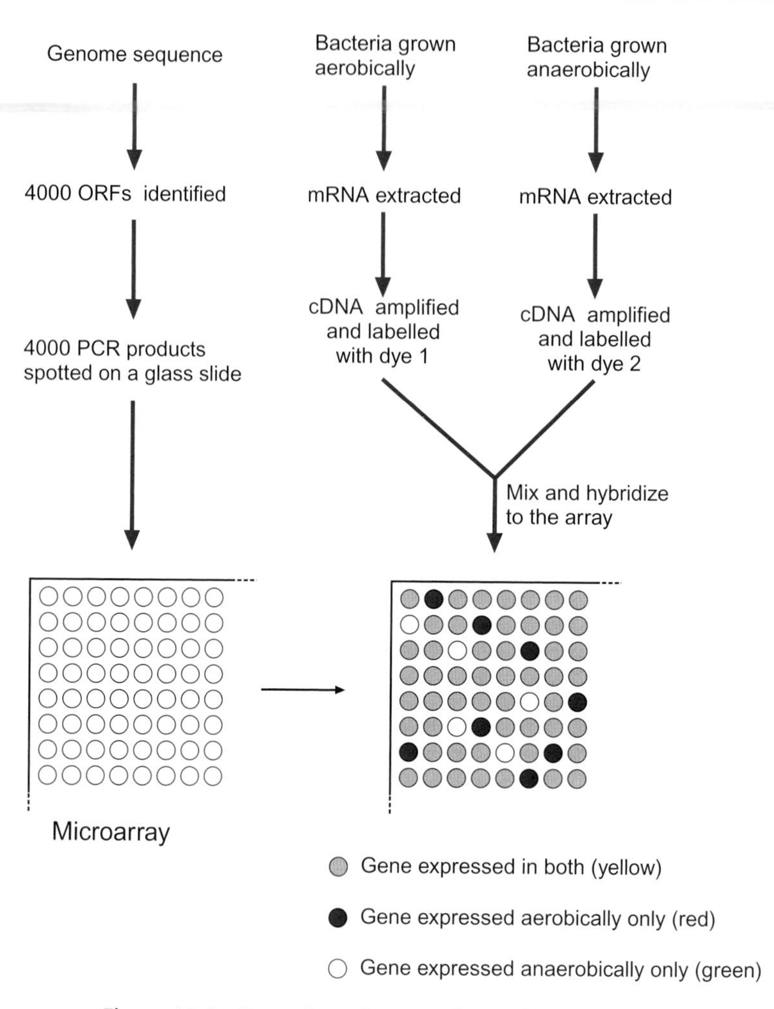

Figure 14.4 Transcriptomic comparison using a microarray

making possible another even more efficient approach for array production (Figure 14.4). Computer methods are used first of all to identify all the open reading frames above a certain size; these are the predicted genes. For each predicted gene, we use the computer to design a pair of primers that will amplify a suitably sized region of that gene. Automated DNA synthesizers are used to make each of these primers, and robotic methods are used to carry out the thousands of PCR reactions needed to make the DNA fragments, which are then robotically spotted onto macro- or microarrays.

Figure 14.4 shows an example of how such an array could be used. In this example, we want to compare the bacterial genes that are expressed under aerobic and anaerobic conditions. From bacteria grown under the different conditions, mRNA extracts are made, and reverse transcribed to produce a mixture of cDNA molecules. These are amplified by PCR and labelled with different fluorescent dyes. The two probes are then mixed and hybridized to the microarray. A microarray reader is used to measure the different fluorescence of the two dyes, for each spot on the microarray. The results are presented visually in different colours. If the probe from the aerobically grown bacteria, labelled with dye 1, hybridizes preferentially (i.e. that gene is expressed at a higher level in aerobic culture), you will see a red spot. Conversely, if the gene is expressed more when grown anaerobically, the spot will be green, because there is more hybridization with the probe labelled with dye 2. If both probes hybridize equally (i.e. there is no difference in expression of that gene) you will see a yellow spot.

Although the visual display is attractive, the real comparison demands a quantitative approach. The microarray reader actually compares the relative levels of the two fluorescent signals and records that data. At a simple level, you might expect that for genes that are expressed to the same extent under both conditions you would get a ratio of 1, and that values greater than, or less than, 1 would indicate differential expression. However, there will be a certain amount of noise in the system, so a cut-off value is commonly employed, taking for example values greater than 2, or less than 0.5, as indicating a difference in expression. More reliably, you could use a microarray that contained a number of spots representing each gene (randomly arranged on the array), and/or do the whole experiment several times, and then use statistical analyses to estimate whether any differences are statistically significant. Note however that a simple statistical test still leaves a problem. If you choose to regard $p < 0.01$ as indicating significance, this means that the event is expected to occur by chance 1 in 100 times. If you are looking at 4000 genes, you would expect to get 40 of them showing such a difference, even if the two samples are identical.

Furthermore, there are a host of assumptions behind this procedure. Most importantly, it assumes that every step in the process (mRNA extraction, cDNA production, labelling, hybridization) is equally efficient for each gene in both samples. This is unlikely to be true, and carrying out repeat experiments will only eliminate random bias, not systematic errors. For example, if a specific gene does not amplify well in one sample (perhaps because of competition with another gene), then no amount of repetition will solve the problem. You also need to remember that, as with most of these techniques, what you are measuring is the *amount* of each mRNA, not the level of transcription of that gene. The amount of each mRNA will be influenced by its stability as well as its rate of production.

Experiments such as these have generated an enormous amount of extremely valuable data, but it is important not to regard them as definitive. They produce a list of candidate genes that are worth looking at further, but the data needs to be verified by more reliable techniques, such as quantitative

RT-PCR (see Chapter 10), applied to each of the candidate genes individually.

Mammalian genomes are much larger, of course, but the availability of complete genome sequences makes it possible to produce a complete array of PCR-amplified products in the same way. However the larger number of genes involved makes it desirable to use even higher density arrays (see below).

14.2.3 Synthetic oligonucleotide arrays

Microarrays, as described above, can equally well be produced with a set of oligonucleotides, synthesized in the conventional manner and then attached to the glass slide. Such oligonucleotide microarrays are used in the same way as the PCR product arrays.

However oligonucleotide arrays can also be made, at a much higher density, by synthesizing the oligonucleotides *in situ*, using photolithographic techniques analogous to those used in the manufacture of computer chips. For this reason, such arrays are also known as *gene chips*. (The term is sometimes used for microarrays in general, but it should be restricted to those produced by this method; the term GeneChip is a trademark of the company that produces them, Affymetrix. Hence these arrays are also known as Affymetrix arrays.)

The procedure, as illustrated in simplified form in Figure 14.5, involves the sequential addition of nucleotides to each position in the array. The use of photolabile blocking groups prevents the addition of nucleotides to positions where they are not wanted. The blocking group is destroyed by the action of a laser to activate selected positions.

Initially, after attachment of linker molecules, with a protective blocking group, to the substrate, a mask is positioned over the array, so that the laser will only be allowed to reach the positions where the first nucleotide is to be attached. After the selected positions have been activated, the first nucleotide, in this case G (with a blocking group), is attached. A second mask is then positioned, to allow activation of positions where the next nucleotide is an A, followed by laser exposure, attachment of A residues (with a blocking group) – and so on.

Such an array can contain millions of oligonucleotides on a single chip, and they are therefore of great benefit for screening organisms with larger genomes – either for expression studies, or for screening for nucleotide polymorphisms.

The principle of the use of a gene chip is similar to that described for microarrays in general, involving hybridization with fluorescently labelled probes.

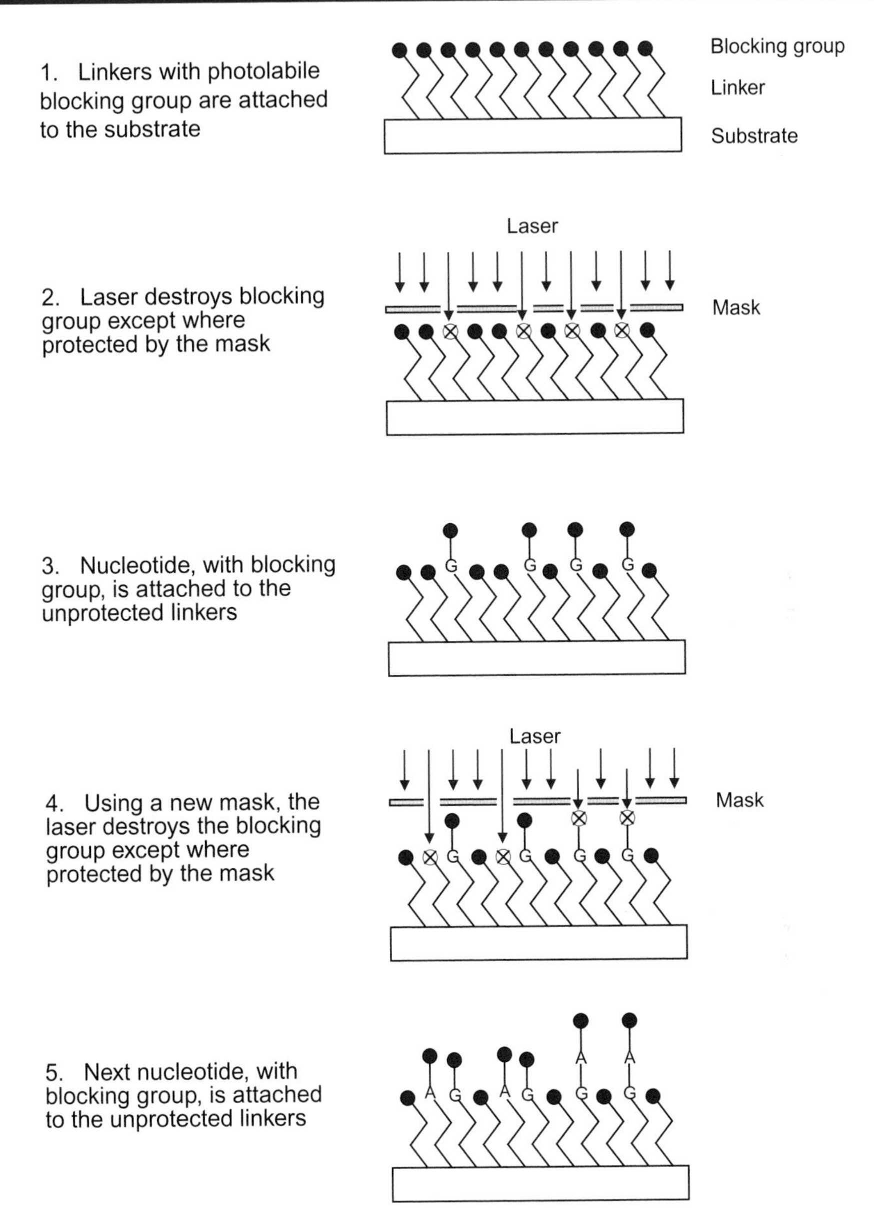

Figure 14.5 Synthesis of oligonulceotides in an array

14.2.4 Important factors in array hybridization

Note that array hybridization is in effect the reverse of conventional hybridization. The conventional techniques involve attaching the target DNA or RNA to a filter (e.g. by southern or northern blots, or using dot blots), and then hybridizing that filter with a specific labelled probe. With macro- or micro-arrays, the specific fragments are attached, unlabelled, to the solid phase, and they are hybridized with the labelled target molecule.

The production of arrays requires a high initial investment. Not only does it require expensive equipment, but also the synthesis of the large number of PCR primers, and the carrying out of large numbers of PCR reactions, is costly. This is especially true for the gene chips, which are only practicable where a substantial level of use of a specific chip is envisaged. However once you (or someone else) has got to that stage, large numbers of arrays can be made; the subsequent hybridization and reading of the results is much less expensive. A growing number of centres are capable of producing micro-arrays, but a much larger number of laboratories are capable of using the microarrays for their own projects.

All the variants of differential and subtractive hybridization and screening are subject to unreliability and artefacts. They should therefore be regarded as rapid screening methods that identify a large number of candidate genes. Other methods, such as quantitative RT-PCR, are needed for unambiguous confirmation of array data, and for providing information about the corresponding transcripts.

It is worth re-iterating that these methods only provide information, at best, about the *levels* of the relevant mRNA species. This is a function of the rate of transcription of that gene and of the stability of the message. Therefore a poorly transcribed but highly stable mRNA may be present at a higher level than that of a more strongly expressed gene where the mRNA is very unstable. The stability of the mRNA is an especially important parameter if you are considering conditions under which specific genes are turned on or off. It is also important to remember the next step of gene expression – translation of the message into protein. The level of mRNA is not necessarily a foolproof predictor of the amount of protein that is being made. In the next section we will look at *proteomics*, which is the study of the spectrum of proteins that are produced by the organism.

14.3 Translational analysis; proteomics

As with the analysis of transcripts, the study of the levels of proteins in different samples can be divided into those methods that are applicable to specific proteins, such as western blots, as described in Chapter 10, and other methods that are intended to demonstrate the full spectrum of proteins that are present; this latter is known as *proteomics*. It should be emphasized that, while the genome of an organism is a fixed entity (give or take a bit of variation), the profile of expressed proteins will be different according to factors such as growth conditions and cellular differentiation.

14.3.1 Two-dimensional electrophoresis

The oldest established method of attempting to identify all the proteins in a sample is *two-dimensional gel electrophoresis*, which involves isoelectric focussing (IEF) in one direction and SDS-PAGE in the other (Figure 14.6). Isoelectric focussing consists of applying an electric field across a stable pH gradient. All proteins have a characteristic *isoelectric point* (or pI), that is a pH value at which they have no net negative or positive charge. At pH values above the pI, the protein will be negatively charged; below the pI the net charge will be positive. Negatively charged proteins will therefore move through the pH gradient towards the positively charged electrode (anode) until they reach a pH equal to their pI, at which point they are no longer charged. Similarly the positively charged proteins will move towards the cathode until they reach their pI. The end result is that each protein will focus at a characteristic point on the pH gradient. The second dimension (SDS-PAGE) separates the proteins according to their size.

After two-dimensional gel electrophoresis, you will get a pattern of spots that can be compared with other patterns to enable the identification of proteins that are present in one strain but not another, or proteins that are differentially expressed according to changes in environmental conditions or phase of growth. Web-based resources are available that can help you with

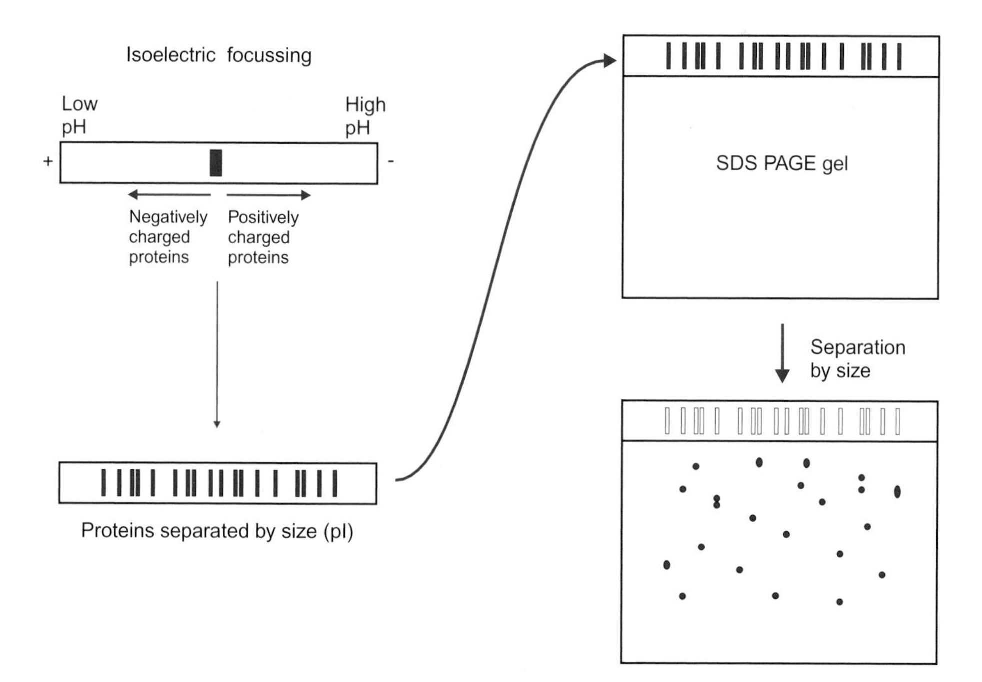

Figure 14.6 Two-dimensional gel electrophoresis

the interpretation (www.mpiib-berlin.mpg.de/2D-PAGE/). However it can be extremely difficult to obtain reproducible results from two-dimensional gel electrophoresis. It is necessary to control the conditions used very carefully to enable a reliable comparison to be made.

It should also be noted that a single protein can occur at different positions in a two-dimensional gel, for example as a consequence of differences in post-translational modification – so the apparent disappearance of a spot, and the appearance of another one in a different place, may not indicate a difference in production of specific proteins, but may be a consequence of altered post-translational modification.

14.3.2 Mass spectrometry

Initially, a two-dimensional gel merely shows us a pattern of spots, some of which may be more interesting than others as they vary according to the nature of the sample. However, that does not tell us what the proteins are. It is possible, using microsequencing techniques, to obtain protein sequences from individual spots. However, this would be a rather cumbersome way of identifying all the spots on a two-dimensional protein gel. Fortunately, if the genome sequence is available, we can use quicker methods. One commonly used procedure is known as matrix-assisted laser desorption ionization–time of flight mass spectrometry, or MALDI-TOF for short. This involves firstly eluting the protein from an individual spot on the gel and digesting it with a proteolytic enzyme such as trypsin. The peptide mixture is added to a matrix which is then subjected to a laser pulse to vapourize and ionize the peptides. The machine then accelerates the ionized peptides towards a detector, which records the time of arrival of peptide ions. The time taken to reach the detector is a function of the mass-to-charge ratio of each ion. The result is a peptide mass fingerprint with a very precise measurement of the relative mass of each peptide. If we know the genome sequence of the organism in question, it is a simple matter for a computer to predict all the tryptic peptides that would be obtained from each protein (since the cleavage site of trypsin is known; it cuts after lysine and arginine residues). Therefore we can compare the experimentally determined peptide mass fingerprint with the peptides predicted from each protein, and thus determine which of those proteins is present in the spot on the two-dimensional gel. MALDI-TOF can even be applied to individual bands on a simple SDS-PAGE gel, provided the band concerned is sufficiently separated from any others.

An alternative, even more powerful procedure, is known as electrospray ionization mass spectrometry. In this case, the sample is introduced into the machine in liquid form to produce charged droplets; as the solvent evaporates, ions are produced which are analysed by the mass spectrometer. The main advantage of this is that it can be readily automated to analyse peaks eluted from HPLC, and so bypass the need for two-dimensional gel electrophoresis.

Mass spectrometry can also be used even if we do not know the genome sequence. Under suitable conditions, peptides can be randomly fragmented by a procedure known as collision-induced dissociation, resulting in a series of ions differing, sequentially, by the mass of single amino acids. This requires two mass analysers, operating in tandem within the same instrument (known as MS/MS). The first separates the original peptides and selects individual peptides, which are then subjected to collision-induced fragmentation before the second analyser determines the sizes of the fragments produced. From this data, the sequence of the peptide can be determined. This sequence can then be compared with databanks of known protein sequences, resulting in an identification of the likely nature of the protein in question.

14.4 Post-translational analysis: protein interactions

Most of the methods for post-translational analysis fall outside the scope of this book. However, procedures for studying which proteins interact with one another is worth considering, as they provide a useful supplement to proteomic analysis in the genome-wide study of protein function.

14.4.1 Two-hybrid screening

One important clue as to the function of a protein is its ability to interact with other proteins. Two-hybrid screening is a commonly used method to identify such interactions; since this method is normally used with yeast cells as a host organism, it is therefore often called *yeast two-hybrid screening* (although versions are now available for bacterial and mammalian cells as well). Although this procedure is commonly used for studying individual proteins, it can also be used for genome-wide screening.

The basis of this procedure is similar to yeast one-hybrid screening, as described in Chapter 10, in that it depends on the modular nature of transcriptional activators such as GAL4. The activator domain (AD), the part which stimulates transcription, does not have to be covalently attached to the DNA-binding domain (BD). If the activator domain can interact with a second protein, and if that protein is bound to DNA adjacent to a reporter gene, transcriptional activation will occur.

The basis of the technique is illustrated in Figure 14.7. As implied by the name, two recombinant plasmids have to be constructed. The first is designed to express one of the proteins that we wish to study (the *bait*) as a fusion protein with a specific DNA-binding domain, while the second plasmid

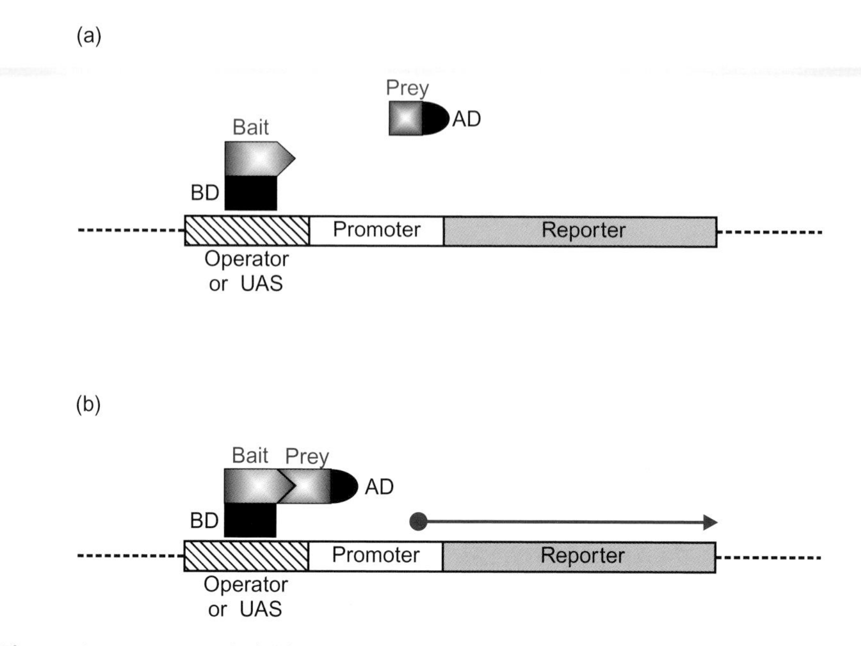

Figure 14.7 Yeast two-hybrid system. Non-interacting proteins: (a) no expression of reporter; (b) expression of reporter

expresses another protein (the *prey*) as a fusion with an activator domain. The host is an engineered yeast strain that contains, inserted into the genome, a reporter gene with a specific operator or upstream activator sequence to which the BD domain will bind. Expression of the reporter gene will only occur if the two fusion proteins interact.

The system can be set up to test the interaction between the products of two specific cloned genes. Alternatively, using a library of DNA fragments in the prey vector, it is possible to identify those proteins that are able to interact with a specific bait.

Although yeast two-hybrid assays originally employed GAL4 AD and BD domains, it is apparent from the description that, since there is no direct physical interaction between them, either or both can be replaced by domains from other proteins. Indeed there are advantages in using heterologous proteins, that is proteins that are not derived from yeast. For example, in some systems the BD domain is derived from the bacterial LexA protein, with the LexA operator, to which LexA binds, attached to the reporter gene.

14.4.2 Phage display libraries

In this method, DNA fragments are inserted into a cloning site within the gene coding for one of the proteins that make up the surface of a filamentous phage such as M13 (see Chapter 5). The fused gene gives rise to a hybrid protein which (we hope) will be incorporated into the phage particle in such a way as

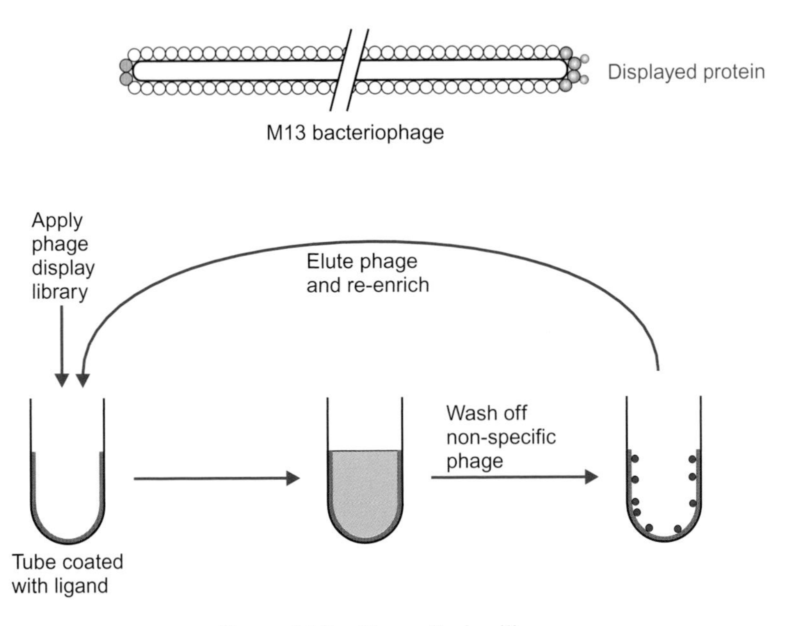

Figure 14.8 Phage display library

to *display* the foreign protein or peptide on the surface of the bacteriophage particles that result from infection of host bacteria. (Figure 14.8). A phage display library can be created either by cloning DNA fragments that code for actual proteins or parts thereof, or by inserting synthetic oligonucleotides designed to give rise to a representative collection of random peptides. The phage particles displaying proteins or peptides with the required properties can be recovered by adsorption to a tube or a well in a microtitre tray coated with the target protein, so as to recover all the phage particles containing different proteins that will bind to the target. In other contexts, other ligands, such as an antibody or a hormone receptor can be used. The non-specific phage are then washed off and the retained phage, enriched for the specific recombinant, are eluted; further rounds of enrichment can be carried out. The power of this technique lies in the ability to test very large numbers of phage particles – M13 preparations can contain 10^{12} phage particles per millilitre – and to carry out repeated rounds of enrichment for the phage displaying proteins or peptides with the properties that you want. Compare this with screening a gene library by hybridization or by antibody screening.

Phage display libraries therefore provide a versatile technique, not only for genome-wide surveys of protein interactions, but also for screening for specific proteins or peptides that can interact with smaller molecules of pharmacological interest. Identification of a peptide sequence that binds to the ligand can be of direct use either in itself (such as in the search for a therapeutic agent), or can serve to provide clues about the partial sequence of a longer ligand-binding protein.

14.5 Integrative studies; systems biology

Although the techniques described in this chapter are designed to look at the complete spectrum of gene expression in an organism, under the chosen conditions, they still do not really reflect the contribution that each gene, or its product, makes to the overall characteristics of that organism. The overall metabolic processes of a cell are defined not only by which enzymes are present, and the extent to which they are produced, but by a complex network of interactions, including physical interactions between the enzymes, interactions with regulatory proteins, and activation or inhibition by small molecules (substrates, products, and messengers such as cyclic AMP). The extent to which we can predict the flow of material through any metabolic pathway by studying the individual enzymes in isolation is severely limited.

However, if bacteria are grown under carefully controlled conditions in a chemostat, it is possible to measure accurately all the input substrates and output products, and, using the information from the genome sequence to predict the possible metabolic pathways, this data can be used to construct a model of the flow of material through each step of the pathways. By analogy with the genome, transcriptome and proteome, this sort of analysis is known by the rather ungainly term *metabolomics*. The details of how this is done are beyond the scope of this book.

We still have to remember that, even under the most carefully controlled conditions, what we usually measure is an average of the situation in all the cells in our sample. Some techniques are now available for measuring gene expression within individual cells – for example inserting a reporter gene (see Chapter 10) coding for a fluorescent protein at a specific point in the genome enables measurement of the fluorescence of individual cells. (This is easier with mammalian cells in culture than with bacteria, as they are bigger and gene expression tends to fluctuate less.) Such experiments show that, even though the average level of gene expression appears to be quite stable and reproducible, there is a significant level of variation from one cell to another, around the mean.

This is still a somewhat artificial situation. In nature, bacteria rarely grow as pure cultures, nor are they homogeneous – there will be a mixture of bacteria in different physiological states, and they will interact with one another, as well as with different bacteria growing in the same niche. A complete model would therefore need to include the interactions between cells of different species within the same niche – and of course for multicellular organisms this would need to be extended to consider the interactions and development of different cells within a multicellular organism. This constitutes the integrated discipline known as *systems biology*, which is still in its infancy but rapidly growing in importance.

15 Modifying Organisms; Transgenics

15.1 Modification of bacteria and viruses: live vaccines

Genetic modification has a relatively long history of application to microorganisms (bacteria and viruses), especially the use of bacteria for the production of novel products. This was covered in Chapter 11. Here, in keeping with our Genes to Genomes theme, we want to focus on modifications that result in an *organism* with novel properties. The main example, in respect of microorganisms, is in the use of genetic modification to produce novel live vaccines. The impact of genetic modification on production of other types of vaccines was covered in Chapter 11.

15.1.1 Live attenuated vaccines

One of the limitations of non-living vaccines, including both conventional killed vaccines and subunit vaccines, is that the protective effect on initial immunization can be relatively poor and short-lived. Boosters are commonly used, but even so the protection is usually inferior to that conferred by a live vaccine.

The conventional route for development of a live attenuated vaccine involves repeated laboratory subculture, especially under conditions unfavourable to the pathogen. For example, a virus may be repeatedly cultured using a cell line that is a poor host for the wild virus, or a bacterium may be repeatedly grown using a medium in which the pathogen grows very weakly. The TB vaccine BCG was developed, early in the twentieth century, by repeated subculture over some twenty years on potato slices soaked in glycerol and ox bile. The principle is that as the pathogen adapts to these unusual conditions it will at the same time be likely to lose the ability to infect human

From Genes to Genomes, Second Edition, Jeremy W. Dale and Malcolm von Schantz

or animal hosts. Although this empirical procedure has been successful in producing several widely used vaccines, it is rather hit or miss. Many mutations are likely to accumulate, apart from those that give rise to the desired attenuation, so there is a good deal of uncertainty in the nature of the resulting strain. Although the sequence can be determined, it is still difficult to know which of these mutations is associated with loss of virulence.

It can be important to understand the nature of the loss of virulence. If it is due to just a single mutation, and especially if that is a simple point mutation, then it is quite possible for that mutation to revert. In other words, a further mutation at that site may restore the original sequence of the gene, and the strain is no longer attenuated. This can happen for example with the live polio vaccine. The vaccine itself is safe, but the attenuated virus persists for some time in the body (which of course contributes to its success in stimulating a good immune response). During that time, there is continued opportunity for it to revert to a virulent form. This does not do any direct damage, as the person 'infected' is by then already protected by the immunity generated by vaccination. However, vaccinated individuals shed, for a time, virus into the environment, and if it has reverted to a virulent form, any non-immunized individuals who come into contact with this revertant are at risk of contracting polio. As the polio eradication campaign became increasingly successful, and the natural occurrence of the virus diminished, reversion of the vaccine strain became a potentially significant cause of the continuation of this disease.

It is also necessary to recognize the possibility that the strain will not only have lost virulence, but also may have lost the ability to induce protection. The structure of the key antigens may have changed, for example. Furthermore, it is possible for a strain to become too attenuated to be useful as a vaccine. The most effective protection is achieved by a micro-organism that is able to undergo at least a limited degree of replication in the tissues before being eliminated by the host defences. If the strain is over-attenuated, the body will eliminate it before it has a chance to evoke a full immune response.

With the classical empirical procedure, it is difficult to control these factors. You have to produce many different variants and test each one for both properties: the loss of virulence, and the ability to produce protective immunity. Genetic technology provides rational ways of achieving these goals in a controlled manner, by site-directed mutagenesis (Chapter 11), or by gene replacement (Chapter 9). The relatively small size of viral genomes makes them especially attractive targets for these techniques.

This requires some knowledge of the genes that are necessary for virulence. If pathogenicity of a bacterium is due to the production of a protein toxin, then you would expect the inactivation of the toxin gene to produce an attenuated derivative. Usually it is not so easy. With most bacteria, pathogenicity is multi-factorial. Much effort is being devoted to attempts to identify genes that are needed for virulence, especially those that are selectively expressed during infection (using for example the techniques described in Chapters 13 and 14) – which would identify targets that could be knocked out in order to attenuate the bacterium.

The most successful approach so far has been to target genes that are involved in certain central biochemical pathways rather than the more obvious virulence genes. This takes advantage of knowledge that the bacterium at the site of infection will not have a ready supply of certain key substrates, such as some amino acids. It will have to make them for itself. Mutants in which key genes in that pathway have been disrupted will therefore be unable to multiply inside the body. Of course we can still grow these strains in the laboratory, simply by adding the required amino acid to the medium. One such pathway is that for aromatic amino acid biosynthesis. *Salmonella* mutants in which the *aroA* gene has been disrupted have reduced virulence.

15.1.2 Live recombinant vaccines

One of the novel advances in vaccine technology that has been made possible by genetic manipulation is the construction of vaccines that confer immunity to several pathogens simultaneously. This can be achieved by inserting the genes for key antigens from different pathogens into a single live vaccine. Much of this work has been carried out using the vaccinia virus (the smallpox vaccine) as the carrier. Vaccinia is a rather large and inconvenient virus to use for genetic manipulation (with a genome size of 187 kb), and the insertion of genes requires the use of *in vivo* recombination. This is shown schematically in Figure 15.1. The required gene is inserted into an *E. coli* vector, adjacent to a promoter derived from vaccinia. This plasmid also carries a defective thymidine kinase (TK) gene from vaccinia, interrupted by the promoter and the cloning site. If an animal cell is infected with vaccinia virus, and at the same time transformed with the recombinant plasmid, the homology between the TK genes on the plasmid and the virus will allow specific homologous recombination. This results in the incorporation of the insert from the recombinant plasmid, together with the adjacent promoter, into the viral DNA. The recombinant virus now lacks a functional TK gene, which provides a way of selecting the recombinants.

Antigens from a wide variety of viruses, and other pathogens, have been expressed in this way as recombinant vaccinia viruses. In one example, a recombinant vaccinia virus expressing a gene from the rabies virus was formulated into a bait for oral immunization of animals in the wild against rabies in parts of Europe. The applications of vaccinia recombinants are not limited to single antigens. It is possible to insert several genes, from different sources,

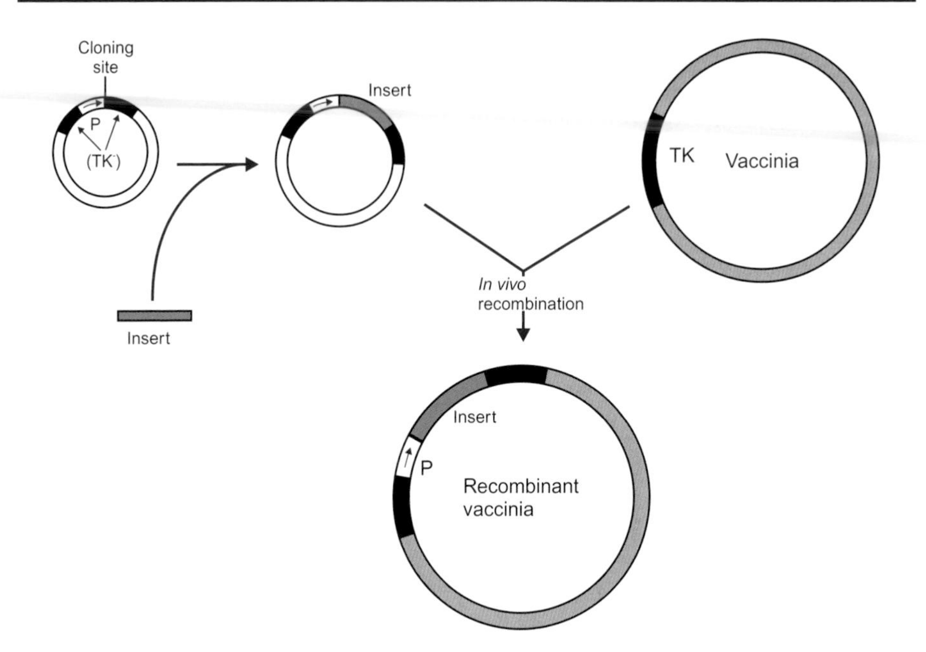

Figure 15.1 Construction of recombinant vaccinia virus. P = a vaccinia promoter; TK = thymidine kinase. The plasmid carries an interrupted vaccinia TK gene. Rcombination occurs between the TK sequences on the plasmid and vaccinia DNA, after co-infecting animal cells. The resulting recombinant is TK, and expresses the insert gene from the vaccinia promoter

into vaccinia, producing candidate vaccines that are capable of immunizing simultaneously against several different diseases.

However, vaccinia is not an ideal candidate vector for human vaccines. Despite its widespread use in the smallpox eradication campaign, it is not a very safe vaccine: for example, it is not suitable for people with skin conditions such as eczema, nor for people who are immunocompromised. However, there are other candidate carriers, notably attenuated strains of *Salmonella* and also the TB vaccine BCG. These are very safe vaccines, and the manipulations involved are relatively straightforward. At a basic level, you simply clone the genes you want into a shuttle vector, including appropriate expression signals, and transform the chosen host. Ultimately there are further considerations. Antibiotic resistance genes are often used as a selectable marker for such manipulations, but for practical use as a vaccine, the inclusion of resistance genes is unlikely to be acceptable, and alternative strategies have to be adopted. Furthermore, plasmids are generally not completely stable, and insertion into the chromosome is preferred. One way of achieving this is to use a suicide plasmid with a bacteriophage integration system. The expression of the phage integrase then recombines the plasmid with the attachment site on the chromosome. There is a wide range of such potential vaccines at different stages of development.

15.2 Transgenesis and cloning

Much of this book deals with the genetic manipulation of individual cells in culture, either by introducing, and expressing, genes from other sources, or in other ways specifically altering the genetic composition of those cells. In this chapter, we will deal with the stable introduction of a gene into another *organism*, which is referred to as *transgenesis*, the result being a *transgenic* organism. For a bacterial or yeast geneticist, there is no distinction between manipulating individual cells and manipulating an organism. In effect, all the previous discussion of genetic modification of bacteria, and unicellular eukaryotes such as *Saccharomyces*, could be labelled as transgenics. On the other hand, for somebody who works with truly multicellular organisms, such as plants or animals, the distinction between manipulating a cell line in culture and genetic modification of a whole plant or animal is very real and important. A gene can be stably transfected into a cell line quite cheaply, and this can be used to provide fast and reasonably useful answers to many questions – but is inevitably limited in scope. In particular, it cannot provide definitive answers to questions about the differentiation of cells in the normal animal, nor about the interaction between different cells. The creation of a genetically manipulated whole organism is much more difficult, but often much more informative. Systems such as the fruit fly *Drosophila*, and the nematode *Caenorhabditis elegans* (*C. elegans*), provide a useful compromise, in that they are reasonably easy to work with but at the same time are capable of addressing some of these difficult questions – assuming that the lessons learned from model systems are also applicable to more complex animals. More controversially than these research-oriented applications, a transgenic organism may be created not to answer a question, but for a specific, usually commercial, purpose. We will discuss both of these uses in some detail later.

It is important to distinguish the production of transgenic plants and animals from cloning of the plants or animals themselves. The former involves the introduction of foreign DNA, in the form of cloned genes. In contrast, the latter means obtaining progeny that are genetically identical to the original plant or animal, which includes methods ranging from the widely publicised procedure that led to the birth of Dolly the sheep to techniques for asexual propagation of plants (such as taking cuttings) that have been in use for hundreds of years.

When you introduce a *transgene* into an organism, you insert into the genome a foreign gene, be it from the same species or from a different one. Owing to the virtual universality of the coding properties of DNA, huge evolutionary distances can be bridged in transgenesis – the example of a fish gene inserted into tomato plants for cold hardiness achieved notoriety. In order to be of any use in a multicellular organism, the transgene must be inserted into germline DNA, so that it can be propagated in subsequent generations. Other applications where genes are inserted into *somatic cells* (cells that are not involved in reproduction of the organism), for example for gene therapy, are considered later in this chapter.

In principle, the technology is available to insert transgenes into any species. In some species, it is also technically possible to do what is conceptually the opposite – to *knock out* a gene (which may involve disrupting the gene rather than deleting it entirely). The use of gene knockouts in bacteria was considered in Chapter 12; as we will see later on, obtaining gene knockouts in whole animals follows a similar procedure. We will also consider another procedure for reducing or removing the effect of a gene without removing the gene itself, using RNA interference (also known as gene knockdown technology).

15.3 Animal transgenesis

In order to ensure expression of the transgene in the entire animal, the transgenic construct must first be introduced into the germline. The techniques for doing this are most advanced, and most commonly used, in mice. The most obvious, and most common, method is to physically inject the transgenic construct into the nucleus of a fertilized egg, which is then cultured *in vitro* for several cell generations before being implanted in a surrogate mother (Figure 15.2). This is conceptually simple, but by no means a trivial affair. Not all eggs survive the injection without damage, not all of them will develop, and not all of them will contain the transgene. Only a proportion of the resulting embryos will be transgenic, so it is necessary to screen the embryos (using gene probes and/or PCR) for the presence of the foreign gene. The result for each embryo is all or nothing: if the gene is present in one cell it is usually present in all cells. *Mosaics* (animals in which the gene is present in some cells or tissues and not in others; also known as *chimaeras*) do not usually occur with this procedure (although mosaics are sometimes created intentionally in other ways) – and so the transgene will be present in the germline cells and transmitted to the progeny. The chromosomal location of the integrated DNA differs from one animal to another, as does the number of copies of the integrated DNA, which may range from a single copy to several hundred. The level of expression of the inserted gene may therefore vary from one animal to another.

One alternative method is to use retroviral vectors, as described in Chapter 5. Technically, this is much simpler. The fertilized egg can be allowed to develop to the eight-cell stage before infecting it with the (defective) recombinant virus (Figure 15.2). This method has distinct limitations, however. Firstly, these vectors can only accommodate approximately 8 kb of DNA, which may not be sufficient for the whole coding region and the necessary promoter elements (see below). Secondly, the modified retrovirus does not

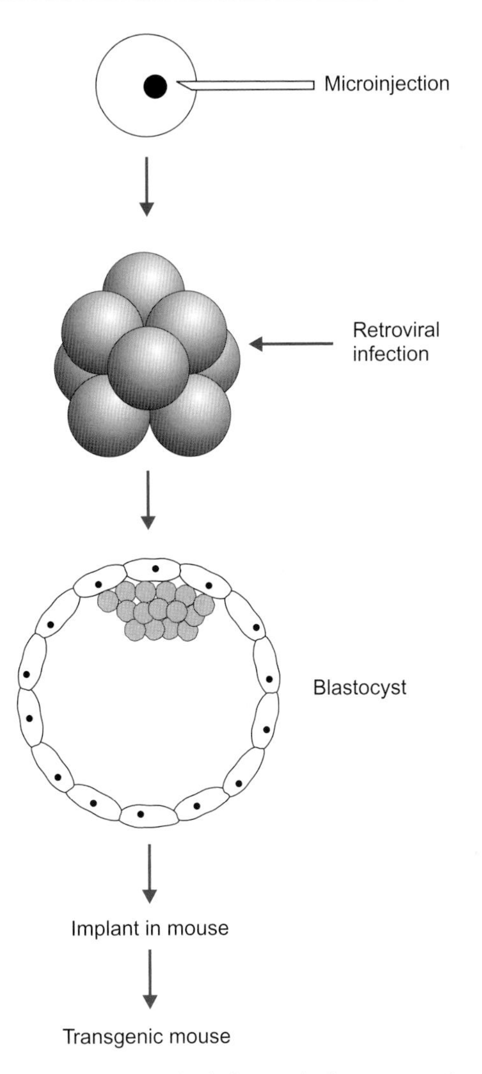

Figure 15.2 Two methods for producing transgenic mice

contain the genes necessary for replication as a virus. It needs assistance from a *helper virus* to be infective. This helper virus could lodge itself into the genome as well and, like a prophage, become infective again at some stage in the future. If the helper virus is also present in the embryo, it can result in the unwanted spread of the recombinant construct to other animals. The method described in Chapter 5 uses *helper cells* as an intermediate stage, to produce infective virus particles that do not need a helper virus, and thus circumvents this problem. However, there is still the possibility that a subsequent retrovirus infection of the transgenic animals might result in mobilization of the recombinant DNA. A further limitation is that, although infection of eight-cell embryos is technically simpler than micro-injection, the fact that cell division has occurred before infection takes place means that there is a possibility that not all the cells in the embryo will be infected. Mosaicism (see above) is therefore a potential problem, but non-chimaeric animals, containing the cloned gene in all their cells, can be obtained by selection from the progeny of the chimaera – since each animal is, of course, derived from only one germline cell that either does or does not contain the transgene. Retroviruses are thus limited in their application for transgenic animals, but do play a significant role in the development of potential techniques for gene therapy (see later in this chapter).

A third method is the transfection of *embryonic stem cells*, and their subsequent incorporation into mouse embryos This method allows selection and enrichment of transgenic cells, which permits an even richer range of potential manipulations. These are discussed in a separate section below.

15.3.1 Expression of transgenes

Even when successfully integrated in the genome, the positioning of the transgene is more or less completely random. The objective is obviously that the gene lodges itself into a region where it does not disrupt another gene. Because most of the DNA in plants and animals does not actually form part of a gene, the odds are actually stacked against any such problems occurring. Nonetheless, occasionally they do, either by the transgene lodging itself directly into a gene or into a region that is important for gene regulation. In some instances, this is lethal. In others, there is an obvious risk that the disruption of a different gene creates a phenotype which has nothing to do with the transgene itself.

The random localization of the transgene will also affect its level of expression. This is because there are large-scale regional differences on the chromosomes – the environment surrounding the transgene may be more or less conducive to its expression.

We also need to consider the regulation of the expression of the inserted transgene. In Chapter 11, we discussed the expression of cloned genes in mammalian cells in culture. In that context, the emphasis was on promoters that were constitutively active, or could be turned on and off by altering the culture conditions. Neither approach is really suitable for a transgenic animal, where we would normally want to target gene expression to specific tissues.

For some purposes, the ideal objective would be to introduce the native transcription unit into a genotype that lacks it. This would require a cloned sequence covering the whole gene and its associated regulators, including activator sequences and promoter elements as well as the introns and exons. The conceptually simplest, and often preferable, method of constructing such a transgene exploits the large coding capacity of yeast artificial chromosome (YAC) vectors, as described in Chapter 5.

One alternative is to mix and match promoter and coding sequences. The closest to a YAC transgene would be to combine a genomic clone of the promoter region of the native gene with a cDNA clone of the same gene. The absence of introns may have some effect, but introns are not *necessary* for mammalian gene expression – some mammalian genes do not have them in the first place. What may make more of a difference, however, is the promoter. You would certainly have pinpointed it using reporter gene constructs before spending a minor fortune on producing a transgenic organism. However, the conditions in the cell line where this was done do not necessarily apply to the conditions in the native cells. Moreover, there may be important elements – positive or negative – that are located some considerable distance away from the transcription start site.

Apart from the obvious combination of a cDNA with its own promoter, there are many inventive and informative combinations that can be – and have been – made, either to obtain tissue-specific expression of the transgene or to identify the tissues or cell types in which that gene would naturally be expressed. Many of these examples are related to topics that have been discussed in earlier chapters in relation to the manipulation of unicellular microorganisms or of cell lines in culture. Here are a few examples:

(1) If the promoter region of the gene you want has not been characterized, a cDNA encoding your protein may be combined with a different promoter that has the same cell or tissue specificity.

(2) In order to investigate the exact cell/tissue specificity of a promoter, or its activity under different conditions, it may be combined with a reporter gene such as *lacZ*, luciferase or green fluorescent protein (which are easy to detect *in vivo*). Similar applications of reporter genes for individual cells in culture were described in Chapter 11.

(3) A promoter may be combined with a gene, such as the gene for the A subunit of diphtheria toxin (DTA) which will specifically damage any cell type in which it is expressed. (The inserted gene does not include the part of the toxin, the B subunit, that is needed for cell entry, so the toxin subunit will only affect the cells in which it is produced.)

(4) A promoter may be combined with a gene that adds a particular functionality to a particular cell type. A common variant of this is the rescue of a recessive mutated phenotype by introducing the wild-type gene, which will provide conclusive proof of the identity of the gene. This can obviously be accomplished with a YAC clone as well. See also the discussion of complementation in Chapter 9.

15.3.2 Embryonic stem cell technology

This technology is based on embryo-derived cell lines that can be grown in culture. Here, we will only discuss the use of mouse-derived embryonic stem cells as a means of generating transgenic animals. The current debate over the use of human embryonic stem cells refers only to their use for research and potential gene therapy (see later in this chapter), not to the generation of transgenic humans. See also the section on genetic diagnosis in Chapter 13.

Some types of cells taken from an adult animal can be grown in culture, for a short time. In some cases, these can give rise to established *cell lines* which can in principle be maintained indefinitely through serial subculture. Manipulation of animal cells in culture (as considered in previous chapters) is considerably easier than introducing genes into a fertilized egg. However, the differentiation of such cells is not usually reversible; they cannot be used to produce a whole animal. This is a marked contrast to the situation in many plants, where cells from various parts of the plant, although differentiated, are still capable of producing whole plants; such cells are said to be *totipotent.* This is why it is so easy to propagate many plants asexually, e.g. by taking cuttings, or even by starting from *in vitro* cell cultures. It is also why there was so much excitement over Dolly the sheep, which was a demonstration that it *is* actually possible, through some intricate techniques, to reverse the differentiation of some cells from an adult animal.

Cells at an early stage in the development of an embryo still have the potential to develop into many types of cell in the adult animal; these are known as *embryonic stem cells* (ES cells). If cells are collected at the blastocyst stage, they can be grown in culture for a limited period of time; if established ES cell lines are used it is not necessary to harvest cells from embryos. The availability of cells in culture, whether they are fresh ES cells or a cell line, means that many of the techniques described in earlier chapters can be employed – in particular, the investigator can include selectable markers in the construct. These convey resistance to toxic compounds that are then added to the growth medium. This is similar to antibiotic selection for plasmids in bacteria: only cells that have successfully integrated the transgene will survive. These selected cells are then injected into blastocysts and implanted into a surrogate mother (Figure 15.3). The progeny will be chimaeric, since the blastocyst will still contain some of the non-manipulated cells, and heterozygous, since normally only one of the chromosomes will be altered. Subsequent breeding and selection of the progeny will give rise to stable homozygous mutant mice.

A further advantage of ES technology is that it allows the introduction of the construct in an exactly specified site in the host genome. This is done by including two sequences that are identical to two particular adjacent regions in a host chromosome. Recombination at these sites will occur with a certain frequency, leading to the specific insertion of the transgene exactly in the desired place, by *homologous recombination* (Figure 15.4). The inserted gene would normally be accompanied by a selectable marker gene, so that we can

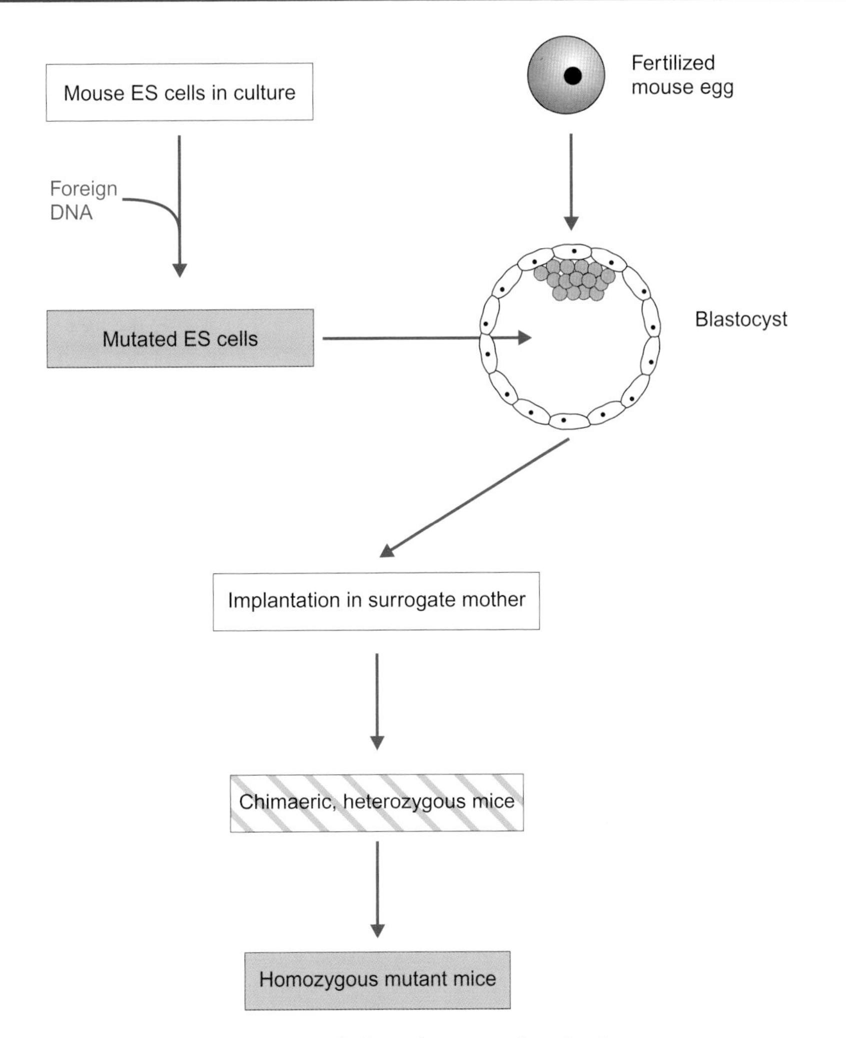

Figure 15.3 Embryonic stem-cell technology

select the cells in which the insert has become integrated into the chromosome. However, it has to be remembered that in animal cells random integration of transfecting DNA occurs quite frequently, so it may be necessary to screen a number of clones in order to find one that has the inset at the right place. This can be made easier by including a second marker gene on the construct, outside the region flanked by the homologous sequences. Random integration of the transfecting DNA would lead to integration of the whole construct, while integration by homologous recombination will only insert the region between the two recombination sites (see Figure 15.4). The cells we are looking for will therefore lack the second marker from the construct, and can thus be easily identified.

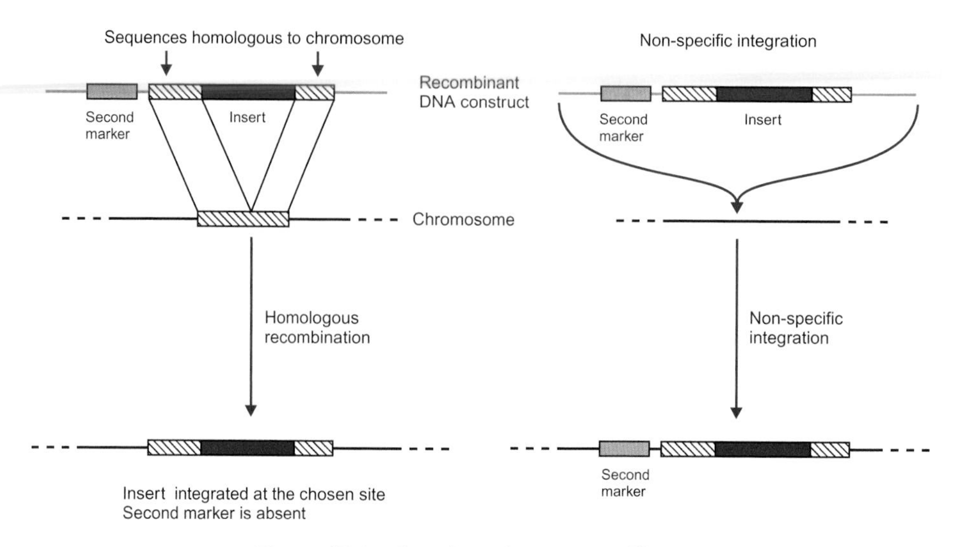

Figure 15.4 Gene insertion at a specific site

15.3.3 Gene knockouts

ES cell technology with homologous recombination also makes it possible to not only *add* a gene, but also to *remove* (or *modify*) an existing one. In the above description of introducing a transgene into a specific site, our choice of homologous sequences would be designed so that integration will take place in a part of the genome that allows the gene to be freely expressed without disrupting any other genes. However, disrupting the gene may be exactly what we want to do, and we can achieve it in a very specific way. A transgenic construct is produced that contains resistance markers, and a part of the gene we wish to manipulate with a specific manipulation made. This could be the disruption of the promoter, the introduction of a stop codon early on in the coding region, or the deletion of part of the gene. In the version shown in Figure 15.5, the target gene is disrupted by the incorporation of a *neo* gene. This gene gets its name because it conveys resistance to neomycin in bacteria, but in animal cells another aminoglycoside antibiotic, known as G418 or Geneticin, is used. It simultaneously fulfils the roles of disrupting the target gene and providing a selectable marker. The resulting offspring will therefore not express the target gene – we have created a *knockout mouse*. This procedure is basically the same as that described in Chapter 9 for allelic replacement in bacteria. Knockout mice have been absolutely crucial in elucidating the exact function of many genes, by investigating the anatomical, physiological, biochemical and behavioural characteristics of the knockout strain. By studying what happens when a gene is removed, scientists are able to draw important conclusions about its function.

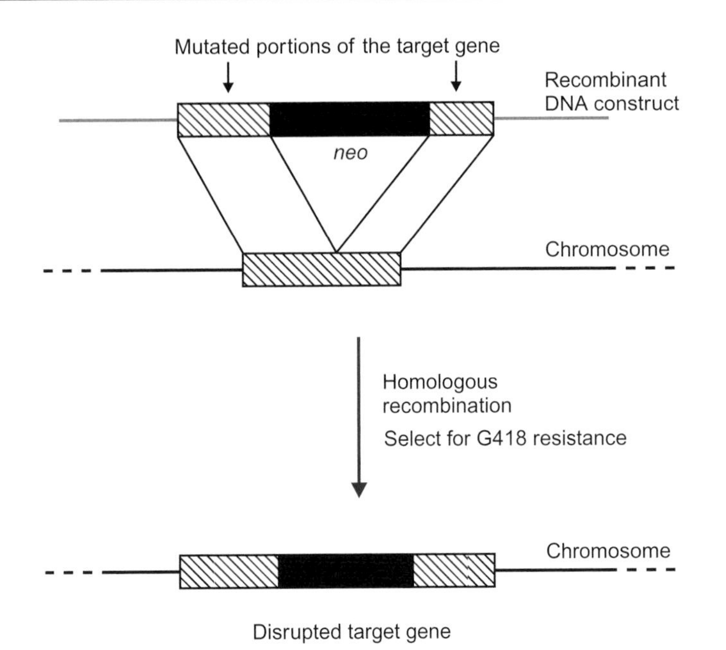

Figure 15.5 Gene knockouts; *neo* codes for an aminoglycoside phosphotransferase that gives resistance to neomycin and to G418 (Geneticin)

The method is not without its pitfalls, however. The absence of the gene may be lethal to the embryo. Alternatively, overexpression of other genes may compensate for the absence of the knocked-out gene, thus obscuring the phenotype.

15.3.4 Gene knockdown technology; RNA interference

In Chapter 12, we introduced the concept of RNA interference (RNAi) or gene silencing, which occurs when dsRNA is introduced into cells. In summary, the dsRNA is first broken into short fragments (siRNA), and the sense strand is degraded by an enzyme complex called RISC. The remaining antisense strand then binds to the corresponding mRNA, which is in turn degraded by RISC. This causes a loss of function of that gene, and is a much more convenient way of achieving that end than knocking out the gene. Although dsRNA cannot be used with mammalian cells, as it induces an effect known as an interferon response, which shuts down protein synthesis, this can be bypassed using either synthetic siRNA, or a recombinant vector designed to express short hairpin RNAs which are processed within the cell to siRNA. The latter strategy allows long-term studies of the effect of silencing specific genes. This method is very versatile, and will basically work with any eukaryotic cell, ranging from unicellular organisms (such as yeast) to multicellular ones (animals or plants). Unlike classical knockout technology, this method has the advantage that you can observe the knockdown in progress. If it is lethal to the cells, you have the opportunity to see what happens, and you are less likely to see compensatory overexpression of other genes.

RNA interference is widely used with cells in culture, and has been more recently extended to germline transmission of siRNA constructs in mice and rats, as well as transfection of mouse organs with siRNA. There is thus considerable potential for gene therapy (see below), although some mouse experiments with high levels of expression of short hairpin RNAs have resulted in serious adverse consequences.

Similar effects can be achieved by using constructs in which the relevant gene (or a part of it) is placed in the reverse orientation, with respect to the promoter. This results in the production of an *antisense RNA*, which reduces the expression of the relevant gene, probably by interfering with translation. We will consider the use of antisense technology further in relation to transgenic plants, later in this chapter.

15.3.5 Gene knockin technology

Most diseases are not caused by the complete disruption of gene function, but by mutations that alter it only in part. These effects are more likely to be subtle and less likely to be lethal. Such mutations can be introduced by *knockin* technology, where one or more exons are replaced with altered ones. Indeed, producing animal models of disease is not the only important application of this technology, or even necessarily the most important one. The concept is important in any situation where we do not want to lose a gene completely, so knockout technology is not appropriate, but rather where we want to introduce specific changes into that gene. One important example of the application of knockin technology is the production of mouse strains in which critical portions of significant genes have been replaced with, or engineered to be more similar to, their human equivalents. The potential value of such strains, for example in the testing of the toxicity and/or action of new drugs, is obvious.

The procedure for producing knockin mice is very similar to that described above for gene knockouts, or to the procedure in Chapter 9 for allelic replacement in bacteria. The desirable mutation(s) is introduced into the exon(s) in question, using site-directed mutagenesis or other procedures described in Chapter 11, and a selectable marker such as the *neo* gene (see above) is inserted into an intervening intron. Homologous recombination will then result in the replacement of the chromosomal gene with the modified sequence.

From this description of the procedures that can be used, it should be clear that transgenic technology is extremely valuable for research into the role and regulation of specific gene products in the development and characteristics of a variety of animals. We now want to turn to a few examples of practical applications of these techniques.

15.4 Applications of transgenic animals

The use of transgenic technology for commercially important animals such as cattle and sheep is highly controversial. This should be put in perspective. Conventional properties of farm animals, such as their rate of growth or milk production, have been manipulated by selected breeding for thousands of years, resulting in the specialized breeds we see today that bear little resemblance to their original ancestors. Against this background, the novel aspect of transgenic technology is the ability to introduce additional genes from other sources. This means that the animals can be used as a source of a variety of useful products.

One approach involves attaching the foreign genes to a promoter that will be active in mammary tissue, such as the casein or β-lactoglobulin promoters. Consequently, the protein concerned will be produced in large quantities in milk. This opens the field not only for enriching the milk itself with various proteins, but also for the use of cattle or sheep as bioreactors for producing large amounts of proteins with various uses. Another example is driving expression of a gene for a protein that you wish to express with the lysozyme promoter in a hen. The transgenic animal will then lay 'bioreactor' eggs which are full of the protein you wish to express. The production of pharmaceutically useful proteins by farm animals has been termed 'pharming'. This can potentially offer substantial advantages over the use of recombinant bacterial cultures. In the first place, it avoids the need to build and run expensive industrial-scale fermenters. In addition, using animal hosts means that the product is likely to have the appropriate post-translational modifications, which is often not the case for products obtained from bacteria.

Recombinant plants also offer considerable potential in this way, as described subsequently.

15.4.1 Studies of development

Multicellular organisms, including humans, are not just a collection of different cells each acting independently. There is a substantial amount of interaction between them, affecting, amongst other things, their development and differentiation. Many of the techniques covered in this book have played, and continue to play, a major role in deciphering the subtleties and complexities of these interactions – including the use of microarray studies to characterize gene expression in different cells, the use of reporter genes to locate cells, *in situ*, in which specific promoters are active, and the use of gene knockouts and RNA interference to switch off specific genes or sets of genes.

Space precludes a thorough coverage of these topics, but some of the model organisms used deserve a brief mention. Amongst the simplest of these are the slime moulds, such as *Dictyostelium discoideum*. When food is plentiful, these behave as unicellular organisms, but they switch to an essentially multicellular organization when food becomes limiting. Under these circumstances they aggregate to form a tiny 'slug' that crawls around and eventually develops into a structure containing a fruiting body and a stalk. Many of the signals that control the differentiation of the initial cells into stalk or spore cells, and the genes responsible, have been identified by the techniques described.

A second important example is the nematode *C. elegans*. This is small (about 1 mm long), easily maintained in the laboratory and has the attraction for developmental biologists that it contains a precise number of cells (959), and the entire ancestry of every cell has been elucidated. RNA interference has proved to be an especially valuable technique with *C. elegans*, since it can be achieved by feeding the nematodes with bacterial clones expressing dsRNA targeted to specific genes. Genome-wide RNAi screens have successfully identified the function of many of the genes of the *C. elegans* genome. Furthermore, the organism is transparent, which greatly facilitates the use of reporter genes (such as that coding for Green Fluorescent Protein, GFP) to study *in situ* gene expression and localization of the product.

Amongst more complex organisms, the zebrafish *Danio rerio* is popular due to its ease of use in the laboratory. This is a small (4 cm) freshwater fish, native to South Asia. The embryos develop rapidly (within a few days), and are transparent – both features being useful for studies of their development. The roles of many of the genes involved in embryonic development have now been identified. In addition to developmental studies, zebrafish are an attractive model for studying microbial infections, as they possess innate and adaptive immune systems with many of the features of mammalian immunity.

The study of these, and other, model organisms is providing many clues as to how development and differentiation occur in other vertebrates including mammals.

15.4.2 Gene therapy

Knowing which mutation, in which gene, causes a disease is not a prerequisite for treating it. Many genetic diseases have been successfully treated for decades or even centuries thanks to insights in the pathophysiology of the conditions, or indeed because of the discovery of treatments by trial and error or even sheer coincidence. Nonetheless, although it is neither the beginning nor the end, discovering the exact genetic nature of a condition is a very important step. Firstly, as described above, it makes it possible to devise precise diagnostic methods to confirm who has, or is predisposed to, a specific condition. Secondly, knowing which gene is defective in the disease means you know which protein may be targeted for the development of new drugs to treat the disease. It also allows us to express the resulting protein (if any) and characterize it in the laboratory (Chapter 11). Furthermore, it puts us in the position of being able to produce a mouse engineered to carry the same mutation, so that we have an animal model that closely approximates the human condition we wish to study (elsewhere in this chapter). All of this will be, potentially, of great help in the development of improved treatments for the condition. However, there will probably always be conditions where treatment is not enough to alleviate the symptoms. This is the rationale for developing treatments involving gene transfer (*gene therapy*).

Gene therapy treatments could potentially take place by targeting the specifically diseased organs in an affected individual (*somatic gene therapy*), or by correcting the gene defect in gametes or fertilized eggs (*germline gene therapy*). There are strong reasons why most gene therapy research focuses on the former. Firstly, the genetic manipulation of embryos is inherently a very contentious issue, and the controversy is exacerbated by the fact that germline therapy would lead to an inheritable alteration in the genetic material. Even the prenatal diagnosis and specific abortion of embryos carrying an affected gene has a far greater acceptance in society as a whole.

Secondly, there is an obvious need to treat people who are alive and suffering now, and indeed there will always be people born suffering from genetic disease. However, somatic gene therapy would only lead to a cure for that patient, and would not remove the underlying genetic cause – so there would still be the risk of the individual concerned passing the defective gene on to his/her offspring.

15.4.3 Viral vectors for gene therapy

The principle of delivering genes to the cells of a human being is not different from that of delivering genes to any other cells. Gene expression can be either *transient* or *permanent*. Delivery can be either directly into the affected tissues – *in vivo* – or into cells transiently removed from the patient's body – *ex vivo*. For *ex vivo* delivery, the methodology is virtually the same as for transfecting mammalian cell lines in culture – genes can be delivered using liposomes, calcium phosphate precipitation or electroporation. However, the methodology that is attracting the greatest interest by far, both for the development of *in vivo* and *ex vivo* methods, is the use of viral vectors.

Amongst these are the *adenoviruses*. These are common causes of respiratory tract infections, especially pharyngitis (sore throat), but it is possible to produce defective viruses that do not cause disease, and to replace the deleted regions with a cloned gene. The recombinant DNA will persist for some time within the infected cell, but not indefinitely. This limitation has both advantages and disadvantages. The advantage is the fact that any negative effects are likely to be reversible. The disadvantage is that re-infection is needed to sustain the effect. Nonetheless, trials with adenovirus vectors have been conducted. One case where they are particularly promising is in the therapy of cystic fibrosis, a recessive disorder caused by a mutation in a chloride transporter, and which manifests itself most strongly in respiratory problems. Naturally, a viral vector based on a respiratory virus is potentially a very promising vehicle for reintroducing the correct form of the gene into these tissues. However the use of adenoviruses as carriers received a major setback in 1999 when a patient in a trial of gene therapy for ornithine transcarboxylase deficiency died from multiple organ failure believed to have been caused by a severe immune response to the adenovirus.

Another DNA virus that can be used for gene therapy is herpes simplex virus I (HSV-I). Unlike adenovirus infections, HSV-I infections persist throughout the lifetime of the patient, although they remain latent through most of it. The preference of this virus for a particular cell type – nerve cells – makes it especially interesting for correcting mutations affecting these cells.

Retroviruses are the natural choice of vector for stable expression. The biology of these viruses, and their use as cloning vectors, was described in Chapter 5. The main feature that is important here is that after infection the RNA genome of the virus is copied into DNA by reverse transcriptase and integrated efficiently into the host genome. By replacing the genes needed to form new viral particles, two birds are killed with one stone – space is created for the introduction of the gene that is to be transferred, and the recombinant virus is prevented from re-infecting other cells.

Integration into the host genome is an effective way of ensuring reliable and lasting expression, that is also transferred to the progeny of the infected cell, and we therefore encountered retroviruses earlier in this chapter as a vector for the production of germline transgenics. Disadvantages include the limited space available for the gene that is to be inserted, and the fact that retroviruses are not very efficient in cells that do not divide, such as nerve cells. In addition, we have to remember that this class of viruses includes some notable pathogens, not only HIV but also other viruses that cause some forms of cancer in humans and animals. Although the vectors described are defective, their widespread use for gene therapy would require extensive checks to ensure that untoward effects do not occur by interaction with other naturally occurring retroviruses. However, the major problem is that random retrovirus insertion can lead to activation of neighbouring genes, which can include genes with oncogenic potential (i.e. they can give rise to the development of cancer). In one trial, two children who had been successfully treated for severe combined immunodeficiency disease, by gene therapy using retroviral vectors, subsequently developed leukaemia.

Despite the problems encountered with both adenovirus and retrovirus vectors, gene therapy has enormous potential for the treatment of life-threatening genetic diseases. The potential does not stop there. In particular, there is considerable potential for novel cancer treatments by suppressing the activity of oncogenes, for example by using *RNAi technology* (see Chapter 12 and earlier in this chapter), or by targeting toxic genes to cancerous tissue. The serious nature of these diseases provides a powerful argument for employing an approach that has such potential. However, there are concerns, apart from the safety aspects. Partly these rest on a feeling that altering the DNA of a human being in this way is somehow unnatural, and that by developing this technology we are in some way in danger of letting the genie out of the bottle and paving the way for altering human genes that do not cause disease. Could the techniques developed for the praiseworthy objective of eliminating the suffering caused by diseases such as cystic fibrosis be exploited for other ends? These are serious questions that demand a wider debate, and scientists have a role in trying to ensure that a broader audience is sufficiently educated about the issues to achieve this on a rational level.

15.5 Transgenic plants and their applications

15.5.1 Introducing foreign genes

Producing transgenic plants is much simpler than producing transgenic animals, as plant cells, unlike animal cells, are *pluripotent*. In many cases, plant cells can be grown in culture, and manipulated, and whole plants regenerated from these individual cells. There are a variety of ways of introducing DNA into plant cells, including protoplast transformation/electroporation, microinjection, biolistics, and transfer from *Agrobacterium tumefaciens*.

For protoplast transformation, the plant cell wall is digested with suitable enzymes to obtain protoplasts. These will take up DNA quite readily, and the efficiency can be improved by the use of electroporation (see Chapter 5), or by fusion with DNA-containing liposomes. Microinjection involves the direct injection of DNA into the nucleus of the plant cell, in much the same way as was described above for animal cells.

Biolistics involves coating tiny beads of gold with DNA and bombarding the plant cells at high velocity; this technique has also been used with other types of cell (see also the discussion of DNA vaccines in Chapter 11). For plants it has the advantage of being also applicable to embryonic plants, and so avoids the difficulty of regenerating some types of plants from individual cells.

The use of the Ti plasmid of *Agrobacterium tumefaciens* as a vector was described in Chapter 5. This bacterium produces a type of tumour in susceptible plants through the transfer of part of the plasmid DNA into the plant cells. Incorporating foreign DNA into the plasmid allows that DNA to be taken up by the plant cells and incorporated into the chromosomal DNA. Because many plants can be grown from cells in culture, the manipulation, and selection of transformants, can be carried out using cell culture. Until fairly recently, it was thought that this was only possible for the group of plants called dicotyledons, which was a serious limitation as the most important food crops (cereals, rice, maize etc.) belong to the other group, monocotyledons. However, recent discoveries have made it possible to use *Agrobacterium* for gene transfer to monocotyledons as well.

Just as in animal transgenesis, a suitable combination of promoter and gene is chosen to express the gene in the desired location and quantity. One promoter that is used for this purpose is derived from the cauliflower mosaic virus. Clearly, producing genetically modified (GM) plants is much simpler than GM animals, and unsurprisingly, the commercial exploitation of this fact has gone much further, and is already a reality.

15.5.2 Gene subtraction

In addition to adding genes to plants, it is possible to inactivate specific plant genes, by the use of *antisense RNA*. This involves cloning all or part of the coding region of a relevant gene in an expression vector, but in the 'wrong' orientation, so that stimulation of the promoter leads to the production of an RNA molecule that is complementary to that which codes for the required protein. Binding of the antisense RNA to the genuine mRNA probably interferes with translation, or leads to degradation by ribonuclease, so resulting in reduced production of the corresponding protein (Figure 15.6), although effects on transcription or processing of the mRNA are also possible.

One example of an application of this is in delaying the ripening, and spoilage, of tomatoes. In the later stages of ripening, an enzyme (polygalacturonase) is produced. This enzyme breaks down polygalacturonic acid in the cell walls of the tomato and thus softens the tomato. Softening is desirable, but if it goes too far, you get a spoilt, squishy tomato. Introduction of part of the polygalacturonase gene, under the control of a cauliflower mosaic virus vector but in the 'wrong' orientation, produces an antisense RNA which binds to the normal polygalacturonase mRNA and greatly reduces the enzyme levels in the fruit. Enough enzyme is made to achieve a gradual softening, but the fruit can be stored for a much longer period before they spoil.

There is some similarity between this description of antisense RNA and the model presented earlier for RNA interference (see Chapter 12, and earlier in this chapter) in that both act via dsRNA to cause a reduction or

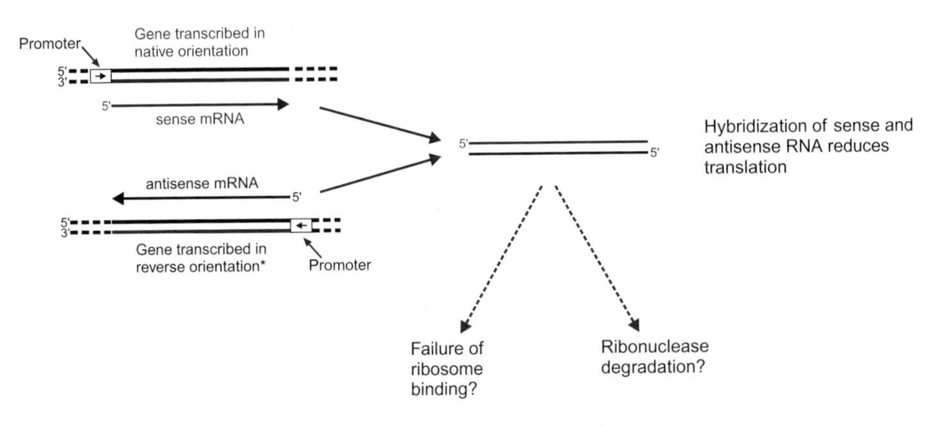

Figure 15.6 Antisense RNA

loss of gene expression. However, this model for antisense RNA is stoichiometric – in other words, one molecule of antisense RNA will inhibit one molecule of mRNA – whereas the RNAi model is catalytic, so a small amount of the siRNA can result in complete removal of all the relevant mRNA. However, the distinction is not totally clear-cut.

RNA interference, as described for mammalian systems earlier in this chapter, can also be used with plants. Sequence-specific gene silencing can be achieved through the use of inverted repeat transgenes, so that the transgene mRNA folds up into a long stable dsRNA. As described for other systems (Chapter 12), this dsRNA is cleaved by Dicer into siRNA which guides specific mRNA cleavage. One important difference in plants is that they possess an RNA-dependent RNA polymerase which can copy the antisense strand of an siRNA, making more dsRNA. This amplifies the effect, and the dsRNA can spread throughout the plant by cell-to-cell transfer. The natural significance of this is that it plays a role in the development of plant resistance to viral infection.

15.5.3 Applications

There are an extremely wide variety of modifications that have been made, and the list will grow very rapidly. One of the applications that has received a lot of publicity is the production of plants that are resistant to insect attack. This most commonly involves the insertion of a gene from the bacterium *Bacillus thuringiensis* coding for an insecticidal toxin which is highly poisonous to certain groups of insects. Expression of this gene by a plant (e.g. cotton) makes it resistant to insect attack, and thus substantially reduces the need for spraying the crop with insecticides.

The second widely publicised application is that of resistance to herbicides (principally glyphosate). This is often confused with insecticide resistance, but the cases are quite different. Expression of a herbicide-resistance gene in a crop plant enables the farmer to spray the crop with a broad-spectrum herbicide to eliminate weeds.

Other examples include genes that convey tolerance to environmental factors such as cold or salt, or influence texture, taste or colour (the latter important both in agricultural and horticultural crops). We have mentioned previously the potential of using plants for the production of pharmaceutically important proteins. Of particular importance are modifications that improve the nutritional content of the product. Many people in poorer countries rely extensively on a single food crop, such as rice, which often does not provide an adequate level of all the essential nutrients and vitamins in their diet. For example, vitamin A deficiency affects up to 250 million children worldwide, with some 500 000 per year becoming blind. The development of so-called 'Golden Rice', which accumulates provitamin A due to the incorporation of genes from various sources, could make a major contribution to alleviating vitamin A deficiency.

In this chapter, as in the rest of the book, we have introduced the techniques and some examples of their applications. Although this only scratches the surface of a rapidly expanding subject, we hope it will have given you some understanding of the underlying concepts involved, of the advances in knowledge that have already been attained, and of their future possibilities. Some of these possibilities may be undesirable, but in many ways these techniques will play a major role in improvements in health, to use just one example. From your study of this book, you should be better placed to understand, and hopefully contribute to, the debate over the use and control of genetic modification.

Glossary

Adaptor: a short double-stranded oligonucleotide used to add sticky ends to a blunt-ended fragment, or to change one sticky end to a different one (see *linker*)

Adenovirus(es): medium-sized DNA viruses, widespread in animals and birds. Derivatives are used as vectors in animal cells, and for gene therapy

Affinity chromatography: use of a specific *ligand*, attached to an insoluble matrix, to bind the required protein and enable its purification

Agrobacterium tumefaciens: a bacterium that produces tumour-like growths (crown galls) on certain plants; used for transfer of DNA into plant cells

Alkaline phosphatase: enzyme that removes the terminal 5′ phosphate from DNA, and thus can be used to prevent ligation

Allelic replacement: see *gene replacement*

Allosteric effect: alteration of the conformation of a protein, through the binding of a *ligand*, resulting in a change of the activity of a different site on the protein

Amplicon: the amplified product of a PCR reaction

Amplification: (a) increase in plasmid copy number under changed growth conditions; (b) multiplication of the number of bacteria or phages in a library (without increasing the number of independent clones); (c) production of a large number of copies of a specific DNA fragment by PCR

Annealing: formation of double-stranded nucleic acid from two single-stranded nucleic acid molecules; see *hybridization*

Anticodon: the region of tRNA that pairs with the codon

Antisense RNA: RNA that is complementary to a specific mRNA, and which can interfere with translation

Autoradiography: detecting radioactively labelled material using X-ray film or a phosphoimager

From Genes to Genomes, Second Edition, Jeremy W. Dale and Malcolm von Schantz

Auxotroph: a mutant that requires the addition of one or more special supplements to its growth medium (cf. *prototroph*)
Avidin: a protein with high affinity for biotin; used for detecting biotinylated probes

Back-translation (reverse translation): predicting all possible nucleic acid sequences coding for a specific amino acid sequence
Bacterial artificial chromosome (BAC): a vector based on the *F plasmid*, used to clone very large DNA fragments
Bacteriophage: a virus that infects bacteria (often shortened to 'phage')
Baculovirus: an insect virus used as a vector for eukaryotic gene expression
Bioinformatics: computer-based analysis of biomolecular data, especially large-scale datasets derived from genome sequencing
Biolistics: the use of high-velocity microprojectiles for introducing DNA into cells
Biotin: a small molecule that can be attached to dUTP and incorporated into a nucleic acid as a non-radioactive label. Detected with avidin
BLAST: basic local alignment sequencing tool; a computer programme for searching databases for sequences similar to a query sequence
Blunt end: the end of a double-stranded DNA molecule in which both strands terminate at the same position, without any single-strand extension (as opposed to a *sticky* end); also known as a *flush* end
Bootstrap: a technique in molecular phylogeny for assessing the robustness of a phylogenetic tree
Box: a short sequence of bases in DNA conforming (more or less) to a consensus for that particular type of box; usually has a regulatory function
Broad host-range plasmid: a plasmid that is capable of replicating in a wide range of bacteria

Caenorhabditis elegans: a nematode species widely used for studies of genetics and cell differentiation
Capsid: viral component that surrounds and contains the nucleic acid
Cassette: a DNA sequence containing several genes and/or regulatory components which can be inserted as a unit into a vector, or which moves naturally as a unit. The term is used with similar meanings in other contexts.
cDNA: complementary (or copy) DNA, synthesized using mRNA as a template, a reaction carried out by reverse transcriptase
cDNA library: a collection of cDNA clones which together represent all the mRNA present in a sample at a particular time (cf. *genomic library*)
Chaperone: a protein that affects the folding of other proteins or the assembly of complex structures

Chromosome: DNA-containing structure in the nucleus of a eukaryote. Also used to refer to the main genetic component of a prokaryote, especially when distinguishing it from a *plasmid*

Chimaera: an animal or plant containing a mixture of cells with different genotypes

Chromosome walking: a technique for identification of DNA regions adjacent to a known marker by sequential hybridization of clones

***cis*-acting**: a control region that influences genes on a more or less adjacent region of the same DNA molecule only, and has no effect on other DNA molecules (cf. *trans-acting*)

Clone: one of an identical set of organisms, or cells, descended asexually from a single ancestor.

Cloning: obtaining a homogeneous population of cells by repeated single colony isolation. Also used to refer to obtaining copies of recombinant DNA carried by such a cell ('gene cloning') and, by further extension, for the reactions used to make such recombinants

CLUSTAL: a computer programme for multiple alignment of protein or DNA sequences

Codon: a group of three bases in mRNA that codes for a single amino acid

Codon bias: difference in the frequency of occurrence of *synonymous codons*

Codon usage: a measure of the relative use of *synonymous codons*

Cohesive end: see *sticky end*

Colony/plaque lift: transfer of colonies or plaques from a plate onto a membrane for hybridization

Competent: a bacterial cell that is able to take up added DNA

Complementary strand: a nucleic acid strand that will pair with a given single strand of DNA (or RNA)

Complementation: restoration of the wild-type phenotype by the introduction of a second DNA molecule, without recombination

Complexity: the number of independent clones in a library

Conjugation: transfer of genetic material from one bacterial cell to another by means of cell-to-cell contact

Consensus sequence: a sequence derived from a number of related, non-identical sequences, representing the nucleotides that are most commonly present at each position

Constitutive: the gene product is always formed irrespective of the presence of inducers or repressors (see *induction*, *repression*)

Contig: composite DNA sequence built up from a number of smaller overlapping sequences during a sequencing project

Convergent evolution: the development of organisms, structures, or proteins, with similar functions but descended from different sources

Copy number: the number of molecules of a plasmid present in a cell

***cos* site**: the sequence of bases of bacteriophage lambda that is cut asymmetrically during packaging, generating an unpaired sequence of 12 bases at each end of the phage DNA

Cosmid: a plasmid that contains the *cos* site of bacteriophage lambda. After introducing a large insert, the recombinant cosmid forms a substrate for *in vitro* packaging

CpG island: a chromosomal region rich in CG dinucleotides

ddNTP: any one of the dideoxynucleotides ddATP, ddCTP, ddGTP, ddTTP

Denaturation: (a) reversible separation of the two strands of DNA by disruption of the hydrogen bonds (usually by heat or high pH); (b) disruption of the secondary and tertiary structure of proteins

Deoxyribonuclease (DNase): an enzyme that degrades DNA

Dideoxynucleotide: a nucleotide lacking an OH at both 2′ and 3′ positions, which therefore terminates DNA synthesis. Used in DNA sequencing

Direct repeat: two identical or very similar DNA sequences, reading in the same direction (see also *inverted repeat*)

DNA ligase: enzyme for joining DNA strands by formation of phosphodiester bonds

dNTP: any one of dATP, dCTP, dGTP, dTTP

Domain: a region of a protein that folds into a semi-autonomous structure

Domain shuffling: a natural process in which various domains of a protein appear to have originated, evolutionarily, from different sources (see *mosaic protein*)

Dot (or slot) blot: a hybridization technique where the samples containing the target nucleic acid are applied directly to the membrane in a regular pattern, usually defined by a manifold

Double cross-over: integration of a gene into the chromosome by two recombinational events, either side of the gene (cf. *single cross-over*)

dsDNA, dsRNA: double-stranded DNA or RNA

Electrophoresis: separation of macromolecules (DNA, RNA, or protein) by application of an electric field, usually across an agarose or acrylamide gel

Electroporation: inducing cells to take up DNA by subjecting them to brief electric pulses

Embryonic stem cell: a *totipotent* cell derived from the embryo of an animal. Used in transgenics

End filling: converting a sticky end to a blunt one by enzymatic synthesis of the complementary strand (cf. *trimming*)

Endonuclease: an enzyme that cuts a DNA molecule at internal sites (cf. *exonuclease*)

Enhancer: a *cis*-acting sequence that increases the utilization of a promoter

Episome: a plasmid that is able to integrate into the chromosome. In eukaryotic systems, often used to emphasize the extrachromosomal state

Epitope: portion of a protein that is recognized by a specific antibody

EST: expressed sequence tag. A partial cDNA molecule, picked at random from a library and sequenced

Eukaryote: a cell that has a discrete nucleus, bounded by a membrane; e.g. fungal, protozoan, plant and animal cells (cf. *prokaryote*)

Exon: Coding sequence, part of a single gene and flanked by *introns*

Exonuclease: an enzyme that removes nucleotides from the ends of a DNA molecule (cf. *endonuclease*)

Expression vector: a cloning vector designed for expression of the cloned insert using regulatory sequences present on the vector

Fingerprint: polymorphic pattern of bands on a gel for differentiating individual genomes

FISH: fluorescent *in situ* hybridization. Using a probe labelled with a fluorescent dye to locate specific genes on a chromosome, or within a cell or tissue

Flush end: see *blunt end*

Footprinting: identification of sequences that bind a specific protein, by visualization of the protection of positions that are protected from attack by DNase I

F plasmid: a natural, low-copy, plasmid of *E. coli*; see *bacterial artificial chromosome*

Frameshift: insertion or deletion of bases, other than in multiples of three. This changes the reading frame of protein synthesis beyond that point

Gel retardation: reduction in the electrophoretic mobility of a DNA fragment due to protein binding (also known as gel shift or band shift)

Gene chip: a high-density array, produced by synthesis of oligonucleotides on a substrate

Gene knockdown: reduction of gene expression by antisense RNA

Gene knockout: inactivation of a gene by homologous recombination (see also *gene replacement*)

Gene replacement (gene knockin): replacement of a chromosomal gene by recombination with a homologous sequence, inactivated or otherwise modified *in vitro*

Gene subtraction: use of antisense RNA for partial or controllable reduction in the activity of a specific gene

Gene therapy: use of specific DNA to treat a disease, by correcting the genetic lesion responsible or alleviating its effects

Genetic map: a map of the genetic structure of a genome, showing the relative position of known genes (cf. *physical map*)

Genome: the entire genetic material of an organism. Sometimes (e.g. 'human genome') used in a more restrictive sense, to refer to the nuclear material only

Genome plasticity: larger-scale variations in the organization of the genome, by rearrangements, inversions etc., and by incorporation of DNA from other sources (see *horizontal gene transfer*)

Genomic library: a collection of recombinant clones which together represent the entire genome of an organism; see also *cDNA library*

Genomics: analysis of data derived from the complete DNA sequence of an organism

Genotype: the genetic make-up of an organism (cf. *phenotype*)

Guessmer: a short synthetic oligonucleotide (oligomer) based on the most likely DNA sequence deduced from amino acid sequence data

Hairpin: a region of DNA or RNA that contains a short inverted repeat, which can form a base-paired structure resembling a hairpin (similar to a *stem–loop structure*)

Heteroduplex: a double-stranded nucleic acid molecule formed by base-pairing between two similar but not identical strands. Since the two strands are not identical, some regions will remain single-stranded

Heterologous probe: a probe that is similar but not identical to the target

Heterozygote: a diploid organism that carries two different versions of a specific gene (cf. *homozygote*)

HMM (hidden Markov model): a statistical technique with many applications in bioinformatics, such as predicting open reading frames in genome sequences

Homologous recombination: a natural process involving the pairing of similar DNA molecules followed by breakage, crossing over and rejoining of DNA strands

Homology: similarity in the sequence of two genes, from different organisms, that share a common evolutionary origin. Often used, more loosely (as in *homologous recombination*), to describe DNA molecules with a sequence that is sufficiently similar for complementary strands to hybridize, without evidence of common evolutionary origin or function

Homopolymer tailing: see *tailing*

Homozygous: a diploid organism in which the two copies of a specific gene are the same (cf. *heterozygote*)

Horizontal gene transfer: transfer of genetic material from one organism to another

Hybridization: The formation of double-stranded nucleic acid molecules by the production of hydrogen bonds between wholly or partially complementary sequences

Hybridoma: cell line producing a *monoclonal antibody*
Hydrophilic: 'water-loving'. Substances, or parts of a structure, that are polar, and interact with water, and therefore tend to be exposed to water
Hydrophobic: 'water-hating'. Substances, or parts of a structure, that are non-polar, and do not interact with water, and therefore tend to remove themselves from an aqueous environment

Inclusion body: an insoluble intracellular protein aggregate formed during high-level expression of a recombinant protein
***In situ* hybridization**: a form of hybridization where the immobilized target is part of a whole chromosome, or in some cases a whole cell
***In vitro* mutagenesis**: see *site-directed mutagenesis*
***In vitro* packaging**: the assembly of mature bacteriophage particles *in vitro* by mixing suitable DNA with cell extracts that contain bacteriophage heads and tails and the enzymes needed for packaging
Indel: insertion/deletion. Indicates, in a comparison of two genome sequences, that there has been either an insertion in one or a deletion in the other
Induction: (a) increasing the synthesis of a gene product through a specific environmental change; (b) applying a treatment to a lysogenic bacterium that results in the bacteriophage entering the lytic cycle:
Insertion sequence: a DNA sequence that is able to insert itself, or a copy of itself, into another DNA molecule; carries no information other than that required for transposition (see also *transposon*)
Insertion vector: a lambda cloning vector into which DNA can be inserted at a single site (cf. *replacement vector*)
Insertional inactivation: destruction of the function of a gene by insertion of a foreign DNA fragment, either by transposition or by gene cloning
Intermolecular: interaction between two different molecules
Intramolecular: interaction between different parts of the same molecule
Intron: intervening sequence in a eukaryotic gene; removed by splicing; see also *exon*
Inverted repeat: two identical or very similar DNA sequences, reading in opposite directions (see also *direct repeat*)
IPTG: iso-propylthiogalactoside. An inducer of β-galactosidase that is not hydrolysed by the enzyme
Island: a region of the genome with different base composition from that of the overall genome (see also *CpG island, pathogenicity island*)
Isoelectric focussing: separation of proteins by electrophoresis in a stable pH gradient, so that each protein will move to its *isoelectric point*
Isoelectric point: the pH at which a specific protein has no net overall charge
Isogenic: strains that are identical in their genetic composition; normally used to mean identical in all genes except the one being studied

Isoschizomers: restriction endonucleases that recognize the same nucleotide sequence (but do not necessarily cut in the same fashion)

IVET: *in vivo* expression technology; a procedure for identifying bacterial genes that are expressed during infection rather than during growth in the laboratory

kilobase: a nucleic acid region that is 1000 bases long (abbreviated to kb)

Labelling: adding a detectable signal to a probe

Lambda: temperate bacteriophage of *E. coli*, used as a cloning vector

Leader: nucleotide sequence at the 5′ end of mRNA, before the start point for translation of the first structural gene. Often involved in regulating gene expression

Ligand: a (usually small) molecule that binds non-covalently to specific site(s) on a protein

Ligation: joining two DNA molecules using *DNA ligase*

Linkage: the degree to which two genes are inherited together

Linkage analysis: mapping the relative position of two genes (or other markers) by determining the extent to which they are co-inherited

Linker: a short double-stranded oligonucleotide with blunt ends and an internal restriction site, used to add sticky ends to a blunt-ended fragment (see *adaptor*)

Lysogeny: a (more or less) stable relationship between a bacteriophage (*prophage*) and a host bacterium (lysogen)

Lytic cycle: multiplication of a bacteriophage within a host cell, leading to lysis of the cell and infection of other sensitive bacteria

M13: a filamentous bacteriophage of *E. coli*, with a single-stranded DNA genome. Used as a cloning vector

Macroarray: a large set of DNA spots immobilized on a membrane. Used for comparative and differential studies of genomes and transcriptomes (cf. *microarray*)

MALDI-TOF: matrix-assisted laser desorption ionization time-of-flight. Mass spectrometry technique used in proteomics for identification of peptides

Mapping: determination of the position of genes (*genetic map*), or of physical features such as restriction endonuclease sites (*physical map*)

Melting: separation of double-stranded DNA into single strands (see *denaturation*)

Melting temperature (T_m): the temperature at which the two strands of a DNA or a DNA/RNA molecule separate (denature)

Messenger RNA (mRNA): RNA molecule used by ribosomes for translation into a protein

Metagenomics: study of genome sequences of a mixture of organisms

Microarray: a large set of DNA spots immobilized on a glass slide. Used for comparative and differential studies of genomes and transcriptomes (cf. *macroarray*)

Microinjection: direct injection of DNA into the nucleus of a cell

Microsatellite: tandem repeats of a short sequence of nucleotides. Variation in the number of repeats causes *polymorphism*

Mobilization: transfer by conjugation of a non-conjugative plasmid in the presence of a conjugative plasmid

Modification: alteration of the structure of DNA (usually by methylation of specific residues) so that it is no longer a substrate for the corresponding *restriction endonuclease*

Molecular beacons: a technique for real-time PCR in which a probe shows fluorescence only when annealed to a PCR product

Molecular clock: rate of variation of a gene or an organism

Monoclonal antibody: a homogeneous population of identical antibody molecules produced by an immortalized lymphocyte cell line

Mosaic: a transgenic animal in which the cloned gene is present in only a proportion of the cells or tissues (cf. *chimaera*)

Mosaic protein: a protein composed of domains from different sources (see *domain shuffling*)

Motif: conserved sequence within a family of proteins indicating a specific function

Multiple cloning site: a short region of a vector containing a number of unique restriction sites into which DNA can be inserted (see *polylinker*)

Multiplex PCR: a PCR reaction carried out with several pairs of primers, amplifying different sequences

Mutagenesis: treatment of an organism with chemical or physical agents so as to induce alterations in the genetic material (see also *site-directed mutagenesis*)

Mutant: a cell (or virus) with a change in its genetic material (cf. *mutation*)

Mutation: an alteration in the genetic material (cf. *mutant*)

Neighbour joining (NJ): a technique for constructing phylogenetic trees, e.g. from sequence data

Nested PCR: a technique for increasing the sensitivity and/or specificity of PCR, by using a second set of primers internal to the first pair

Nick: a break in one strand of a double-stranded DNA molecule

Nonsense mutation: base substitution creating a *stop codon* within the coding sequence, causing premature termination of translation

Northern blot: a membrane with RNA molecules transferred from an electrophoresis gel for hybridization

Oligonucleotide: a short nucleic acid sequence (usually synthetic)
Open reading frame (ORF): a nucleic acid sequence with a reading frame that contains no stop codons; it therefore defines a potentially translated polypeptide
Operational Taxonomic Unit (OTU): in the construction of phylogenetic trees, a group of organisms, or sequences, that are treated as a single organism or sequence
Operator: a region of DNA to which a repressor protein binds to switch off expression of the associated gene. Usually found adjacent to, or overlapping with, the promoter
Operon: a group of contiguous genes (in bacteria) that are transcribed into a single mRNA, and hence are subject to coordinated induction/repression
Ordered library: a collection of clones containing overlapping fragments, in which the order of the fragments has been determined
Origin of replication: position on a DNA molecule at which replication starts. Most commonly used to mean that part of a plasmid that is necessary for replication
Outgroup: in phylogenetic analysis, a sequence (or organism) that is distantly related to those being studied. Enables the production of a *rooted tree*

P element: a transposable element in *Drosophila*, used in the construction of cloning vectors, and in transposon mutagenesis
P1: a bacteriophage that infects *E. coli*. Used as the basis for some cloning vectors
Packaging: the process of incorporating DNA into a bacteriophage particle (see also *In vitro packaging*)
Packaging limits: the range of DNA sizes that can be packaged into a specific bacteriophage particle
Palindrome: a sequence that reads the same in both directions (on the complementary strands)
Partial digest: cutting DNA with a restriction endonuclease using conditions under which only a fraction of available sites are cleaved
Partitioning: distribution of copies of a plasmid between daughter cells at cell division
Pathogenicity island: a DNA region (in bacteria) carrying virulence determinants; often with a different base composition from the remainder of the chromosome
Phage: see *bacteriophage*
Phage display: a technique for expressing cloned proteins on the surface of a bacteriophage
Pharming: producing a recombinant protein from a genetically modified farm animal
Phenotype: the observable characteristics of an organism (cf. *genotype*)

Phylogenetic tree: graphical display of the relatedness between organisms or sequences, used to infer possible evolutionary relationships

Phylogeny: study of the evolutionary relationships amongst groups of organisms or sequences

Physical map: a map of the physical structure of a genome, e.g. showing restriction sites, position of specific clones, or ultimately the complete sequence (cf. *genetic map*)

Pilus: in bacteria, a filamentous appendage on the surface, often involved in *conjugation*

Plaque: a region of clearing, or reduced growth, in a bacterial lawn, as a result of phage infection

Plasmid: an extrachromosomal genetic element, capable of autonomous replication

Plus and minus strands: mRNA is defined as the plus (sense) strand, and the complementary sequence as the minus (antisense) strand. DNA sequences maintain the same convention, so it is the minus strand of DNA that is transcribed to yield the mRNA (plus strand)

Point mutation: an alteration (or deletion/insertion) of a single base in the DNA

Polar mutation: a mutation in one gene that affects the expression of others (e.g. genes downstream in an operon). The phenotypic effect may not be directly caused by the original mutation

Polyadenylation : a natural process (mainly in eukaryotes) which produces a long string of adenyl residues at the 3′ end of the mRNA (should strictly be 'polyadenylylation' but 'polyadenylation' is commonly used)

Polycistronic mRNA: messenger RNA coding for several proteins (see *operon*)

Polylinker: synthetic oligonucleotide containing several restriction sites (see *multiple cloning site*)

Polymerase chain reaction (PCR): enzymatic amplification of a specific DNA fragment, using repeated cycles of denaturation, primer annealing and chain extension

Polymorphism: detectable variation in genome structure between individuals in a population

Positional cloning: using the mapped position of a gene to obtain a clone carrying it

Post-genomics: studies that go beyond the genome sequence; genome-wide studies of the products of transcription (*transcriptomics*) and translation (*proteomics*)

Post-translational modification: modification of the structure of a polypeptide after synthesis, e.g. by phosphorylation, glycosylation, proteolytic cleavage

Primary structure: the base sequence of a nucleic acid, or the amino acid sequence of a protein

Primer: a specific oligonucleotide, complementary to a defined region of the template strand, from which new DNA synthesis will occur

Primer walking: a technique in DNA sequencing whereby information from one sequence run is used to design another primer to extend the sequence determined

Probe: a nucleic acid molecule that will hybridize to a specific target sequence

Prokaryote: a cell that does not have a discrete nucleus bounded by a membrane (cf. *eukaryote*). Often used as synonymous with bacteria but also includes the *Archaea* (formerly termed *Archaebacteria*, but now recognized as evolutionarily distinct)

Promoter: region of DNA to which RNA polymerase binds in order to initiate transcription

Promoter probe vector: a vector carrying a promoterless reporter gene, so that inserts carrying a promoter can be detected

Proof-reading: the ability of DNA polymerase to check, and correct, the accuracy of the newly made sequence

Prophage: the repressed form of bacteriophage DNA in a lysogen; it may be integrated into the chromosome or exist as a plasmid

Protein engineering: altering a gene so as to produce defined changes in the properties of the encoded protein

Proteome: the complete content of different proteins in a cell (cf. *genome, transcriptome*)

Proteomics: global study of protein expression in an organism

Protoplast: formed by complete removal of the cell wall, using osmotically stabilized conditions

Prototroph: a nutritionally wild-type organism that does not need any additional growth supplement (cf. *auxotroph*)

Pseudogene: a gene, usually recogniable as a copy of another gene, that does not produce a protein product

Pulsed-field gel electrophoresis: separation of large DNA molecules by application of an intermittently varying electric field

Purine: one of the two types of bases in nucleic acids (adenine, guanine); see *pyrimidine*

Pyrimidine: one of the two types of bases in nucleic acids (cytosine and thymine in DNA; cytosine and uracil in RNA); see *purine*

RACE (rapid amplification of cDNA ends): a PCR-based method for obtaining the full length of a cDNA

Random primers: synthetic oligomeric nucleotides (usually hexamers) designed to act as primers for DNA synthesis at multiple sites

Reading frame: a nucleic acid sequence is translated in groups of three bases (*codons*). There are three possible ways of reading the sequence (in one direction), depending on where you start. These are the three reading frames

Real-time PCR: a PCR technique, using fluorescent dyes, that makes it possible to monitor the progress of the amplification as it occurs

Recombinant: (a) an organism containing genes from different sources. either by natural horizontal gene transfer, or as a result of gene cloning; (b) a DNA molecule formed by joining two different pieces of DNA

Recombination: (a) the production of new strains by mating two genetically distinct parents; (b) the generation of new DNA molecules by breaking and re-joining the original molecules:

Relaxation: conversion of supercoiled circular plasmid DNA to an open circular form

Replacement vector: a lambda cloning vector in which a piece of DNA (the *stuffer fragment*) can be removed and replaced by the cloned fragment (cf. *insertion vector*)

Replica plating: transfer of colonies from one plate to others, in the same position, for differential screening

Replication: synthesis of a copy of a DNA molecule

Replicon: a DNA molecule that can be replicated; also used to refer to the replication control region of a plasmid (cf. *origin of replication*)

Reporter gene: a gene that codes for a readily detected protein for study of the regulation of gene expression

Repression: (a) reduction in transcription of a gene, usually due to the action of a repressor protein; (b) establishment of lysogeny with temperate bacteriophages

Repressor: a protein that binds to a specific DNA site to switch off transcription of the associated gene

Restriction: reduction or prevention of phage infection through the production of *restriction endonucleases* which degrade foreign DNA; see also *modification*

Restriction endonuclease: an enzyme that recognizes specific DNA sequences and cuts the DNA, usually at the recognition site

Restriction fragment length polymorphism (RFLP): variation between individuals or strains in the size of specific restriction fragments; used for strain typing, and for locating particular genes

Restriction mapping: determination of the position of restriction endonuclease recognition sites on a DNA molecule

Retrotransposon: a transposon that transposes by means of an RNA intermediate (cf. *retrovirus*)

Retrovirus: a virus with an RNA genome that is copied, by reverse transcriptase, into DNA after infection

Reverse genetics: making specific changes to the DNA and then examining the phenotype; contrasts with classical genetics in which you select mutants by their phenotype and then study the nature of the mutation

Reverse hybridization: hybridization in which the specific nucleic acid fragments are fixed to a substrate and hybridized with a labelled mixture of nucleic acids such as a cell extract

Reverse transcription: production of cDNA from an RNA template, by reverse transcriptase

Reverse translation: see *Back translation*

Ribonuclease: an enzyme that digests RNA

Ribosome binding site: the region on an mRNA molecule to which ribosomes initially attach, in bacteria

RNA interference: reducing expression of a specific gene through the action of dsRNA (see *siRNA*)

RNA polymerase: enzyme that synthesizes RNA, generally using a DNA template

Rooted tree: a phylogenetic tree in which the position of the putative common ancestor can be inferred (see *outgroup*, *unrooted tree*)

RT-PCR: reverse transcription PCR. Technique for producing an amplified DNA product from an mRNA template

SDS-PAGE: polyacrylamide gel electrophoresis in which proteins are separated according to molecular weight in the presence of sodium dodecyl sulfate

Secondary structure: the spatial arrangement of amino acids in a protein, or of bases in nucleic acid

Selectable marker: a gene that causes a phenotype (usually antibiotic resistance) that can be readily selected

Shine–Dalgarno sequence: see *ribosome binding site*

Shotgun cloning: insertion of random fragments of DNA into a vector

Shotgun sequencing: genome sequencing strategy involving the sequencing of large numbers of random fragments; the individual sequences are subsequently assembled by a computer

Shuttle vector: a cloning vector that can replicate in two different species, one of which is usually *E. coli*; facilitates cloning genes in *E. coli* initially and subsequently transferring them to an alternative host without needing to re-clone them

Sigma factor: polypeptide that associates with (bacterial) RNA polymerase core enzyme to determine promoter specificity

Signal peptide: amino acid sequence at the amino terminus of a secreted protein; involved in conducting the protein through the membrane, or targeting it to specific cellular locations

Signal transduction: extracellular conditions alter the conformation of a transmembrane protein which in turn alters the regulation of metabolic pathways within the cell

Silent mutation: a change in the DNA structure that has no effect on the phenotype of the cell

Single cross-over: recombination at a single position, e.g. between a plasmid and chromosome, leading to integration of the whole plasmid (cf. *double cross-over*)

Single nucleotide polymorphism (SNP): a single base difference in the DNA sequence of part of a population

siRNA: short interfering RNA. Small dsRNA molecules, responsible for RNA interference

Site-directed mutagenesis: a technique for specifically altering (*in vitro*) the sequence of DNA at a defined point

Southern blot: a membrane with DNA fragments transferred from an electrophoresis gel, preparatory to hybridization

Splicing: removal of introns from RNA and joining together of the exons

Start codon: position at which protein synthesis starts, usually AUG

Stem–loop structure: a nucleic acid strand containing two complementary sequences can fold so that these sequences are paired (stem), with the region between them forming a loop of unpaired bases (cf. *hairpin*)

Sticky end: the end of a DNA molecule where one strand protrudes beyond the other; also known as a *cohesive end*; see *blunt end*

Stop codon: a codon which has no corresponding tRNA, and which signals the end of a region to be translated

Stringency: conditions affecting the *hybridization* of single-stranded DNA molecules. Higher stringency (higher temperature and/or lower salt concentration) demands more accurate pairing between the two molecules

Structural genes: genes coding for enzymes (or sometimes other products), as distinguished from regulatory genes

Stuffer fragment: piece of DNA that is removed from a vector such as a replacement lambda vector, and replaced by the cloned DNA fragment

Suicide plasmid: a vector that is unable to replicate in a specific host. Maintenance of the selected marker requires integration into the chromosome

Supercoiling: coiling of a double-stranded DNA helix around itself

Superinfection immunity: resistance of a lysogen to infection by the same (or related) bacteriophage

Suppression: the occurrence of a second mutation which negates the effect of the first without actually reversing it

SV40: Simian virus 40. Small virus isolated from monkeys; used as a vector, and also as a source of expression signals in mammalian expression vectors

SYBRGreen: dye which fluoresces when bound to dsDNA. Use for *real-time PCR*

Synonymous codons: different codons that code for the same amino acid

Synteny: comparison of the overall organization of genes in a genome

T7: Lytic bacteriophage of *E. coli*. The requirement of T7 RNA polymerase for a highly specific promoter is used in several contexts

TA cloning: a method for cloning PCR products, exploiting the tendency of *Taq* polymerase to add a non-specific adenyl residue to the 3′ end of the new DNA strand

Tagging: constructing a recombinant so that the protein formed has additional amino acids at one end, facilitating purification by affinity chromatography, or targeting the protein to specific destinations

Tailing: adding a not exactly defined number of nucleotides to the 3′ end of DNA, using *terminal transferase*

Tandem repeat: occurrence of the same sequence two or more times, directly following one another

***Taq* polymerase**: a thermostable DNA polymerase, commonly used for PCR

TaqMan: a technique for real-time PCR in which fluorescence develops as a consequence of destruction of the labelled probe during DNA synthesis

T DNA: the part of the *Ti plasmid* that is transferred into plant cells

Telomere: end region of a eukaryotic chromosome containing sequences that are replicated by a special process, counteracting the tendency of linear molecules to be shortened during replication

Temperate: describes a bacteriophage that is able to enter *lysogeny*

Template: a single strand of nucleic acid used for directing the synthesis of a complementary strand

Terminal transferase: an enzyme that adds nucleotides, non-specifically, to the 3′ ends of DNA (see *tailing*)

Terminator: site at which transcription stops

Tertiary structure: folding of secondary structure components of a protein

Ti plasmid: tumour-inducing plasmid, in *Agrobacterium tumefaciens*. Used as a vector in genetic manipulation of plants (see *T DNA*)

Topoisomerase: an enzyme that alters the supercoiling of DNA by breaking and rejoining DNA strands

Totipotent: describes a cell that is capable of giving rise to all types of cell within a whole animal or plant

***trans*-acting**: a gene that influences non-adjacent regulatory DNA sequences, through the production of a diffusible protein (cf. *cis-acting*)

Transcription: synthesis of RNA according to a DNA template

Transcriptional fusion: a recombinant construct in which a promoterless insert is transcribed from a promoter on the vector (cf. *translational fusion*)

Transcriptome: the complete mRNA content of a cell (cf. *genome, proteome*)

Transduction: bacteriophage-mediated transfer of genes from one bacterium to another

Transfection: introduction of viral nucleic acids into a cell

Transformation: introduction of extraneous DNA into a cell. Also used to mean the conversion of an animal cell into an immortalized, tumour-like, cell

Transgenic: describes an animal or plant possessing a cloned gene in all its cells, so that the introduced gene is inherited by the progeny of that animal or plant

Translation: synthesis of proteins/polypeptides by ribosomes acting on a mRNA template

Translational fusion: a recombinant construct in which an insert lacking a translation start site is joined (in frame) to a fragment carrying translational signals (cf. *transcriptional fusion*)

Transposon: a DNA element carrying recognizable genes (e.g. antibiotic resistance) that is capable of inserting itself into the chromosome or a plasmid, independently of the normal host cell recombination machinery (cf. *insertion sequence*)

Transposon mutagenesis: disruption of genes by insertion of a transposon

Trimming: converting a sticky end to a blunt end by removing the unpaired nucleotides (cf. *end filling*)

Two-dimensional gel electrophoresis: separation of a complex mixture of proteins by a combination of *isoelectric focussing* and *SDS-PAGE*

Ty element: retrotransposon, used for transposon mutagenesis in yeast

Universal primers: sequencing primers derived from the sequence of the vector (pUC series); any insert can be sequenced with the same primers

Unrooted tree: a phylogenetic tree in which the position of the putative common ancestor cannot be inferred (see *rooted tree*)

UPGMA (Unweighted Pair Group Method with Arithmetic means): a technique for constructing phylogenetic trees, e.g. from sequence data

Upstream activator sequence: a sequence, upstream from the promoter, which is required for efficient promoter activity

Vaccinia: smallpox vaccine virus, used as a vector for recombinant vaccine construction

Vector: a replicon (plasmid or phage) into which extraneous DNA fragments can be inserted, forming a recombinant molecule that can be replicated in the host cell

Western blot: a membrane with proteins transferred from an electrophoresis gel, usually for detection by means of antibodies

X-gal: 5-bromo-4-chloro-3-indolyl-β-D-galactoside. Chromogenic substrate for β-galactosidase

Yeast artificial chromosome (YAC): a vector that mimics the structure of a yeast chromosome, used to clone very large DNA fragments

Yeast centromere plasmid (YCp): a yeast vector containing a centromere; replicates at low copy number

Yeast episomal plasmid (YEp): an autonomously replicating vector based on a yeast plasmid (the '2 μm circle')

Yeast integrating plasmid (YIp): a yeast vector that relies on integration into the yeast chromosome

Bibliography

General books

Brown, T.A. (2006) *Gene Cloning and DNA Analysis – an Introduction*, 5th edn. Blackwell Science: Oxford.

Dale, J.W. and Park, S.F. (2004) *Molecular Genetics of Bacteria*, 4th edn. Wiley: Chichester.

Gibson G. and Muse S.V. (2004) *A Primer of Genome Science*, 2nd edn. Sinauer: New York.

Lewin, B. (2006). *Essential Genes.* Pearson Education: London.

Lodge, L., Lund, P. and Minchin, S. (2007) *Gene Cloning – Principles and Applications.* Taylor & Francis: Abingdon.

Primrose, S.B. and Twyman, R.M. (2003) *Principles of Genome Analysis and Genomics*, 3rd edn. Blackwell: Oxford.

Primrose, S.B. and Twyman, R. (2006) *Principles of Gene Manipulation*, 7th edn. Blackwell: Oxford.

Watson, J.D., Caudy, A.A., Myers, R.M. and Witkowski, J.A. (2007) *Recombinant DNA: Genes and Genomes – a Short Course*, 3rd edn. W.H. Freeman: Philadelphia, PA.

Laboratory manuals

Sambrook, J. and Russell, D.W. (2001) *Molecular Cloning: a Laboratory Manual*, 3rd edn. Cold Spring Harbor Laboratory Press: Cold Spring Harbor, NY.

Also, many commercial suppliers of molecular biological materials produce very informative catalogues and methods sheets, usually available on-line.

Special topics

Higgs, P.G. and Attwood, T.K. (2005) *Bioinformatics and Molecular Evolution.* Blackwell: Oxford.

Lesk, A.M. (2005) *Introduction to Bioinformatics.* Oxford University Press: Oxford.

Salemi, M. and Vandamme, A.-M. (eds) (2003) *The Phylogenetic Handbook – a Practical Approach to DNA and Protein Phylogeny.* Cambridge University Press: Cambridge.

Strachan, T. and Read, A.P. (2004) *Human Molecular Genetics 3.* Garland Publishing: Abingdon.
Sudbery, P. (2002) *Human Molecular Genetics*, 2nd edn. Prentice Hall: Harlow.
Xiong, J. (2006) *Essential Bioinformatics.* Cambridge University Press: Cambridge.

Web sites

Note that this is only a small selection of the available sites. Many other sites not mentioned are also valuable resources. Note also that web site addresses may change, although the ones selected are likely to be reasonably stable, at least in their home pages. Within each site, the structure is likely to be more fluid, so only the home pages are listed. You will need to explore the site from that point to find the facilities that you want.

See Box 9.1 for further and more specific sites.

Address	Organization	Facilities include
http://www.ebi.ac.uk/	European Bioinformatics Institute	Databases: EMBL, UniProt, TrEMBL Tools: BLAST, FASTA, MPsrch, CLUSTAL
http://www.ncbi.nlm.nih.gov/	National Center for Biotechnology Information (NCBI)	GenBank database plus search and comparison tools
http://www.ddbj.nig.ac.jp/	DNA Data Bank of Japan	DDBJ database plus search and comparison tools
http://www.isb-sib.ch/ http://www.expasy.org/	Swiss Institute of Bioinformatics (SIB)	UniProt database Prosite; SWISS-2DPAGE ExPASy (Expert Protein Analysis System) proteomics server
http://www.sanger.ac.uk/	Wellcome Trust Sanger Institute (WTSI)	Genome sequence data
http://www.ensembl.org http://genome.ucsc.edu	EMBL, EBI, WTSI University of California Santa Cruz	Genome browsers
http://www.tigr.org/	The Institute for Genomic Research	Genome sequencing projects
www.genomesonline.org		List of sequenced genomes

Index

From Genes to Genomes, Second Edition, Jeremy W. Dale and Malcolm von Schantz